£ 20/-

Early Hominid Behavioural Ecology

Early Hominid Behavioural Ecology

Edited by

James S. Oliver, Nancy E. Sikes and Kathlyn M. Stewart

ACADEMIC PRESS

London San Diego New York Boston

Sydney Tokyo Toronto

ACADEMIC PRESS LIMITED
24–28 Oval Road
London NW1 7DX

U.S. Edition Published by
ACADEMIC PRESS INC.
San Diego, CA 92101

This book is printed on acid free paper

A catalogue record for this book is available from the British Library.

(Reprinted from *Journal of Human Evolution*, Vol. 27, Nos 1–3, 1994).

ISBN 0-12-5256604

Printed in Great Britain by Henry Ling Ltd, Dorchester, Dorset.

Contents

List of Contributors

R. V. Bellomo, *Department of Anthropology, University of South Florida, 4202 East Fowler Avenue, SOC 107, Tampa, Florida, FL 33620-8100, U.S.A.*

L. C. Bishop, *Department of Anthropology, Yale University, PO Box 208277, New Haven, CT 06520, U.S.A.*

R. J. Blumenschine, *Department of Anthropology, Rutgers University, New Brunswick, NJ 08903, U.S.A.*

H. T. Bunn, *Department of Anthropology, University of Wisconsin, Madison, WI 53706, U.S.A.*

S. D. Capaldo, *Department of Anthropology, Rutgers University, New Brunswick, NJ 08903, U.S.A.*

J. A. Cavallo, *Department of Anthropology, Rutgers University, New Brunswick, NJ 08903, U.S.A.*

C. S. Feibel, *Department of Geology and Geophysics, University of Utah, UT 84112-1183, U.S.A.*

J. W. K. Harris, *Department of Anthropology, Rutgers University, Douglass Campus, New Brunswick, New Jersey, NJ 08903-0270, U.S.A.*

A. Hill, *Department of Anthropology, Yale University, Box 208277, New Haven, CT 06520, U.S.A.*

M. Kibunjia, *Division of Archaeology, National Museums of Kenya, PO Box 40658, Nairobi, Kenya* and *Department of Anthropology, Rutgers University, New Brunswick, New Jersey, NJ 08903, U.S.A.*

E. M. Kroll, *Department of Anthropology, University of Wisconsin, Madison, WI 53706, U.S.A.*

H. M. McHenry, *Department of Anthropology, University of California, Davis, CA 95616, U.S.A.*

J. S. Oliver, *Anthropology Section, Research & Collections Center, Illinois State Museum, 1011 East Ash Street, Springfield, IL 62703, U.S.A.*

T. W. Plummer, *Department of Anthropology, National Museum of Natural History, Smithsonian Institution, Washington DC 20560, U.S.A.*

R. Potts, *Department of Anthropology, National Museum of Natural History, Smithsonian Institution, Washington DC 20560, U.S.A.* and *National Museums of Kenya, PO Box 40658, Nairobi, Kenya*

M. J. Rogers, *Department of Anthropology, Rutgers University, Douglass Campus, New Brunswick, New Jersey, NJ 08903-0270, U.S.A.*

M. M. Selvaggio, *Southern Connecticut State University, Department of Sociology and Anthropology, 501 Crescent Street, New Haven, CT 06515, U.S.A.*

J. M. Sept, *Anthropology Department, Indiana University, Bloomington, IN 47405, U.S.A.*

N. E. Sikes, *Department of Anthropology, University of Illinois, 109 Davenport Hall, 607 South Mathews Avenue, Urbana, IL 61801, U.S.A.*

N. Stern, *Department of Archaeology, La Trobe University, Bundoora, Victoria 3083, Australia*

K. M. Stewart, *Canadian Museum of Nature, PO Box 3443, Station D, Ottawa, Ontario, K1P 6P4, Canada*

J. S. Oliver
Anthropology Section, Research and Collections Center, Illinois State Museum, 1011 East Ash Street, Springfield, IL, 62703, U.S.A.

N. E. Sikes
Department of Anthropology, University of Illinois, 109 Davenport Hall, 607 South Mathews Avenue, Urbana, IL, 61801, U.S.A.

K. M. Stewart
Canadian Museum of Nature, PO Box 3443, Station D, Ottawa, Ontario, K1P 6P4 Canada

Introduction to "Early Hominid Behavioural Ecology": new looks at old questions

Journal of Human Evolution (1994) **27**, 1–5

Describing and understanding the evolution of hominid behaviour is one of the primary goals of palaeoanthropology. A crucial, initial part of this process is being able to tap the archaeological and palaeontological remains for primary evidence of early hominids and their activities. Interpretation of this evidence requires the utilization of principles and models derived from studies of the behaviour and ecology of modern organisms. Until now, little melding of primary fossil and artifactual data with behavioural and ecological interpretation has been undertaken. Earlier studies of Plio-Pleistocene hominids were necessarily tool- and site-specific, and only the most speculative statements on hominid behaviour could be made. As the focus of palaeoanthropological research has shifted away from the level of the site to study of a more dynamic relationship between hominid activities and their ecological contexts, so have our methods for examining such variables expanded.

As the papers in this volume reveal, researchers are now able to formulate and carry out innovative methodologies for documenting specific palaeogeographical and palaeoecological contexts of early hominids. Further, the continuing development of taphonomic and middle-range research allows us to make linkages between the fossil and artifactual remains, and the principles and models of modern evolutionary theory and behavioural ecology. We believe we should emphasize not only the archaeological behaviours defined by specific data, but also the palaeoecological and biological context. In this way we will learn more about the evolutionary ecology of early hominids, and will thereby come to a better understanding of the many novel hominid behaviours that appeared in the Plio-Pleistocene.

Considering the new methods and approaches now being practised, it is time to present in one volume both the new methodologies and the new archaeological and palaeoecological data, in order to better understand early hominid behavioural ecology. Most of the contributions in this volume have evolved from papers presented in the "Early Hominid Behavioural Ecology" symposium at the 62nd Annual Meeting of the American Association of Physical Anthropologists, held in Toronto in 1993. The impetus for the symposium and subsequently for this volume came from its contributors, many of whom were conducting research in East Africa in the summer and fall of 1990. Many were embarking on dissertation research (Bishop, Capaldo, Kibunjia, Oliver, Rogers, Selvaggio, Sikes); others had recently finished their doctoral work and were initiating new research projects (Plummer, Stewart) and still others were continuing long-standing research projects in East

0047–2484/94/010001+05 $08.00/0

Africa (Bunn, Feibel, Harris, Hill, Potts). It was the frequent interaction and mixing of these and other researchers as we came and went from Nairobi and the National Museums of Kenya that contributed most to this volume. Discussions over breakfast, lunch or dinner and beer ranged from the mundane to the academically inane. Somewhere in between came comments, evaluations of recent papers and hypotheses or positions on issues related to Plio-Pleistocene hominids. Through it all, however, discussions were grounded in new findings from ongoing research that sought to further our understanding of the ecological position and behaviour of early hominids. These discussions were continued back in North America, and incorporated research by other scientists who were also taking new approaches to early hominid behaviour and ecology. Our goal in assembling this volume was to have the contributors present some of these new archaeological and palaeoecological data, and what these data tell us about early hominid behavioural ecology.

New methods, models and data on early hominid behavioural ecology

The paper by Potts provides a stimulating lead-off discussion for the volume. He summarizes the current models on early hominid behaviour, but states that these have become static reconstructions with which we have rigidly defined hominid activities. Instead, Potts suggests that these models are best viewed through the interactions of their component behavioural and environmental variables. By focusing on these variables, we can view hominid behaviour in terms of varied responses to environmental, habitat and resource change, rather than as part of an inflexible model. The relationships of some of these variables are then examined by Potts in the context of data from the landscape archaeology project at Olorgesailie.

Sikes, and Plummer and Bishop utilize different and innovative methodologies to provide some of the environmental data necessary to "flesh out" Potts' variables. Sikes suggests that the integration of stable carbon isotopic analyses of paleosols with landscape approaches to Oldowan archaeology is a robust method for determining the floral microhabitat context of hominid behaviour and resource use. This position is illustrated by geochemical analyses of a basal Bed II paleosol from an area near the HWK and FLK site complexes that, based on modern soil and vegetation analogs, supported a riparian forest to grassy woodland. This environmental setting at Olduvai can be compared with palaeoenvironmental data from other hominid sites to examine variability in hominid land use patterns. In their paper, Plummer and Bishop undertake a taxon-free analysis of the measurements of metapodials from 37 extant African antelope species, in order to reconstruct bovid habitat preference. They classify the modern metapodials into three habitat groups, and then apply these classification criteria to fossil elements. Based on analysis of elements from four Bed I Olduvai Gorge sites, Plummer and Bishop suggest that the palaeoenvironments during Bed I deposition were more closed than has been indicated in previous studies.

The focus of the volume is turned to hominids as biological organisms with the paper by McHenry. By predicting changes in brain and body size based on analyses of hominid postcranial samples dating from 4 to 1·3 my, his anatomical study allows inferences to be made on early hominid ranging patterns and mating systems. McHenry suggests that the apparent substantial body size increase and reduction in sexual dimorphism present by 1·7 my may be related to "a significant expansion in ranging area".

Investigation of early hominid land use patterns has in particular benefited by new field and laboratory methodologies, and the following three papers examine archaeological remains and patterns of intra- and inter-site distribution in order to make inferences about land use patterns. Stern examines laterally extensive sedimentary horizons of the lower Okote member at Koobi Fora for archaeological traces, and emphasizes the importance of understanding time-averaging in making behavioural inferences. Specifically, she addresses the problem of differences in temporal scale and scale of resolution between the geological record, the entombed archaeological traces, and the ecological and behavioural questions being asked. Next, Kroll examines the vertical and horizontal distributions of lithics and fauna at archaeological sites at Koobi Fora, as an indicator of site use over the ancient landscape. Her data indicates that hominids repeatedly exploited large areas of the landscape, and suggests that areas under trees may have been preferentially utilized by early hominids.

In their paper, Rogers, Feibel and Harris examine known archaeological traces, particularly lithic discard patterns, and their palaeogeographical contexts in the Turkana Basin at successive time intervals between 2·3 and 1·5 my. The archaeological traces dating to ca. 1·6 my are found in a variety of geographic settings, rather than the more restricted contexts at about 2·3 my. The 1·6 my traces coincide with the emergence of *Homo erectus*, and may imply certain behavioural adaptations, including greater foraging range, greater diet quality and breadth, and possibly larger group size with inherently greater social interactions.

Bellomo and Kibunjia separately use archaeological data to indicate how hominids directly manipulated certain aspects of their environment. The control and use of fire would have dramatically changed the ability of early hominids to interact with the environment and associated fauna, and Bellomo's paper outlines a methodology by which evidence of fire from natural processes can be distinguished from that resulting from human activities. He then presents a series of artifact distribution analyses that seek to discern and evaluate patterns of association between the fired features and artifacts at FxJj20 Main at Koobi Fora. Bellomo suggests that, based on a variety of indicators, early hominids probably used fire primarily as protection against predators, as a source of light and/or as a source of heat. Kibunjia's paper analyses technological characteristics of lithic artifacts from the Pliocene-aged Lokalalei site, west Turkana, dated at 2·36 my. His analysis indicates that Lokalalei inhabitants were not well skilled at striking off whole flakes from parent forms. Their resulting implements may therefore not have performed well at cutting or slicing, thus possibly limiting the size of animals hunted or scavenged. Kibunjia suggests the Lokalalei technology represents a step in a continuum from simpler to more complex lithic tool-making.

The next six papers discuss early hominid subsistence. The authors utilize actualistic studies to aid in reconstruction of scavenging, hunting and other foraging behaviours, and infer from these a wider range of early hominid social and ecological behaviours. Blumenschine, Capaldo and Cavallo's paper presents a thought-provoking conceptual framework for conducting behavioural-ecological analyses of extinct hominid species. The authors argue for a need to integrate middle-range research in its broadest sense with behavioural-ecological modelling in order to generate testable hypotheses on early hominid behavioural ecology. Based on years of experimental and actualistic observations of carnivore behaviour in northern Tanzania, they model the foraging and carcass utilization strategies of Oldowan hominids by combining the ecologically-based concept of competition for carcasses with the "ecological taphonomy" of bones.

Selvaggio combines observations on modern carnivore feeding behaviour with experiments to quantify the incidence of butchery marks and tooth marks on long bones following carnivore defleshing, in order to evaluate cut and tooth marks on archaeological assemblages.

She found that the incidence of butchery marks and carnivore tooth marks is related to the condition of bones upon carnivore abandonment, which in turn is dependent on the number of carnivores involved in defleshing the limbs.

Stewart's paper provides culinary relief from the archaeological diet of mammals, by focusing on fish as an alternative early hominid food source. She discusses fish as a nutritional and seasonal alternative food source for early hominids, and documents its importance as a seasonal resource in the Late Pleistocene archaeological record. Stewart then reviews the evidence for fish procurement by early hominids, with an emphasis on fish remains from Beds I and II Olduvai Gorge sites.

The relationship between archaeological sites in east Turkana and their proximity to a large ancestral Omo River is re-evaluated in Bunn's paper, in contrast to previous hominid land-use patterns based on a lake-dominated model of the Koobi Fora landscape. Bunn also documents single-event butchery and/or carcass consumption sites in ephemeral lake margin localities, where butchery marks attest to the presence of hominids in the near absence of stone tools. He then suggests hominid foraging behaviour at the lake margin sites may best be described as a "feed as you go strategy".

Oliver's analysis of hammerstone and carnivore-induced damage on bones from the FLK *Zinjanthropus* site indicates that early *Homo* was the predominant agent of modification, and may have had regular access to meat-rich carcasses. Based on an assessment of food transport and processing behaviour in other mammals, notably carnivores, Oliver infers that early *Homo* may have employed a similar dual-unit foraging pattern whereby a caregiver/infant unit forages near a secure central place while other group members forage farther afield.

Sept's paper takes on the bone people by stating that a bone-biased archaeological record does not necessarily mean that meat and/or marrow were critical foods in the early hominid diet. She states that plant foods would have been low-risk, dependable foods with predictable locations and known processing times, and suggests that archaeologists should investigate possible hominid feeding locations, in order to reconstruct local habitats and their edible vegetation. Sept examines the distribution and quality of modern plant foods available along the Semliki River in Zaire, and estimates the costs and benefits of foraging for plant foods in savanna riverine habitats.

In the last contribution to the volume, Hill provides a personal insight into earlier problems and current issues in researching early hominid behavioural ecology. In particular Hill examines the methodologies used by researchers in this volume to investigate early hominid behavioural ecology, with a view to how their methods have improved compared to earlier studies.

Summary

Many of the contributors to this volume crossed paths in Nairobi, Kenya in 1990. The new research and lively discussions permitted by the interaction of so many was the impetus for the resulting symposium and this volume. The challenging diversity of papers present new archaeological, palaeoecological and palaeoanthropological data on early hominid behaviour and ecology. We believe that both by emphasizing the archaeological behaviours defined by specific data and by providing a palaeoecological and biological context we will come to a better understanding of the behavioural and evolutionary ecology of Plio-Pleistocene hominids. The papers in this volume do this by opening up new perspectives on old questions.

Acknowledgements

We first express our gratitude to the co-editors of the *Journal of Human Evolution*, in particular Leslie Aiello, for her support and encouragement in putting together this volume. We also thank our contributors for their inspiration in assembling this volume, and for their participation in the AAPA symposium in Toronto in 1993. Desmond Clark provided an overview of the symposium in Toronto in 1993, and has continued to be supportive of the volume. We are very grateful to our many reviewers, several of whom went beyond the call of duty to provide timely and constructive comments. We thank the National Museums of Kenya and its staff for providing offices and research space for many of us in 1990. We gratefully acknowledge the support and administrative assistance provided by our parent institutions, the Illinois State Museum in Springfield, Illinois, and the Canadian Museum of Nature in Ottawa, Canada. Many people contributed along the way to publishing the volume, and in particular we would like to thank Melinda Aiello (ISM) and Donna Naughton (CMN) for their help in assembling papers, Julianne Snider (ISM) for the cover artwork and other figures, and Genny Early (Academic Press) for her efforts and patience. Finally, we thank the governments and people of Kenya, Tanzania, Ethiopia and Zaire for graciously providing us the opportunity to conduct research on our shared past.

Richard Potts

Department of Anthropology, National Museum of Natural History, Smithsonian Institution, Washington, DC 20560, U.S.A.; and National Museums of Kenya, P.O. Box 40658, Nairobi, Kenya

Received 24 September 1993
Revision received 3 February 1994 and accepted 20 February 1994

Keywords: Early hominid behavior, paleolandscape, central place, stone cache, routed foraging, riparian woodland, paleoenvironment, stone tools, animal bones, Olorgesailie, Olduvai, East Turkana, natural selection, variability, paleoecology.

Variables versus models of early Pleistocene hominid land use

Alternative reconstructions, or models, have been advanced to explain the archeological behavior patterns of early toolmakers in East Africa. Current models include central place foraging (home bases), multiple place foraging at stone caches, routed foraging, and riparian woodland scavenging. These models imply that certain socioecological variables affected hominid attraction to specific places. Such factors include social cohesion, predator avoidance, costs of stone transport, habitat patch choice, and tethering of toolmakers to fixed resources. When viewed as alternative hypotheses, however, these models become static and miss possible simultaneous effects of variables on early hominid behaviors. A paleolandscape study of excavated artefacts and habitat indicators in Member 1 (ca. 0·99 Ma) of the Olorgesailie Formation, Kenya, illustrates the interplay of these variables in comparison with other Plio-Pleistocene contexts: (1) resource tethering was important for some raw materials and possibly handaxes, but not for other components of the artefact assemblage; (2) stone transport varied greatly among different basin contexts; (3) overlap with predators was minimal at Olorgesailie but large in Bed I Olduvai; (4) no specific correlation between hominid traces and microhabitat is evident at Olorgesailie; (5) no clear spatial indications of human social aggregation (e.g., shelters, hearths, activity areas) have been found.

Diverse sedimentary basins can be expected to differ in the strength of these factors. Growing awareness of variability in Plio-Pleistocene contexts changes the pursuit of archeology away from the single best reconstruction toward an understanding of environment-behavior covariation. Geographic and temporal variations in hominid land use suggest that natural selection may have entailed responses to habitat and resource variability, not merely consistent selection pressure or adaptation to a model environment.

Journal of Human Evolution (1994) **27**, 7–24

Introduction

Assessment of early hominid behavioral ecology requires that we pinpoint in the geologic record environmental variables, behavioral variables, and the likely interaction between the two. Studies of living organisms have identified general linkages among a number of such factors (e.g., Krebs & Davies, 1991; Rubinstein & Wrangham, 1986; Smuts *et al.*, 1987); these include resource distribution (in time and space), pattern of social grouping, predation risk, food choice, and ranging distance. Yet the fossil record also has something to offer—its long temporal span in which evolutionary questions may predominate (Figure 1). With this in mind, the ambitious aim of bringing behavioral ecology to bear on evolutionary questions is to portray the character of natural selection over time, specifically the factors that affected hominid survival and reproduction.

The best source, really the only source, of data about the ranging patterns and resources used by earlier hominids comes from excavated archeological evidence. These data obviously pertain to toolmakers but whose exact species identity often remains unknown, especially in Plio-Pleistocene settings (see, e.g., Wood, 1991; Susman, 1991).

A series of models about hominid resources and ranging patterns has been inspired by late Pliocene and early Pleistocene studies at Olduvai and Koobi Fora. Several variations on these models exist, but four have dominated attention so far (Table 1): central place foraging (or the home base model); the routed foraging model; the stone cache model; and various behavioral reconstructions based on habitat-specific resources, including the riparian

0047–2484/94/010007+18 $08.00/0

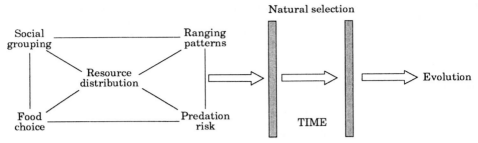

Figure 1. Studies in modern behavioral ecology have established certain generalized relationships among factors that affect the survival and reproductive success of animals. A few of these factors are depicted on the left side of the diagram. Over time the type and strength of interaction among these factors may be altered by natural selection (middle part of the diagram). For example, a change in food distribution may favor larger social groups, which could enhance defense against predation and may thus affect future choice of food patches. As interactions among these factors varied, certain combinations of behaviors (e.g., ranging distances, degree of group cohesion) proved more successful on a consistent basis than others, resulting in evolutionary change (right side of diagram). Some of these factors and their interactions may be measured or otherwise interpreted from the fossil record (see text).

woodland scavenging model. Since the early 1980s, debates have ensued about which model is correct in that it best accounts for the composition and clusterings of excavated stone artefacts and fauna. Hence these models are usually treated as rival interpretations or competing hypotheses about the behavior of early toolmakers (see Binford, 1981; Blumenschine, 1987; Isaac, 1981, 1983, 1984:23; Potts, 1988, 1991).

The aim of this paper is to illustrate why this research approach may not be as sound as commonly presumed. In treating these behavioral models as competing interpretations, the interesting behavioral and ecological variables become isolated from one another, and the simultaneous interplay among them is masked. The interplay, however, is what enables one to discern (a) the variability in hominid activities (at least those related to tool use), (b) the contexts that elicited certain behavioral responses, and (c) the shifts in both context and response over time. These issues are more important to an understanding of hominid evolution than is any search for the single best model of hominid behavior or environmental setting.

Models and variables

According to Isaac (1984), the "pivotal question" in the archeology of early humans is to explain how high-density clusters of stone artefacts and animal bones were formed. In addition to essential taphonomic considerations (e.g., Binford, 1981; Bunn *et al.*, 1980; Isaac, 1984; Potts, 1988), a number of behavioral hypotheses, or models, have been developed to answer this question. The key distinction between models lies in the main attraction, or magnet, that stimulated the concentration of debris in certain places of the landscape and created lower density scatters in others (Table 1).

The central place model, or central place foraging, was proposed by Isaac (1983, 1984) in lieu of the terms home base and food sharing to explain dense artefact-bone accumulations (e.g., Isaac, 1978). Zoologically, a central place is a single, highly delimited living site (e.g., den, nest) from which an animal starts to forage and then returns (Hamilton & Watt, 1970). The central place is typically associated with certain social interactions such as provisioning or sharing food. As a model applied to hominids, it is assumed that members of the social group

Table 1 **Four current models of early hominid behavior and land use, and the factors explaining repeated return of toolmakers to specific places on the landscape**

Model	Magnets and factors
Central place foraging	Campsite-type social behaviors; social focus attracts the accumulation of debris
Stone cache	Transport of stones defines multiple points to which food transport is drawn; predation risk too high to attract central place social behaviors
Routed foraging	Activity tethered to fixed resources (outcrops, water, shelter, shade trees)
Riparian woodland	Reliable resources (e.g., scavenging opportunities) in specific habitat zones attracted toolmakers and created conditions for debris accumulation

converged daily at one fixed point. Because foragers varied in the types and quality of foods they could find, sharing at the central place was advantageous to each individual. Animal tissues became an integral part of sharing, which was responsible for the accumulation of stones and animal parts on the landscape. Isaac (e.g., 1984:67) used the terms central place and home base interchangeably; reaggregation of the group for the purpose of vital social activity was the driving force behind the repeated use of one site at a time.

According to the stone cache model (e.g., Potts, 1988), modified and unmodified rocks served as significant secondary sources of stone. This model stands alone in considering the costs involved in carrying and using stones in the foraging process. The premise is that toolmakers had to have stones and certain kinds of food together simultaneously. This requirement led hominids to redistribute *both* resources to common ground (to multiple, not central, places on the landscape). The critical magnet consisted of stones dropped previously at loci where stone tool transport and use may originally have been incidental, e.g., presence of a tree, a carcass, or some unknown resource (Potts, 1988:280–1). Over time, debris accreted around these drop points as they became remembered focal spots of transported rocks (see also Schick, 1987). The stone cache idea was developed from a computer optimality model of energy costs; it thus reflects a behavioral *strategy* and, like all evolutionary strategies, need not represent an intentional or conscious activity—in this case, purposeful stockpiling of stone prior to food acquisition. Rather, in this model, the accumulation of stone and food refuse was concurrent, and could occur where even small amounts of excess (unused) stone had previously been dropped (Potts, 1988). A second factor—predation risk—enters into the picture. Where this was high (i.e., where large carnivores and hominids were attracted to the same specific places and carcass parts), social activity was low and removed from these places.

The routed foraging model (Binford, 1984) focuses on stationary features of the landscape and how these determine the ranging patterns of animals, including hominid toolmakers. According to this model, hominid activities were tethered to fixed resources on the landscape, such as lithic outcrops, water, and trees for shade or shelter. Movements of toolmaking hominids were constrained, or routed, by the location of these resources. Regular stopping at these points led to the deposition and concentration of carried artefacts and bones.

Alternative models have been developed that map hominid behavior on to specific types of habitat. A general rendition of this idea is: concentrated traces of hominid activity occurred in particular vegetation zones where resources useful to toolmakers could reliably be found. According to the riparian woodland model (e.g., Blumenschine, 1986, 1987; see also Cavallo & Blumenschine, 1989; Marean, 1989), consistent availability of scavengeable carcasses in woodlands near water represents one such situation. This model is based on observations that, in the modern Serengeti, scavengeable carcasses offer a regular food source in woodland habitats particularly during the dry season. Thus, toolmakers regularly visited or inhabited this vegetation zone (or riparian patches of trees) because they could obtain marrow and meat from dead animals there. Tool and bone debris accumulated as hominids were attracted repeatedly to scavenging opportunities in wooded places.

These models are typically forwarded as reconstructions of what early hominid behavior was like. There are more important issues at stake, however, than which model provides the most appropriate reconstruction. The critical issues are evolutionary, i.e., relationships between ecological and behavioral variables that over time resulted in interesting evolutionary phenomena: e.g., variables that stimulated the use of stone tools, that prompted the transport of rocks around the environment, that affected hominid overlap with large carnivores, and that led complex social behaviors to become linked with the movement of resources.

The methodological challenge has been to discern unambiguous evidence for each of the proposed magnets. One approach is to dissolve the models into variables that allow at least relative measures of these key factors (see also Blumenschine, 1991).

Five such variables may be defined as follows:

(1) Resource transport: The tendency of hominids to carry resources from one place to another. This is measured in terms of quantity and type of transported material.

(2) Tethering: The proximity of debris to known stationary resources. Stones linked to known outcrops offer a good opportunity for measuring this variable. Archeological materials situated closer to fixed sources were more tightly tethered.

(3) Habitat heterogeneity: The degree to which debris concentrations correspond with distinctive habitats or vegetation zones.

(4) Predation potential: The overlap between carnivores and hominids in their use of carcasses and space.

(5) Strength of social aggregation: In hominids this means the tendency to form a primary social focus to which kin and other group members return. The only material evidence known to apply to hominids consists of shelters, hearths, or other distinctive features or spatial patterns indicative of social structuring of a delimited place on the landscape.

Rather than choosing one model or reconstruction over another, I suggest that *all* of these variables at certain times would have been critical and potentially simultaneous in their effect on toolmaker behaviors.

Methodology

To illustrate the interplay among these variables, I call upon continuing research at Olorgesailie, southern Kenya, and develop comparisons with sedimentary basins of different age, place, and paleoenvironmental context. Since 1985, a collaborative project of the Smithsonian Institution and the National Museums of Kenya has studied a lake-margin paleosol exposed extensively at the top of the oldest geologic unit in the Olorgesailie sequence (Member 1) (Isaac, 1977). The target paleosol is bracketed by ash and pumice layers dated ca. 0·992 Ma and ca. 0·974 Ma respectively (Deino & Potts, 1990, 1992).

The purpose of this project has been to document lateral variation in the presence of stone artefacts, faunal remains, sediments, taphonomic contexts, and other indications of ancient habitat. Erosion of upper Member 1 has cut a varied topography ranging from steep hill-sides of diatomaceous siltstones to small plains immediately overlying the target stratum. We

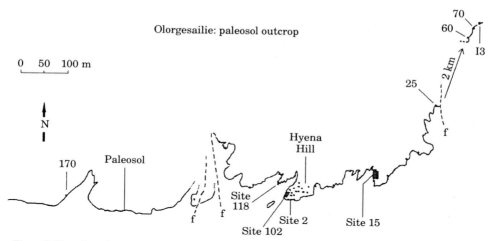

Figure 2. Plan view of the paleosol outcrop in upper Member 1, Olorgesailie Formation. Two faults provide a minor and a major interruption of the lateral extent of the layer 4 paleosol. A sample of excavation locations ("sites") up through 1992 are shown; these excavations sample the erosional transect through the ancient lake margin.

have been able to sample this paleosol by excavation over a total outcrop distance of about 5 km.

A plan view of the paleosol exposure is shown in Figure 2. Excavations have been conducted according to sampling schemes that have varied over the past eight years. At first excavations were distributed randomly along the outcrop. Two areas, termed Hyena Hill and Site 15, were exceptions. Because of the richness and type of finds, large excavations were made in these two places. Over the past year as our sampling has reached its final phase, we have excavated in areas where gaps existed in the lateral spacing between trenches, thus creating a more regular pattern of sampling. The I3 site, excavated by Isaac (1977) in the same paleocatena, is at the easternmost extent of the exposure.

A 3-dimensional coordinate grid throughout the Olorgesailie basin was set up using a Topcon EDM (Potts *et al.*, in press). Any artefact or locus in the region can thereby be tied into the same coordinate system with a precision of 2–5 mm, based on replicate measures. The advantage of this system is that the paleolandscape represented by a given stratigraphic interval, rather than each individual trench, can be conceived as the unit of analysis.

The basic sedimentary sequence is as follows (Potts, 1989a). The target layer, rich in fossilized bone and stone artefacts, consists primarily of fine silt-sized grains, clay-sized particles in cutans of a weakly-modified soil, and fine sand. This brown-colored stratum, termed layer 4, is underlain by white laminated to massive diatomaceous siltstone (layer 5), the top of which variably grades upward into the rootmarked zone or forms a sharper erosional contact with layer 4. The target paleosol (10–45 cm thick) is overlain by a complex sequence of interbedded yellow and grey siltstones and sandy siltstones, some of which are shallow channels fills. This unit is capped by a grey reworked tuff, except to the east where white diatomaceous siltstone overlies the target stratum and contains a series of discontinuous airfall ash horizons. Details of the upper Member 1 sediments will be published based on field observations and laboratory studies by A. K. Behrensmeyer and T. F. Jorstad.

The paleolandscape refers to strata encompassing an interval of time during which modification of sediments by vegetation and other processes indicative of a stable land surface took place. It encompasses any associated lateral variations in erosional or depositional episodes. The paleolandscape of upper Member 1 integrates the time and events from the initial exposure of the diatomite lake flats (layer 5), which in many places provided the parent material modified in layer 4, and extends through the development of a stable vegetated surface over most of the lateral exposure of layer 4 until its final burial.

In many places this period of time was uneventful beyond the establishment of vegetation on freshly exposed lake flats and the incorporation of faunal remains and artefacts into the modified zone. In other places, however, the original lake flats were eroded and other sediment, particularly tuffaceous material, was deposited prior to soil formation. In the lowest topographic zone (near site 15, see below) several strata were deposited, and a vegetation mat was probably established on at least two occasions, during the period of soil formation equivalent to a single layer 4 elsewhere (e.g., site 2). As noted above, Isaac's I3 site occurred within sediments laterally equivalent to layer 4. Soil modification at this locus is very weak, however, and suggests only minor development of vegetation in proximity to a lava peninsula that formed a locally high topographic area.

Paleolandscape analysis recognizes from the outset that no paleosol, especially no package of multiple strata, represents an "isochronous surface" (see Stern, 1994). Hence, the distribution of *in situ* materials in the target paleosol at Olorgesailie reflects both constant factors and specific episodes over an extended period of time. Although episodic traces may inform about specific events of intense interest (e.g., the death of hyenas at site 102, see below), the paleolandscape also registers the spatial dispersion of archeological remains and environmental features over an interval greatly exceeding the duration of such events.

Narrowing of the stratigraphic interval and prudent choice of depositional environment (e.g., paleosol *vs.* fluvial cut-and-fill deposits), nonetheless, may greatly reduce the amount of time averaging represented by the study. In the case of upper Member 1 Olorgesailie, the paleolandscape integrates processes over a period of 10^2–10^3 yrs (on the basis of argon age brackets and extremely weak clay development). In a situation such as this where taphonomic re-distribution of materials appears to have been minimal, study of *in situ*, time-integrated information may help to determine factors that affected hominid behavior over periods perhaps more pertinent to evolution. Although archeology has made an ideal of Pompeii and the search for other "instants in time", paleolandscapes contain significant information of a different sort (see also Blumenschine & Masao, 1991). Mapping of behavioral and other biotic traces in relation to environmental indications—both stable and episodic—is the proximate goal of paleolandscape excavation.

Detailed sedimentary and taphonomic studies of all excavations in the Member 1 paleosol are under way. On a broader spatial scale, the sources of rock utilized by Olorgesailie toolmakers are the subject of geochemical analysis and field mapping by W. G. Melson. Although the definitive results of this research will be published later, the basic types of lava rock used in toolmaking have been identified in the laboratory and field. A small, representative sample of artefacts from the paleosol have been classified as to rock type; these include the handaxes from site I3 (identified by Melson and the author) and all stone artefacts from sites 70, 60, 25, 15, 2, 118, and 170 (identified by the author). These sites span the east to west range of excavations across the paleosol outcrop (Figure 2).

The lithic sources crop out on the slopes and foothills of Mt Olorgesailie, south of the exposures. Variation in rock type is distributed on an east-west axis, which parallels the eroded lateral extent of the target paleosol. The major rock types employed in this study are:

1. Tabular Trachyte: The previously unrecognized source of the majority of stone handaxes manufactured at Olorgesailie. This important lava source, found by W. G. Melson, is aligned with the easternmost exposure of Member 1 sediments on the Legemunge Plain.

2. Lava Tongue: An outcrop, situated near site 118, from which a distinctive trachyte and basalt were available.

3. Basanite: A spatially constrained lava flow visible slightly further west of the Lava Tongue and higher up on Mt Olorgesailie.

4. Lava Ridge: A prominent horst partly consisting of distinctive basalt, immediately west of the westernmost excavation in the Member 1 paleosol.

Identification of these stone sources allows an assessment of variables such as transport distance and resource tethering.

For the purposes of this study all craniodental faunal remains were identified to test for possible variation in species representation over the lateral extent of the target outcrop; and other faunal remains were examined for evidence of carnivore and hominid modification.

Comparisons with other localities, particularly Bed I Olduvai (Hay, 1976; Leakey, 1971; Potts, 1988), play a crucial role in illustrating the diversity in Plio-Pleistocene settings and the variability in hominid behaviors.

Results

Correspondence of artefact and bone densities

The dominant mechanisms of preservation in the target paleosol involved (1) incorporation of stones and bones into layer 4 during soil formation at original loci of deposition (e.g., site 2); (2) incorporation of more complete remains related to biotic activity (e.g., hyena burrowing, site 102); and (3) burial of materials by a combination of downward movement in wet substrates and low-energy introduction of sediment in localized depressions (e.g., artefacts and fossil elephant skeleton at site 15).

Densities of *in situ* stone artefacts and faunal remains are displayed in Figure 3, measured per unit volume of sediment. It is evident that toolmakers deposited stone artefacts throughout the lake margin zone and in certain places contributed to dense accumulations. Peak concentrations of bones and stone artefacts coincide at site 2 (in the area called Hyena Hill), site 15, and I3. I3 is unusual for its number of handaxes (*n*=37, 18% of the flaked pieces); only two complete handaxes derive from all other excavations in the paleosol. At site 102 a dense concentration of bones occurred consisting of four complete skeletons of hyena (*Crocuta crocuta*) preserved in a burrow system. The burrows emanate from the top of layer 4 and descend into the underlying diatomite. Very few stone artefacts were associated with the burrow system, and apparently were introduced with layer 4 sediments and paleosol chunks that filled the burrows. As reported previously, at site 15 a skeleton of *Elephas recki* was found intimately and virtually exclusively associated with about 400 stone artefacts, suggestive of an episode in which toolmakers were attracted to a single carcass (Potts, 1989*a,b*).

Other observations are organized according to behavioral ecological variables that may have influenced toolmakers.

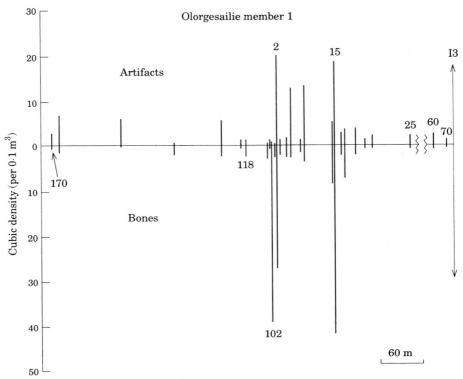

Figure 3. Concentration of stone artefacts (top) and fossilized bones (bottom), as measured per one-tenth cubic meter of sediment. Specific sites mentioned in the text are labelled. Three sites (2, 15, I3) manifested especially dense clusters of both artefacts and bones, whereas site 102 preserved a dense concentration of bones only.

Resource tethering

Figures 4–6 depict the relative east-to-west location of rock types compared against the percentage weight or number of specimens in a sample of excavations along the paleosol. The dataset from I3 consists only of handaxes. By both weight and number of specimens, tabular trachyte drops off very quickly and evenly, and is essentially unknown west of Hyena Hill (Figure 4). Artefacts made out of rock from the Lava Tongue depict an even more dramatic dropoff away from the source (Figure 5). These data indicate that tabular trachyte handaxes, their flake products, and stone from the Lava Tongue were minimally transported. Discard of these rock types by hominids was restricted to the immediate vicinity of the sources.

Yet other rock types (Figure 6: Basanite and Lava Ridge Basalt) show no clear spatial relationship between the fixed location of the source and the discard of lithic material over the paleolandscape. This suggests that the toolmakers were less strictly tethered in the use of these rock types.

The great concentration of handaxes in the eastern sector (in the upper Member 1 paleosol, and also in Member 7, ca. 0·78 Ma) raises the question of whether tethering to raw material source may have governed handaxe manufacture and deposition. The prevalence of tabular trachyte bifaces at I3 hints that this factor might have played a role. Given that the I3 data

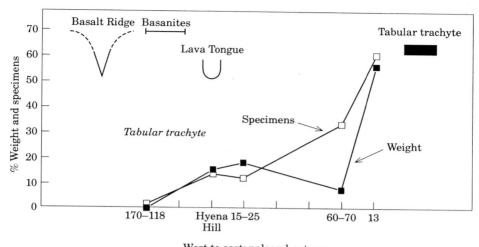

Figure 4. Percentage representation (weight and number of specimens) of tabular trachyte rock in a sample of artefact assemblages stretching from west to east across the paleosol. The relative east-west location of each rock source is depicted at top. Excavated stone artefacts are grouped as follows (from west to east): sites 170 and 118; Hyena Hill excavations, including site 2; sites 15 and 25; sites 60 and 70; and site I3.

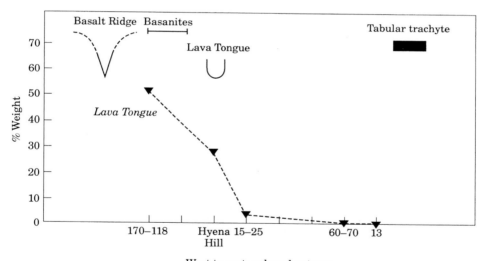

Figure 5. Percentage representation (weight) of rocks from the Lava Tongue in a sample of artefact assemblages stretching from west to east across the paleosol. See Figure 4 caption.

include handaxes only, however, it is clear that the biased spatial distribution of this artefact type also involved rock sources to the west. It appears, therefore, that tethering to something other than raw material source caused hominids to drop handaxes mainly in the eastern part of the basin, at least during upper Member 1 times. That different types of raw material and different components of the artefact record (handaxes *vs.* non-handaxes) might yield different measures of resource tethering was unexpected.

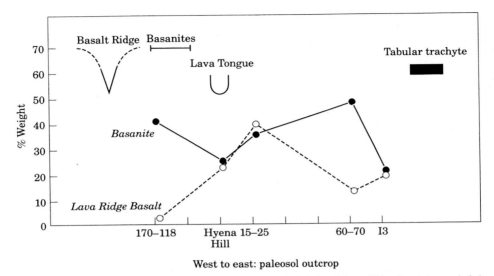

Figure 6. Percentage representation (weight) of Basanite (solid circles) and Lava Ridge Basalt (open circles) in the same sample of artefact assemblages portrayed in Figures 4 and 5.

Resource transport

Cumulative transport distances of rare stone materials, such as obsidian and quartzite, were 16–45 km. Yet 99% of the artefacts from Olorgesailie were manufactured from local lavas. The prevailing transport distances of these materials were ≤1–2 km. As noted above, these distances apparently varied depending on the type and/or location of the lava source.

It is worth noting that manuports are rare in the upper Member 1 paleosol (range: 0–4% of stones), and very rarely are they of rock types that match the raw materials of the flakes and chipped stones found within the same excavations.

This situation contrasts greatly with that in Bed I Olduvai. Carried but unmodified stones were recorded in every excavation; mean representation was 12% (range: 1·4–32%) of the specimens in the Bed I lithic assemblages. Manuport representation by weight was even greater. Modified pieces of a particular raw material, furthermore, were almost always associated with manuports of the same rock type. This spatial association holds especially well for lava rocks (Potts, 1988). Even though very few quartzite manuports are known in Bed I, the site (FLK "Zinj") that preserves the greatest quantity of chipped quartzite also had abundant manuports of this material. The comparison indicates that the transport of stone across the paleolandscape at Olorgesailie was curiously different from that of Bed I Olduvai.

Habitat heterogeneity

There are several methods by which to assess spatial heterogeneity. Variations in substrate (wet-dry; pliable-indurated) can be discerned by evidence of footprints and of bones turned vertically by trampling. Both kinds of evidence are abundant in the vicinity of site 15 and decrease in frequency to the east and west. Variations in original topography, moreover, can be calculated by correcting for present displacements and dips due to faulting, according to a method developed by A. K. Behrensmeyer (Potts & Behrensmeyer, in preparation). Hyena Hill

Table 2 **Taxonomic representation of large mammals in three areas of the paleolandscape of upper Member 1 Olorgesailie–Hyena Hill, Site 15, and I3**

Family	NISP range (%)[1]	MNI range (%)[1]	Species	Unique to Area[2]
Equidae	46–50	29–38	*Equus oldowayensis*	
			E. grevyi	
			Hipparion sp.	
Rhinocerotidae	3–15	6–7	*Ceratotherium simum*	
Bovidae	16–29	21–23	*Megalotragus* sp.	
			Connochaetes sp.	
			Alcelaphus sp.	
			Pelorovis sp.	
			Redunca sp.	I3
			Taurotragus oryx	
			Tragelaphini, medium	
			Hippotragini, medium	
Giraffidae	0–1	0–1	*Giraffa* sp.	I3
Suidae	3–7	4–7	*Metridiochoerus* cf. *andrewsi*	
			Phacocheorus sp.	
Hippopotamidae	7–25	8–14	*Hippopotamus gorgops*	
			Hippopotamus amphibius	
Cercopithecidae	1–6	4–7	*Theropithecus oswaldi*	
			Cercopithecus sp.	HH
Hyaenidae	0–5	0–12	*Crocuta crocuta*	HH
Viverridae	0–2	0–7	*Mungos* sp.	HH
			Herpestes ichneumon	Site 15
Elephantidae	0–99	0–7	*Elephas recki*	

Sample sizes are as follows: Hyena Hill: 17 taxa, 117+ Specimens, 48 MNI. Site 15: 10 taxa, 35+ Specimens, 14 MNI. I3: 12 taxa, 145+ Specimens, 13 MNI.

[1]Range of family-level representation (relative to all large mammals) over the areas of the paleosol, based on percentage of specimens (NISP) and minimum number of individuals (MNI).

[2]Species unique to a particular area of the paleolandscape are identified by either HH, Site 15, or I3 after the name. All other taxa are found in all three segments.

and site 15 were topographically distinct. The elephant site occurred in a localized low, from which there was a gradual slope to a local high at site 2. Thick rootmarks (3–6 mm diameter) typical of reeds or other marsh-type plants were abundant at site 15, while a fine network of grass-type rootlets left traces in the sediments at site 2. Research in progress on the stable isotopic composition of layer 4 organic matter and carbonates aims to further distinguish variations in vegetation across the paleolandscape (Sikes, in preparation).

Based on the lateral distribution of large mammals, there is little to attest to habitat heterogeneity. From Hyena Hill to site 15 to I3 (about 3 km away from Hyena Hill), a very consistent set of grazing species is preserved (Table 2). One consistent feature of the fauna across the paleolandscape is the oddly high representation of equids relative to bovids. This peculiarity is unknown in Olduvai Beds I and II; in the Koobi Fora Formation; in the later middle Pleistocene fauna from Lainyamok, just 42 km southwest of Olorgesailie (Potts *et al.*, 1988); or in present savanna habitats of East Africa. Trees are suggested by the presence of giraffe and reduncine bovids at I3. Yet these are very minor components in an otherwise grazing fauna, probably a very specialized array of grazing animals. Peaks in the distribution of hominid artefacts crosscut these minor faunal differences and changes in substrate and local topography.

Predation risk
Site 102 makes it clear that large, bone-seeking carnivores were present on the paleolandscape. A hyena occupation was situated close to a place where hominids deposited an unusually dense concentration of artefacts and fauna (site 2). Beyond this suggestion of overlap in habitat, however, there is no indication of temporal coincidence and minimal evidence of predation risk to the toolmakers in Member 1 Olorgesailie, especially in comparison with Bed I Olduvai. Taphonomic differences exist between the assemblages from Olorgesailie and Olduvai, but at Olorgesailie there is yet to be observed a single gnaw mark in places where both tools and bones were accumulated. At site 2, for instance, stone percussion and possible cut marks are present, but there is no sign of carnivore attraction to this place of bone concentration. This differs strikingly from Bed I Olduvai, where abundant evidence exists of extensive overlap in hominid and carnivore use of space and carcasses (Potts, 1988, 1989*a*; Bunn, 1981).

Spatial focus of social activity
This is perhaps the most difficult variable to assess. In the spatial arrangement of excavated materials there is nothing to suggest the occurrence of hearths, shelters, or traces of distinctive social units within the area of transported resources. If they had occurred, there is reason to expect the preservation of such traces, since delicate features such as footprint trails and the hyena burrow entrance are preserved at the top surface of layer 4. Spatial artefacts of social cohesion, however, undoubtedly would be altered, possibly due to repeated visits by hominids themselves. Our excavations, particularly at Hyena Hill, continue to search diligently for this kind of spatial evidence.

Discussion

No single behavioral model applies
If our main objective were to find the correct model of hominid land use, a variety of conflicting positions could be advocated.

It might be stressed, for instance, that despite spatial clustering of archeological materials in parts of the Olorgesailie paleolandscape, evidence is lacking for any distinctive point of social aggregation. Hence a magnet of social cohesion (e.g., food sharing at a campsite) evident in the home bases of modern foragers was absent.

On the other hand, predation potential—essential to the argument against home bases in Bed I Olduvai—appears to have been negligible in Member 1 Olorgesailie. I have urged that overlap with large and small carnivores and attendant predation risks held important influence on the ranging, transport, and social behaviors of toolmakers in Bed I Olduvai (Potts, 1988). But such risks (measured by spatial overlap between carnivore and hominid modification of bones) may have been substantially less at Olorgesailie. The same may be true in the lower Okote Member of East Turkana, though carnivore activity is apparent at sites where bone preservation is adequate to observe surface modifications (Bunn *et al.*, 1980; Bunn, 1982). If we paid attention only to those situations where carnivore activity overlapped minimally with hominid activity, such as at Olorgesailie, we might advocate that foragers did converge daily upon a safe social refuge.

We might also insist that the stone cache hypothesis is wrong, because evidence of secondary stone sources and predation risk measure so weakly at Olorgesailie. Manuports were rare and unrelated to toolmaking; and low predation risk meant that there was no advantage in keeping intense aspects of social behavior and grouping away from the stone-bone transport sites.

Alternatively, we could conclude that the tethering aspect of behavior in the routed foraging model did exist at Olorgesailie, but only for certain kinds of stone materials and not others. Or we could focus on the fact that no evidence exists of riparian woodland faunas, and that habitat variation does not appear to explain the biased distribution of stone artefacts.

However, from a behavioral ecological standpoint, none of the key models currently discussed in the literature is necessarily right or wrong. Each one defines one or more important variables linking behavior and environmental conditions. The fact that these conditions varied from basin to basin, and through time, suggests that *behavioral variation* and its relationship with environmental variation is what we are attempting to define rather than a test of mutually exclusive reconstructions.

Contextual variation between basins

Certain behaviors of Plio-Pleistocene toolmakers were rather rigid, prone to replication over hundreds of thousands of years in regionally diverse settings. The practices of Oldowan stone-on-stone percussion—even allowing a distinction between older (>2 Ma) and younger manifestations (Kibunjia, 1994)—have been shown to fit this description well (Leakey, 1971; Toth, 1985; Toth & Schick, 1986; Potts, 1991).

Transport of stones and animal parts, on the other hand, appears to have varied between basins according to paleogeographic context and resource distribution. Bed I Olduvai and the lower Okote Member of the Turkana-Omo system presented rather different settings in which to acquire stone raw material (compare Hay, 1976; Harris, 1978; Toth, 1982; Brown & Feibel, 1991). Hominids consistently transported manuports at Olduvai but not at East Turkana. The toolmakers at East Turkana were evidently constrained by the availability of suitable stones along channels leading to the central axis of the basin. For some reason Olduvai toolmakers carried unmodified rocks and chipped stones often several kilometers away from sources. In the different situation at Turkana, core size decreases toward the basin axis (Toth, 1985). The presence of cutmarked bone near this central axis where stone tools are very rare (Bunn, 1994) illustrates the point, that in different contexts—even within the same sedimentary basin—the archeological record may reveal variation in hominid transport, placement, and discard of resources.

It is pertinent to observe that sites in Beds I and II Olduvai may be unique in several ways. For instance, Olduvai exhibits stratigraphically stacked series of archeological remains (e.g., FLK North-I, MNK Main-II), which seem to transcend shifts in fauna, climate, and landscape (for FLK North-I see, e.g., Bonnefille & Riollet, 1980; Butler & Greenwood, 1976; Jaeger, 1976). Leakey (1971) was able to find large numbers of artefacts and associated fossil bones by digging downward (even in small areas) through many stratigraphic layers (see Figure 7). This approach would not work at Olorgesailie nor apparently at East Turkana. Different strata produce abundant archeological remains mainly in spatially distinct loci; digging laterally from high-density sites may raise the chances of finding good samples while digging downward does not. Finally, Olduvai may be unique, so far, in its comparatively high measures on variables that may reflect repeated use of secondary sources of transported rock unconfined to a central place (see also Blumenschine & Masao, 1991).

Member 1 Olorgesailie, later in time, presents still a different setting. There is some evidence to show that handaxes sometimes served as sources of sharp flakes, and may have rarely been dropped during foraging because of their utility as a portable stone source (Potts,

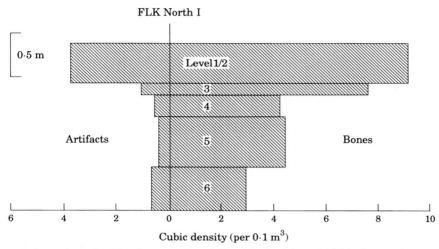

Figure 7. Approximate densities of stone artefacts (left) and fossilized bones (right) in five superjacent strata at FLK North, Bed I Olduvai. Densities are given as number of specimens per one-tenth cubic meter of sediment (estimated). Unlike Olorgesailie, consecutive strata in Beds I and II Olduvai yield assemblages of stone tools and animal bones. Except for level 1/2, it is unlikely that the strata at FLK North I yielded remains that greatly exceed probable background densities. Comparison with Figure 3, for instance, shows that even FLK North level 1/2 had artefact-bone densities similar to those in the background scatter across the paleolandscape of Member 1 Olorgesailie. However, the generally more complete preservation of animal bones at Olduvai suggests that sedimentation and burial rates were much higher at Olduvai, which would yield lower densities of remains. A more accurate picture of contemporaneous background densities in Bed I Olduvai may be derived in the strata above the "Zinj" horizon at FLK-I. This locality is less than 100 meters away from FLK North, and certain levels above the "Zinj" horizon are stratigraphically equivalent with the artefact-bone levels at FLK North. Above the "Zinj" horizon, only nine of 19 strata below Tuff IF yielded any stone artefacts or animal bones. In these strata, including those equivalent with FLK North levels 1–6, bone densities range from 0–0·49 per 0·1 m^3, while stones range from 0–0·03 per 0·1 m^3. The stones and bones at FLK North thus represent significantly higher concentrations than in the upper levels at FLK-I (as do the remains in the "Zinj" horizon [conservatively, 32 artefacts per 0·1 m^3; and >120 bones specimens per 0·1 m^3]). Stratigraphically-stacked occurrences of dense artefact-bone collections at Olduvai (at FLK North and other sites, but not at FLK-I) is peculiar to Olduvai among published Plio-Pleistocene localities.

1989a). Shifts in technology and curation, then, may account for some of the apparent differences between Olorgesailie, Bed I Olduvai, and the lower Okote Member. Still, the particular character of each sedimentary basin is an important determinant of these differences. Even the location of exposed outcrop relative to lithic sources is crucial to assess. The available outcrop at Olorgesailie, for instance, closely parallels the foothills of the mountain, and thus has a unique relationship with lithic source outcrops among the basins I have considered. This may imply that some of the variation between basins in stone-faunal transport and assemblage composition may simply reflect erosional sampling of non-analogous portions of hominid foraging ranges.

   Contextual variation between basins, therefore, may arise from differences in physical features (e.g., mountains, inselbergs), paleoenvironmental history (e.g., fluctuation in lakes, rivers, and vegetation), and erosion. A behavioral ecological approach to hominid land use encourages archeologists to seek out any behavioral variations that may coincide with such contextual differences regardless of how the latter have been caused.

Behavioral evolution

It might be tempting to assign the archeological differences between Bed I Olduvai (1·8–1·7 Ma), the lower Okote Member at Turkana (1·6–1·5 Ma), and Member 1 Olorgesailie (0·99 Ma) to a progressive, evolutionary time series. Indeed archeologists have long sought to define "evolutionary" sequences of artefacts and technology based on the assumption of progressive change. But what about the evolution of hominid behavior and consequent effects on land use? Evidence from Olorgesailie indicates, for instance, that carnivore activity was minimized at sites where hominids discarded stone artefacts and processed animal bones. We might assume that this signals an adaptive improvement related to changes in hominid body size and behavior over time. It is also apparent that, at some point in hominid evolution, strong degrees of kin-based social bonding eventually coincided with predetermined central places of food transport and processing (i.e., the development of home bases). Can we assume, though, that sites later in time necessarily show progressive changes in these directions?

I think the answer must be no; the evolution of behavior cannot adequately be treated by such a simplistic assumption. Behavioral evolution arises from behavioral variation, and in the perspective sought here such variation arises in relation to specific environmental contexts. In other words, behavioral variations comprised the raw material on which natural selection acted to produce behavioral evolution.

Differentiating between specific responses by hominids to particular basinal contexts, on the one hand, and major evolutionary changes, on the other, will eventually require analysis of larger samples of contexts and sites. Differences between Bed I Olduvai and Member 1 Olorgesailie might reflect permanent shifts in hominid behavior. But this can only be told as data become available from other localities permitting comparisons that control for time, technology, and environmental context. A focus on variables rather than static reconstructions, again, better serves the purposes of such comparisons.

Habitat variation over time

At Olorgesailie climates and landscapes varied greatly through time. Based on fossil diatoms and lithologies, a curve of lake levels between ca. 1·0 and 0·5 Ma indicates that large-scale fluctuations occurred (Figure 8). Basinwide paleosols (P), carbonates/calcretes (C), and gravels (G) further show that even this curve of habitat alteration is too simplistic. Such repeated alterations call into question the search for *the* correct ecological analogue or any attempt to construe the environment of hominid foraging as a stable, unvarying state. At Olorgesailie, Olduvai, or any other sedimentary basin that offers a long sequence of environments, we see evidence of significant shifts over time in climates and related landscape features (e.g., Hay, 1976; Bonnefille & Riollet, 1980; Feibel *et al.*, 1991). It therefore seems incorrect that the *evolution* of hominid behavior would be rendered by a single behavioral reconstruction—advocated on the basis of one sedimentary basin, some-times even a single excavated site—or by striving to find the one best model of hominid foraging opportunity.

On the basis of the nascent comparisons presented here, archeological studies of hominid behavior need to specify the local and regional conditions (specific to time and place) to which a particular reconstruction of hominid behavior applies. Much more will be learned as studies compare paleoenvironmental variables from diverse contexts and map out how hominid activities co-varied.

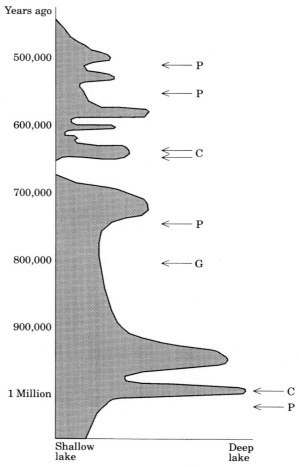

Figure 8. A curve of relative lake level in the Olorgesailie Formation, based on diatom and lithofacies data from Owen & Renaut (1981) displayed on a time column from Deino & Potts (1990). Deep lake levels (e.g., ca. 985 ka) refer to periods when lake diatomites filled the entire basin, delimited evidently by the surrounding volcanic highlands. The approximate stratigraphic positions of basinwide paleosols (P), carbonates and calcretes (C), and gravels (G) are also shown. These represent intervals when lake sediments were not deposited within the area of available sediment exposure.

Conclusion

The main points of this paper are as follows:

1. Behavioral models currently under consideration tend to be treated as rival hypotheses about hominid adaptation. As such, they become static reconstructions and miss the contextual variability in toolmaker behavior (of one or more species). These models, however, can be dissected into a number of salient behavioral ecological variables, which acted simultaneously on hominid toolmakers. Understanding land use may be enhanced by considering spatially diverse components in a single system of artefact production. In Member 1 Olorgesailie, different assemblage components registered different measures of stone resource tethering and transport behavior. This illustrates well the multivariate complexity

operating within a spatially- and temporally-constrained setting. Moreover, evaluation of the entire array of magnets (variables) that attracted hominids to distinct portions of Pliocene and Pleistocene landscapes requires combining analyses of fauna and artefacts.

2. Paleoanthropologists should expect to see geographic and temporal variation in hominid resource use and ranging patterns. What holds in one sedimentary basin, or even one segment of a basin, does not necessarily apply in another. New excavations and reanalyses of old sites are usually justified as ways to assess and choose between particular reconstructions of Plio-Pleistocene hominid behavior. Such studies may take on a new challenge, however, which is to address the variability in archeological traces and hominid land use in different spatial and temporal contexts. This is a critical enterprise in early hominid archeology.

3. Fluctuations in habitats and resources may have driven natural selection. This idea leads us to focus on *heterogeneity* in selective conditions, rather than to look for the single best environmental analogue or behavioral model. By adopting ecological heterogeneity as a critical factor in the origin of hominid behaviors, we may significantly revise the way natural selection is construed to have operated—not merely as *selection pressure* or as *adaptation to* a model environment, but as a response to habitat and resource variability from place to place and over time. The implications regarding natural selection and the processes of faunal, including hominid, evolution during the Pleistocene are profound.

Acknowledgements

Permission and support by the National Museums of Kenya, particularly Dr M. G. Leakey and Dr M. Isahakia, are gratefully acknowledged. The Olorgesailie research was funded by the Smithsonian's Human Origins Program, Scholarly Studies Program, and the National Museum of Natural History. I thank the members of our research team, especially J. M. Nume, A. K. Behrensmeyer, W. G. Melson, T. J. Jorstad, and A. Deino; and J. B. Clark for her assistance with the manuscript, including preparation of figures. The manuscript benefitted from comments by C. Monahan, T. W. Plummer, N. Sikes, and anonymous reviewers. This is a publication of the Smithsonian's Human Origins Program.

References

Binford, L. R. (1981). *Bones: Ancient Men and Modern Myths*. New York: Academic Press.
Binford, L. R. (1984). *Faunal Remains from Klasies River Mouth*. Orlando: Academic Press.
Blumenschine, R. J. (1986). *Early Hominid Scavenging Opportunities*, British Archaeological Reports International Series 283.
Blumenschine, R. J. (1987). Characteristics of an early hominid scavenging niche. *Curr. Anthrop.* **28**, 383–407.
Blumenschine, R. J. (1991). Hominid carnivory and foraging strategies, and the socio-economic function of early archaeological sites. *Phil. Trans. R. Soc. London* B334, 211–222.
Blumenschine, R. J. & Masao, F. T. (1991). Living sites at Olduvai Gorge, Tanzania? Preliminary landscape archaeology results in the basal Bed II lake margin zone. *J. hum. Evol.* **21,** 451–462.
Bonnefille, R. & Riollet, G. (1980). Palynologie, végétation et climats de Bed I et Bed II à Olduvai, Tanzanie. *Proc. Eighth PanAfr. Congr. Prehist. Quat. Stud.*, pp. 123–127. Nairobi.
Brown, F. H. & Feibel, C. S. (1991). Stratigraphy, depositional environments and palaeogeography of the Koobi Fora Formation. In (J. M. Harris, Ed.) *Koobi Fora Research Project*, vol. 3, pp. 1–30. Oxford: Clarendon.
Bunn, H. T. (1981). Archaeological evidence for meat-eating by Plio-Pleistocene hominids from Koobi Fora and Olduvai Gorge. *Nature* **291,** 574–577.
Bunn, H. T. (1982). Meat-eating and human evolution. Ph.D. Dissertation, University of California, Berkeley.
Bunn, H. T. (1994). Early Pleistocene hominid foraging strategies along the ancestral Omo River at Koobi Fora, Kenya. *J. hum. Evol.* **27**, 247–266.
Bunn, H. T., Harris, J. W. K., Isaac, G., Kaufulu, Z., Kroll, E., Schick, K., Toth, N. & Behrensmeyer, A. K. (1980). FxJj 50: An early Pleistocene site in northern Kenya. *World Archaeol.* **12,** 109–136.

Butler, P. M. & Greenwood, M. (1976). Elephant-shrews (Macroscelididae) from Olduvai and Makapansgat. In (R. Savage & S. Coryndon, Eds) *Fossil Vertebrates of Africa*, vol. 4, pp. 1–56. London: Academic Press.

Cavallo, J. A. & Blumenschine, R. J. (1989). Tree-stored leopard kills: expanding the hominid scavenging niche. *J. hum. Evol.* **18**, 393–399.

Deino, A. & Potts, R. (1990). Single crystal ^{40}Ar/^{39}Ar dating of the Olorgesailie Formation, southern Kenya rift. *J. Geophys. Res.* **95**, B6, 8453–8470.

Deino, A. & Potts, R. (1992). Age-probability spectra for examination of single-crystal ^{40}Ar/^{39}Ar dating results: Examples from Olorgesailie, southern Kenya rift. *Quat. Int.* **7/8**, 81–89.

Feibel, C. S., Harris, J. M. & Brown, F. H. (1991). Palaeoenvironmental context for the late Neogene of the Turkana Basin. In (J. M. Harris, Ed.) *Koobi Fora Research Project*, vol. 3, pp. 321–346. Oxford: Clarendon.

Hamilton, W. J. & Watt, K. E. F. (1970). Refuging. *Ann. Rev. Ecol. Systematics* **1**, 263–297.

Hay, R. L. (1976). *Geology of the Olduvai Gorge*. Berkeley: University of California Press.

Harris, J. W. K. (1978). The Karari Industry. Ph.D. Dissertation, University of California, Berkeley.

Isaac, G. L. (1977). *Olorgesailie*. Chicago: University of Chicago Press.

Isaac, G. L. (1978). The food-sharing behavior of protohuman hominids. *Sci. Am.* **238**, 90–108.

Isaac, G. L. (1981). Archaeological tests of alternative models of early hominid behavior: Excavation and experiments. *Phil. Trans. R. Soc. London* **B292**, 177–188.

Isaac, G. L. (1983). Bones in contention: Competing explanations for the juxtaposition of early Pleistocene artifacts and faunal remains. In (J. Clutton-Brock & C. Grigson, Eds) *Animals and Archaeology: 1. Hunters and Their Prey*, pp. 3–19. Oxford: British Archaeological Reports International Series 163.

Isaac, G. L. (1984). The archaeology of human origins: Studies of the Lower Pleistocene in East Africa 1971–1981. *Adv. World Archaeol.* **3**, 1–87.

Jaeger, J.-J. (1976). Les rongeurs (Mammalia, Rodentia) du Pleistocene inferieur d'Olduvai Bed I. In (R. Savage & S. Coryndon, Eds) *Fossil Vertebrates of Africa*, vol. 4, pp. 57–120. London: Academic Press.

Kibunjia, M. (1994). Pliocene archaeological occurrences in the Lake Turkana basin: A review of patterns and gaps in the record. *J. hum. Evol.* **27**, 159–171.

Krebs, J. R. & Davies, N. B., Eds (1991). *Behavioural Ecology*, Oxford: Blackwell.

Leakey, M. D. (1971). *Olduvai Gorge*. Vol. 3. London: Cambridge University Press.

Marean, C. W. (1989). Sabertooth cats and their relevance for early hominid diet and evolution. *J. hum. Evol.* **18**, 559–582.

Owen, R. & Renaut, R. (1981). Palaeoenvironments and sedimentology of the Middle Pleistocene Olorgesailie Formation, southern Kenya rift valley. *Palaeoecol. Afr.* **13**, 147–174.

Potts, R. (1988). *Early Hominid Activities at Olduvai*. New York: Aldine de Gruyter.

Potts, R. (1989a). Olorgesailie: New excavations and findings in Early and Middle Pleistocene contexts, southern Kenya rift valley. *J. hum. Evol.* **18**, 269–276.

Potts, R. (1989b). Ecological context and explanations of hominid evolution. *Ossa* **14**, 99–112.

Potts, R. (1991). Why the Oldowan? Plio-Pleistocene toolmaking and the transport of resources. *J. Anthrop. Res.* **47**, 153–176.

Potts, R., Shipman, P. & Ingall, E. (1988). Taphonomy, paleoecology and hominids of Lainyamok, Kenya. *J. hum. Evol.* **17**, 597–614.

Potts, R., Jorstad, T. & Cole, D. (in press). The role of GIS in interdisciplinary investigations at Olorgesailie, Kenya, a Pleistocene archeological locality. In (H. Maschner & M. Aldenderfer, Eds) *Anthropology through Geographic Information and Spatial Analysis*. New York: Oxford University Press.

Rubinstein, D. I. & Wrangham, R. W. Eds (1986). *Ecological Aspects of Social Evolution*. Princeton: Princeton University Press.

Schick, K. (1987). Modeling the formation of Early Stone Age artifact concentrations. *J. hum. Evol.* **16**, 789–807.

Smuts, B. B., Cheney, D. L. & Seyfarth, R. M. (1987). *Primate Societies*. Chicago: University of Chicago Press.

Stern, N. (1994). The implications of time-averaging for reconstructing the land-use patterns of early tool-using hominids. *J. hum. Evol.* **27**, 89–105.

Susman, R. (1991). Who made the Oldowan Tools? Fossil evidence for tool behavior in Plio-Pleistocene hominids. *J. Anthrop. Res.* **47**, 129–151.

Toth, N. (1982). The stone technologies of early hominids at Koobi Fora, Kenya. Ph.D. Dissertation, University of California, Berkeley.

Toth, N. (1985). The Oldowan reassessed: A close look at early stone artifacts. *J. Archaeol. Sci.* **12**, 101–120.

Toth, N. & Schick, K. (1986). The first million years: The archaeology of protohuman culture. *Adv. Archaeol. Method Theory* **9**, 1–96.

Wood, B. (1991). *Koobi Fora Research Project*, vol. 4. Oxford: Clarendon Press.

Nancy E. Sikes
Department of Anthropology, University of Illinois, 109 Davenport Hall, 607 S. Mathews Avenue, Urbana, Illinois 61801 U.S.A.

Received 10 September 1993
Revision received 11 March 1994 and accepted 12 March 1994

Keywords: Early hominids, habitat preferences, floral microhabitats, stable carbon isotopes, paleosols, Olduvai Gorge, landscape archaeology, Plio-Pleistocene.

Early hominid habitat preferences in East Africa: Paleosol carbon isotopic evidence*

Interpretations of the habitat preferences of sympatric *Homo* and *Australopithecus* have been based mainly on regional and/or macrohabitat reconstructions of the Plio-Pleistocene environment. Microhabitat reconstructions are biased by the preservation of many East African sites in disturbed fluvial contexts and by the rarity of organic remains that may provide site-specific evidence of habitat diversity. In order to determine early hominid foraging behavior and land-use patterns, however, their site-specific floral microhabitat use must be reconstructed across the Plio-Pleistocene landscape. Variability in the distribution and density of excavated archaeological traces over a paleolandscape may be related to hominid activities, resource use, and habitat preference. The stable isotopic composition of paleosol (buried soil) organic and pedogenic carbonate carbon can be used to estimate the original proportion of grasses (C_4) to woody (C_3) vegetation, and is a powerful tool for microhabitat reconstruction. In collaboration with a landscape archaeology project, isotopic values were determined on basal Bed II paleosol samples from trenches excavated at Olduvai Gorge (Tanzania). Interpreted with reference to modern East African soil and vegetation analogs, carbon isotopic ratios indicate a 1 km^2 area near HWK and FLK in the eastern paleolake Olduvai margin supported a riparian forest to grassy woodland ~ 1.74 Myr ago. Stone artifacts and hammerstone-fractured bones are abundant across the waxy claystone paleosol, which corresponds to level 1 at HWK-E. Isotopic evidence from this preliminary study (including FLK *Zinjanthropus*) suggests Plio-Pleistocene hominids in East Africa may have preferred relatively closed woodland habitats that may have offered food, shade, and predator and sleeping refuge.

Journal of Human Evolution (1994) **27**, 25–45

Introduction

"Maybe we left the forest a while ago but the trees only much more recently."
(Isaac, 1983*a*:525)

The fragmentation of the Miocene rain forest coeval with the inception and spread of open savanna grasslands has been considered the prime mover in hominid origins and evolution since Darwin's 1871 *Descent of Man* (Brain, 1981; Isaac, 1983*a*; Vrba, 1985; Hill & Ward, 1988). Yet available paleoenvironmental evidence from Plio-Pleistocene hominid fossil and archaeological localities in Africa, derived from sedimentary geological, geochemical, paleo-botanical, and paleozoological reconstructions (e.g., Hay, 1976; Behrensmeyer, 1975*a*; Bonnefille, 1984*a*; Wesselman, 1984; Dechamps & Maes, 1985; Cerling & Hay, 1986), portrays a diversity of vegetation communities similar to today's tropical savanna mosaic, from swamps to treeless or wooded grassland, woodland, gallery forest, and montane forest. Very few early hominid localities (i.e., Laetoli, Swartkrans) are reconstructed as open grasslands (Harris, 1987; Vrba, 1988). Within this diversity of floral microhabitats, each hominid species, like other mammals, presumably had a different diet, habitat preference, and land-use pattern. Hominid behavior would have depended on the availability, predictability and distribution of resources, such as fresh water, plant and animal foods, predator refuges, sleeping areas, or

---

*This paper was first presented at the symposium "Early Hominid Behavioral Ecology: New Looks at Old Questions", held 15 April, 1993 at the 62nd Annual Meeting of the American Association of Physical Anthropologists in Toronto.

0047–2484/94/010025+21 $08.00/0

Table 1   **Hypotheses of contemporary Plio-Pleistocene hominid habitat, resource, or geographic preference (illustrative cases)**

Habitat, resource, or context	Homo	H.h.	H.e.	A.r./b.	A.b.	Generic
Montane forest				a		
Gallery forest/ riparian woodland		b			c	d
Closed habitat					e	
Closed/mesic			f		g	
Groves of trees	h		i			j
Dambo (wet grassland)						k
Open savanna		l	m			n
Open/xeric				o	p	
More open than A.r./b.	q					
Not open/arid specialists					r	
Both open/arid & closed/wet	s					
Riverine					t	
Stream channel margins						u
Lake margin	v					
"Hints" at lake margin	w					
Fresh water					x	y
Stone sources						z

Homo refers to *H. habilis* and early *H. erectus*, *H.h.* to *H. habilis*, *H.e.* to early *H. erectus*, *A.r./b.* to *Australopithecus robustus/boisei*, and *A.b.* to *A. boisei*. "Generic" includes *H. habilis* and *A. boisei* at a minimum.

References: a: Bonnefille, 1984*b*; b: Marean, 1989; c: Behrensmeyer, 1978*a*, 1982, 1985; d: Blumenschine, 1986, 1987, Cavallo & Blumenschine, 1989; e: Behrensmeyer & Cooke, 1985; f: Coppens, 1980; g: Shipman & Harris, 1988; h: Bunn, 1991; i: Clark, 1987; j: Bunn *et al.*, 1980, Isaac, 1981, 1983*a*, 1983*b*, 1984, Kroll & Isaac, 1984, Blumenschine, 1991; k: Jolly, 1970; l: Coppens, 1980, Bonnefille, 1984*b*; m: Marean, 1989; n: Foley, 1987; o: Boaz, 1977, Vrba, 1975, 1985, Grine, 1981; p: Coppens, 1980; q: Behrensmeyer, 1978*a*, 1982, 1985; r: Grine, 1988, Vrba, 1988; s: Vrba, 1985, Shipman & Harris, 1988; t: Behrensmeyer, 1975*b*, 1978*a*, 1982, 1985; u: Isaac, 1966, 1972, 1976, 1978*a*, 1978*b*, 1978*c*, Sept, 1984, 1986; v: Behrensmeyer, 1975*b*, 1978*a*; w: White, 1988; x: Vrba, 1988; y: Hay, 1976, Isaac, 1978*a*, Binford, 1984, Potts, 1989; z: Binford, 1984, Potts, 1984, 1988, 1991, Schick, 1987, 1991, Toth, 1987.

stone sources, within the diverse Plio-Pleistocene vegetation landscape. Consequently, it is within the *context of microhabitats* that the ecological and behavioral dynamics of hominid resource use occurred. Our knowledge, however, about the floral microhabitat context of hominid behavior during the Plio-Pleistocene is limited.

As widely recognized (e.g., Isaac, 1981, 1984; Behrensmeyer, 1982; Toth & Schick, 1986; White, 1988), taphonomic processes leave us with a biased picture of early hominid geographical and/or depositional contexts. The majority of Plio-Pleistocene hominid finds, for example, are surface collections, subject to differential exposure and recovery techniques. And in East Africa there is a strong bias toward fluvial deposition and preservation in low-lying, sedimentary basin contexts. Nevertheless, understanding hominid ecology and behavior is an important research focus within paleoanthropology, and the literature on habitat preferences of contemporary Plio-Pleistocene hominids continues to grow, as illustrated in Table 1. The examples in the table are based on a variety of methods (analogy, taphonomy, bovid habitat indices) and/or data (contextual, floral, faunal), and range from the general to the specific. Differences also reflect whether the hypothesis is derived from hominid fossil localities and/or archaeological sites where hominid species identity may be less precise. Table 1 shows there is no strong consensus regarding Plio-Pleistocene hominid habitat preferences in East Africa.

Although species identification may be more certain, the depositional context of isolated hominid fossils may not represent habitat preference during life (e.g., Vrba, 1988; White, 1988). In contrast, the residue of hominid activities (stones and/or bones) preserved *in situ* in primary context Plio-Pleistocene archaeological sites attests to a life association between hominids and the paleocommunity. Archaeological research during the Plio-Pleistocene in East Africa, however, has been concentrated on the excavation of large-scale sites (e.g., Leakey, 1971), few of which are contemporaneous. In order to reconstruct early hominid paleoecology and land-use patterns, archaeological survey and/or excavation needs to document isochronous evidence of hominid activity across paleolandscapes. This approach to the Early Stone Age, termed landscape archaeology, has only recently become a research focus in East Africa (e.g., Isaac *et al.*, 1980; Potts, 1989; Blumenschine & Masao, 1991; Stern, 1993). Variability in the distribution and density of excavated archaeological traces across a paleolandscape may be related to hominid resource use, and thus can be interpreted as indicative of hominid habitat preference.

Stable carbon isotopic analysis of paleosol organic matter and co-existing pedogenic carbonates is well established in paleoanthropology as an independent means of reconstructing hominoid and hominid paleoenvironments in East Africa (Cerling & Hay, 1986; Cerling *et al.*, 1988, 1991, 1992; Cerling, 1992; Kingston, 1992; Sikes, in prep.). This robust method of quantifying floral microhabitat diversity is particularly valuable where other classes of paleoenvironmental evidence may be lacking, or equivocal. Biotic remains, such as untransported fossil wood or microfauna, local pollen, or leaf impressions that may provide site-specific evidence of paleoenvironmental context, for example, are rare in the Plio-Pleistocene record. Soil organic matter and co-existing pedogenic carbonates are derived from the immediate vicinity of the sample site, and if unaltered, reflect the original surface plant biomass. As defined here, paleosols also have the taphonomic advantage of being buried in place. With the scale of resolution possible with this technique, the qualitative characterization of tropical habitats as either "open" or "closed" may be turned into a quantitative and precise assessment of the proportion of tropical grass (C_4 plants) to woody vegetation (C_3 plants) in a local floral microhabitat.

By adding the method of stable carbon isotopic analysis of paleosol organic matter and pedogenic carbonates from landscape archaeology excavations to our repertoire of paleoenvironmental reconstruction techniques, we can better understand Plio-Pleistocene hominid behavior and habitat preferences. This position is illustrated by the results of geochemical research on a basal Bed II paleosol conducted by the author in collaboration with a landscape archaeology project at Olduvai Gorge (Blumenschine & Masao, 1991). The results are interpreted against background data on the isotopic ecology of modern floral microhabitats in East Africa. Behavioral hypotheses regarding hominid habitat preferences and ratios of C_3 to C_4 (tree/grass) biomass are then formulated based on a review of current models of archaeological site formation and recent studies relating the abundance and distribution of modern plant and animal resources in sub Saharan Africa to analogous Plio-Pleistocene settings. Last, the behavioral hypotheses are tested with the floral microhabitat evidence from Olduvai and discussed in light of the Plio-Pleistocene isotopic and archaeological record from East Africa.

Isotopic background

Carbon isotopes in tropical savanna ecosystems

Major portions of the modern tropical savanna floral mosaic are comprised of mixtures of two groups of plants with different photosynthetic pathways, termed C_3 and C_4. C_3 plants fix a three-carbon molecule in the first stage of photosynthesis, and include nearly all trees, shrubs and herbs, a few sedges, *Typha* (rushes), and grasses adapted to cool growing seasons and shade (Smith & Epstein, 1971; Smith, 1972). The majority of C_4 plants are tropical grasses and sedges that are better adapted to water stress, high temperature, strong sunlight, and low atmospheric CO_2 concentrations (Smith & Epstein, 1971; Smith, 1972; Björkman & Berry, 1973; Ehleringer *et al.*, 1991). In East Africa, C_4 grasses are replaced by C_3 grasses in shaded areas such as closed canopy forests, and at high altitudes between 2000 to 3000 m (Tieszen *et al.*, 1979). A third group of plants (CAM or Crassulacean acid metabolism) uses both C_3 and C_4 photosynthetic pathways and includes cacti, euphorbias, and agaves. CAM plants, however, are rarely significant components of the floral biomass and are not discussed further.

C_3 and C_4 plants are clearly distinguished by the ratios of stable carbon isotopes ($^{13}C/^{12}C$) incorporated from the atmosphere (Smith & Epstein, 1971; Smith, 1972; Deines, 1980). (Isotopic ratios are referred to in the per mil (‰) notation where $\delta‰ = [R_{sample}/R_{PDB} - 1] \times 1000$. R is the $^{13}C/^{12}C$ or $^{18}O/^{16}O$ ratio, and PDB is the isotopic reference standard.) C_3 plants have $\delta^{13}C$ values that range between -35 and $-22‰$, averaging $-26‰$. The large variation in C_3 plant $\delta^{13}C$ values is primarily due to the extremes of local conditions, from closed canopy recycling of CO_2 (-35 to $-28‰$; Vogel, 1978) to ^{13}C enrichment when water stressed (-26 to $-22‰$; Ehleringer & Cooper, 1988). C_4 plants discriminate less against atmospheric CO_2, which results in $\delta^{13}C$ values between -16 and $-8‰$, averaging $-12‰$.

Soil organic carbon is derived from plant matter, and thus has a $\delta^{13}C$ value which closely reflects that of the standing plant biomass (Nadelhoffer & Fry, 1989; Martin *et al.*, 1990; Ambrose & Sikes, 1991). $\delta^{13}C$ values of soil organic matter may be slightly enriched ($+1–2‰$) relative to those of plants (Nadelhoffer & Fry, 1989), and range between -30 and $-23‰$ in C_3-dominated floral habitats and between -15 and $-10‰$ in C_4 floral habitats. Ecosystems with a mixture of C_3 and C_4 plants have intermediate $\delta^{13}C$ values.

The carbon isotopic composition of soil carbonates is also derived from the local plant biomass (Cerling, 1984; Cerling *et al.*, 1989). Isotopic effects during pedogenic carbonate formation result in a systematic enrichment of ^{13}C by 14 to 17‰ over the $\delta^{13}C$ value of the parent flora. Thus $\delta^{13}C$ values of pedogenic carbonates formed in 100% C_3 and 100% C_4 habitats range between -14 to $-9‰$ and $+1$ to $+4‰$, respectively. The net difference between co-existing organic matter and pedogenic carbonate $\delta^{13}C$ values also ranges from 14 to 17‰. This difference can be reliably used to indicate a lack of diagenesis in paleosols and thus suitability for paleoecologic reconstruction (Cerling *et al.*, 1989). The oxygen isotopic composition of soil carbonate is related to that of local meteoric water (Cerling, 1984). In general, cooler and/or moister regions usually have more negative $\delta^{18}O$ values than hot and dry areas (Cerling, 1992).

Scale of resolution

A very fine scale of floral microhabitat resolution is possible with soil carbon isotopic analysis. Accurate data on the mean isotopic composition of the parent flora is derived from the $\delta^{13}C$ values of the upper layer of modern soils (0–5 or 0–10 cm level below the surface) (Nadelhoffer

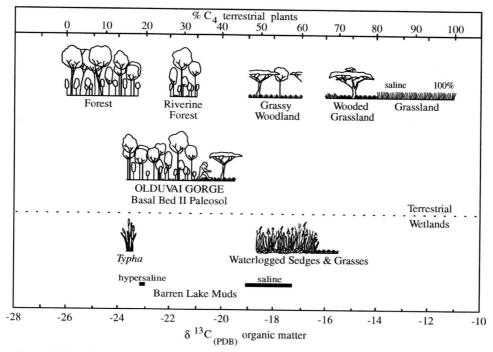

Figure 1. The stable carbon isotopic composition of modern soil and sediment organic matter from East African freshwater and alkaline ecosystems (Ambrose & Sikes, 1991; Sikes, in prep.) plotted by floral microhabitat, as defined in the text. The upper horizontal scale indicates the approximate equivalent percentage of terrestrial C_4 flora (Ambrose & Sikes, 1991). $\delta^{13}C$ values indicate the Olduvai Gorge basal Bed II paleosol supported a relatively closed riverine forest to grassy woodland habitat. This interpretation is supported by the pedogenic carbonate evidence discussed in the text. Vegetation graphics modified after Kingdon (1974) and Harris (1980).

& Fry, 1989; Martin *et al.*, 1990; Ambrose & Sikes, 1991; Sikes, in prep.). In habitats where the climatic regime has remained stable throughout the period of modern soil formation, soil organic and carbonate $\delta^{13}C$ values are relatively constant with depth (Cerling *et al.*, 1989; Martin *et al.*, 1990; Ambrose & Sikes, 1991; Sikes, in prep.). Dramatic shifts in soil organic $\delta^{13}C$ values with depth are detectable within as little as 15–20 cm and, in areas undisturbed by anthropogenic modification, result from current or past changes in the standing plant biomass in concert with climatic variation (Dzurec *et al.*, 1985; Schwartz *et al.*, 1986; Martin *et al.*, 1990; Ambrose & Sikes, 1991).

As illustrated in Figure 1, standard physiognomic types of tropical plant communities, including those with mixtures of C_3 and C_4 plants, can be differentiated according to the proportion of C_4 biomass present, as well as their soil organic $\delta^{13}C$ values. The endpoints for the $\delta^{13}C$ value of soil organic matter beneath 100% C_3 forest and 100% C_4 grassland in Kenya and Tanzania, are − 26‰ and − 11‰, respectively (Ambrose & Sikes, 1991; Sikes, in prep.). The most negative floral habitats are C_3 closed canopy forests and relatively closed canopy woodlands (includes riverine or riparian forest) where herbs and shrubs dominate the discontinuous ground cover. Less negative vegetational "savanna" has an open canopy cover with a continuous herbaceous layer dominated by tropical C_4 grasses and a clear seasonality of development related to water stress (Boulière, 1983). The three structural types of savanna are defined by percent of canopy cover as: (1) grassy woodland (including bushland and shrubland) with an open canopy of >20%; (2) wooded grassland (including bush/shrub grassland) with a canopy cover of <20%; and (3) treeless or sparsely treed grassland (including semi-desert dwarf shrub grassland) with a canopy cover of <2% (Pratt & Gwynne, 1977). High montane grasslands and permanent swamps are distinct physiognomic categories that do not have pronounced seasonality. Montane grasslands are not illustrated.

In East Africa, where the local vegetation mosaic can be complex, lateral change in floral microhabitat across a landscape is paralleled in the soil $\delta^{13}C$ values beneath the surface. $\delta^{13}C$ values for soil organic matter in C_3 *Olea* forest (-25‰) growing less than 20 m from open C_4 grassland (-11‰), for example, reflect that of the parent flora (Sikes, in prep.). Soil organic $\delta^{13}C$ values from open C_4 grasslands less than 100 m from closed montane C_3 forests, on both steep slopes and nearly level ground, are also distinct and derived from the standing plant biomass (Ambrose & Sikes, 1991:fig. 3). These examples from research conducted on modern soils and plants in East Africa demonstrate the power of this technique to distinguish between floral microhabitats within the tropical savanna mosaic on a very fine scale.

The data set partially illustrated in Figure 1 (Sikes, in prep.) substantially expands previous analyses of modern soils from East Africa (Ambrose & Sikes, 1991; Cerling *et al.*, 1991) and includes ∼50 samples from a diversity of terrestrial and wetland settings in freshwater and alkaline ecosystems analogous to paleoenvironments reconstructed for Plio-Pleistocene hominid localities, including lacustrine, riverine, and paludal margins. In lake margin zones where soil development may be only weakly expressed, soils with terrestrial floral isotopic signals can be identified by the presence of soil carbonates and by the carbonate carbon and oxygen isotopic values (e.g., Cerling & Hay, 1986; Sikes, in prep.; and see below). It is thus possible to reconstruct accurate floral microhabitat interpretations in the form of C_3/C_4 (tree/grass) ratios on a similar scale in the Plio-Pleistocene basins occupied by early hominids in East Africa.

Floral microhabitats at Olduvai Gorge

Landscape archaeology in basal Bed II

A landscape approach to Oldowan archaeology at Olduvai Gorge has thus far concentrated on small-scale excavation near the intersection of the Main and Side Gorges of a paleosol at the base of Bed II (Blumenschine & Masao, 1991). The rootmarked paleosol is equivalent to the HWK-E level 1 "occupation floor" excavated by M. D. Leakey (1971) and directly overlies Tuff IF (Hay, 1976). Tuff IF is dated to 1.75 ± 0.01 Myr (Walter *et al.*, 1991) and marks the top of Bed I (Hay, 1976). The ∼1·74 Myr basal Bed II paleosol was exposed in 17 trenches within a 1 km^2 area near the HWK and FLK localities, but locally eroded in an eighteenth trench (Blumenschine & Masao, 1991:fig. 1). The main paleosol is a waxy claystone, which interdigitates with a more earthy claystone variant in the eastern portion of the study area, nearer an alluvial fan (Hay, 1989*b*). A time span of ∼10,000 years may be represented by the average thickness (67 cm) of the waxy claystone (Sikes *et al.*, in prep.). During the time of soil formation, the study area was within the eastern lake margin zone of the Olduvai paleobasin where fresh water from highland streams to the east reduced the alkalinity of the small lake (Hay, 1976:fig. 38). The basal Bed II paleosol was buried by low-energy lacustrine sediments, which were subsequently modified by pedogenesis (the "clay with root casts" of HWK-E level 2 and FLK-N; Leakey, 1971; Hay, 1976, 1989*b*).

Table 2 Stable isotopic composition of organic matter and pedogenic carbonate from paleosols at Olduvai Gorge, Tanzania

Locality[1]			Sample[2]	Organic matter δ13C[3]	Carbonate δ13C	δ18O	Δ[4] (CaCO3-SOM)
Basal Bed II Paleosol							
42b	HWK-EE	1	Earthy	−20·32			
43	HWK-E	3	Waxy	−21·51			
43	HWK-E	4	Waxy	−21·07			
43	HWK-E	5	Waxy	−21·37			
42d	HWK-E	6	Earthy	−23·71			
42b	HWK-EE	7	Waxy	−19·48			
42d	HWK-E	10	Waxy	−21·70 ± 0·7 (2)	−4·26	−5·33	17·4
44	HWK-W	11	Earthy	−20·09	−5·68	−5·67	14·4
44	HWK-W	12	Waxy	−20·46			
44	HWK-W	13	Waxy	−19·94 ± 0·4 (2)	−3·89	−5·34	16·0
45d		15	Waxy	−21·49 ± 1·2 (3)	−5·55	−6·13	15·9
45c		17	Waxy	−21·99 ± 0·4 (2)			
45a	FLK-N	18	Waxy	−20·99 ± 0·8 (4)			
Total earthy and waxy claystone				−21·13 ± 1·1(21)	−4·84 ± 0·9	−5·62 ± 0·4	16·3
Lower Bed II Alluvial Fan[5]							
27			TZ81–86		−3·9	−4·4	
27			TZ81–85		−4·8	−4·8	
Middle Bed I FLK Zinjanthropus level 22							
45	FLK		89-8-6E[6]	−19·77			14·1
45	FLK		FLK-I[5,6]		−5·7	−6·2	

[1]Sample locality as in Hay (1976) and M. D. Leakey (1971). Geological localities 27, 45c, and 45d have no archaeological site designation.

[2]Sample numbers for the basal Bed II paleosol refer to archaeological trench numbers in Blumenschine & Masao (1991) and Hay (1989b).

[3]Average δ13C for trenches with more than one sample; number of samples indicated in parentheses.

[4]Δ (CaCO3-SOM)=δ13C (carbonate)-δ13C (soil organic matter). Calculated with average SOM where there is more than one sample per trench.

[5]Data from Cerling & Hay (1986).

[6]Sample provided by R. L. Hay.

Evidence of hominid activity in the form of varied, usually high densities of flaked stones, cores, hammerstones, and/or hammerstone-fractured bones was found in 16 of the 17 trenches exposing the main waxy claystone (Blumenschine & Masao, 1991). The artifacts appear undisturbed, and artifact density in over half the trenches is higher than calculated for the HWK-E level 1 "living floor". None of the trenches contained hominid fossils.

Carbon isotopic analysis

Stable isotopic analysis of the basal Bed II paleosol organic matter and associated pedogenic carbonates was performed by the author in collaboration with the landscape archaeology project (Blumenschine & Masao, 1991). The δ13C values for soil organic matter from 13 of the 17 trenches exposing the paleosol range from −23·7 to −19·5‰, averaging −21·1‰ (Table 2). Four co-existing pedogenic carbonates have a mean δ13C value of −4·8‰. The average difference of 16·3‰ between the pedogenic carbonate and organic carbon δ13C values falls within the 14 to 17‰ range of difference found in modern soils (Cerling et al., 1989). This agreement is a strong argument for a lack of diagenesis and retention of the original carbon isotopic ratios derived from the local terrestrial plant biomass during pedogenesis.

The stable carbon and/or oxygen isotopic composition of the basal Bed II paleosol can be distinguished from that of paleolake Olduvai basal Bed II diatomite, and Beds I and II lacustrine sediments and calcite crystals (Sikes, in prep.; Sikes *et al.*, in prep.). Such differentiation further supports the independent formation and unaltered condition of the basal Bed II paleosol isotopic ratios. In particular, the similarity of the pedogenic carbonate $\delta^{18}O$ values from the basal Bed II paleosol and lower Bed II alluvial fan (averaging -5.6 and $-4.6‰$, respectively) (see Table 2), argue in favor of a terrestrial and pedogenic, not wetland origin when compared to the enriched $\delta^{18}O$ values (averaging $-0.6‰$ for 15 samples) of Beds I and II lacustrine calcites (Hay, 1976; Cerling & Hay, 1986).

The carbon isotopic ratios of both organic matter and co-existing carbonate nodules from the Olduvai basal Bed II paleosol indicate the small study area of the eastern paleolake margin supported a local floral biomass of 20 to 45% C_4 plants during the period of soil formation ~ 1.74 Myr ago (Figure 1). By analogy with the $\delta^{13}C$ values of modern East African floral microhabitats (Ambrose & Sikes, 1991; Sikes, in prep.), the narrow range of paleosol organic carbon values represents a relatively closed gallery or riparian forest to grassy woodland. Although wetland soils and sediments have similar $\delta^{13}C$ values (Figure 1), the pedogenic carbonates and their $\delta^{18}O$ values support a vegetated terrestrial setting, rather than barren mudflats, a C_3 *Typha* swamp, or waterlogged C_4 sedges or grasses, as noted above. The $\delta^{13}C$ values of lower Bed II alluvial fan pedogenic nodules (Cerling & Hay, 1986) (Table 2) also suggest a grassy woodland was present ~ 6 km east of the study area. None of the carbon isotopic values from the Olduvai basal Bed II paleosol represent wooded or open C_4 grasslands which typify the nearby Serengeti Plain today.

This reconstruction of a riparian forest to grassy woodland contrasts with previous interpretations which depict the lower Bed II Olduvai paleolake margin as mudflats, *Typha* swamp, or open grassland (Hay, 1976, 1989b; Bonnefille, 1984b; Kappelman, 1984; Blumenschine & Masao, 1991; Shipman & Harris, 1988). Some of the discrepancy between interpretations may be attributed to a difference in sampling scale and in paleoenvironmental reconstruction methods. Fossil fauna and pollen, for example, both usually represent a much broader range of habitats than the 1 km^2 basal Bed II study area. Rootmarkings are on site, but interpretations of paleosol macromorphology appear to be less precise than isotopic reconstructions (Cerling *et al.*, 1991, 1992; Kappelman, 1993). Buried in place, unaltered paleosols represent floral microhabitats in the immediate vicinity of the sample site. Silicified wood fragments (Hay 1989b), as well as some woodland fauna (Leakey, 1971; Kappelman, 1984; Blumenschine & Masao, 1991), tend to support a more woody interpretation for the basal Bed II study area (Sikes, in prep.; Sikes *et al.*, in prep.).

The floral mosaic across the basal Bed II paleolandscape may have resembled the modern Lake Nakuru (Kenya) ecosystem. Within the Nakuru basin, there is a succession of plant communities, from alkaline adapted grasses and sedges on exposed mudflats to *Acacia xanthophloea* woodland and groundwater forest on the basin floor near the shallow alkaline lake. Gallery forest lines freshwater streams entering the lake, the vegetation ending only a few meters from the water's edge. *Typha* was recently present at one stream entrance; *Olea* forest grows in the uplands. A lake altitude of 1759 m compares to paleolake Olduvai (~ 1500 m). Annual rainfall (~ 900 mm) and temperature (18°C) at Nakuru are also comparable to that suggested for lower Bed II from pollen spectra ($\sim 800-900$ mm, Bonnefille & Riollet, 1980) and carbonates (13 to 16°C, Cerling and Hay, 1986), respectively. The Nakuru basin soils beneath the stands of *A. xanthophloea* have organic matter $\delta^{13}C$ values of -26 and -18‰ for forest and grassy woodland, respectively. Gallery forest dominated by *Ficus* has a $\delta^{13}C$ value of -22‰.

Rainfall and temperature at Olduvai today average 566 mm and 23°C, respectively. This shift toward greater aridity is reflected in less negative $\delta^{18}O$ values (~ 0.0‰, Hay 1989*a*; Cerling, 1984) for modern grassland pedogenic carbonates near the Gorge compared to basal Bed II carbonates (-5.6‰, Table 2). The comparatively enriched $\delta^{13}C$ values (-12.3‰ for organic matter and $+0.2$‰ for carbonates, Hay, 1989*a*; Cerling *et al.*, 1991; Sikes, in prep.) from nearby modern grasslands with >90% C_4 plant biomass also reflect this climatic change (compare to Table 2 & Figure 1).

The stable isotopic composition of the middle Bed I FLK (level 22) *Zinjanthropus* paleosol organic and pedogenic carbonate fractions suggests a similarity with basal Bed II floral microhabitats (Table 2). The soil organic $\delta^{13}C$ value of -19.8‰ for the ~ 1.77 Myr (Walter *et al.*, 1991) *Zinjanthropus* paleosol is close to the basal Bed II endmember of -19.5‰ and indicates a riparian or grassy woodland was present in the eastern paleolake margin below Tuff IC (Table 2 & Figure 1). Although the isotopic data are very limited, pollen and micro- and macrofaunal evidence (Bonnefille & Riollet, 1980; Kappelman, 1984; Potts 1988; Shipman & Harris, 1988) also suggest more woody C_3 floral habitats, including *Acacia* woodland and gallery forest, were present near the *Zinjanthropus* site. A taxon-free morpho-metric analysis of the bovids from Bed I sites, including the *Zinjanthropus* level (Plummer & Bishop, 1994), indicates floral habitats may have been more closed than previously concluded from a taxon-based approach (Kappelman, 1984). Finally, a rainfall estimate of ~ 900 mm from gastropods (Hay, 1976) matches that for basal Bed II from pollen spectra (Bonnefille & Riollet, 1980).

Ecological variables and hominid habitat selection

Models of site location and formation

During the last two decades, research on Plio-Pleistocene archaeological site formation processes (e.g., Isaac, 1981, 1983*b*, 1984; Binford, 1984; Potts, 1984, 1988; Sept, 1992*b*) has focused on alternative behavioral and ecological explanations for the dense artifact and bone concentrations previously designated as home bases (later central places; cf. Isaac 1983*b*, 1984) (see discussion in Potts, 1994). Trees, for example, are a central part of Isaac's (1981, 1983*b*, 1984) dynamic modeling of archaeological sites as "common amenities" or through-flow systems. This concept may have evolved from early observations of a tendency for lowland archaeological sites (Oldowan and Acheulean) in East Africa to be situated along sandy stream channels, a location that may have been a hominid habitat preference because of the "benefits" of water, plus shade and plant food provided by fringing gallery strips in an otherwise inhospitable landscape (Isaac, 1966, 1972, 1976, 1978*a*, 1978*b*, 1978*c*) (see Table 1). Likewise, together with Isaac (1981, 1983*b*, 1984), the key factors in numerous hypotheses of hominid habitat preferences and/or archaeological site location that focus on behavioral and ecological explanations for the accumulation of dense archaeological assemblages are the shade, plant food, predator refuge, vantage points, and/or sleeping accommodations provided by groves of trees (e.g., Bunn *et al.*, 1980; Kroll & Isaac, 1984; Sept, 1984, 1986, 1992*b*; Clark, 1987; Blumenschine, 1991; Bunn, 1991). Blumenschine (1986, 1987; and see Cavallo & Blumenschine, 1989; Marean, 1989) has suggested that Plio-Pleistocene hominids may have occupied a vacant scavenging "niche" in riparian woodlands.

In addition to groves of trees, fresh water (e.g., Hay, 1976; Isaac, 1978*a*; Binford, 1984; Potts, 1989) or stone sources (including caches and dropped/discarded stone or *de facto* caches) (e.g., Binford, 1984; Potts, 1984, 1988, 1991; Schick, 1987, 1991; Toth, 1987) may have been landscape features or magnets which attracted early hominids to particular resources or locations on the Plio-Pleistocene landscape. Preferred use of these areas may have resulted in the scattered residues and dense time-averaged aggregates of stones and/or bones that archaeologists call sites.

Modern resource distribution

A key to understanding early hominid behavior and habitat preferences resides in comprehension of the ecological variables that shape the abundance, predictability, and distribution of modern plant and animal resources across today's landscape. The same ecological principles and processes that underlie the basic structure of plant and animal communities today are likely to have operated in the past (Behrensmeyer, 1982; Bell, 1982; Foley, 1984; Sept, 1992*a*). The Plio-Pleistocene vegetation mosaic appears to have been broadly similar to that present on today's landscape (e.g., Bonnefille, 1984*a*; Bonnefille & Vincens, 1985; Dechamps & Maes, 1985), and the fossil record suggests modern faunal relationships have relevance for past interactions (e.g., Blumenschine, 1986; Marean, 1989; Blumenschine *et al.*, 1994).

Plant foods were fundamental to the diet of early hominids (see e.g., Peters & O'Brien, 1981; Sept, 1986, 1992*a*) and systematic surveys of above-ground and below-ground plants in East and South Africa indicate the greatest number of potential wild plant food parts likely to have been eaten raw by early hominids (nuts, fruits, flowers, beans), the quality and quantity of which vary seasonally, are from C_3 plants (deciduous trees, shrubs, arborescents, and perennial forbs) (Peters & Maguire, 1981; Peters & O'Brien 1981, 1982, 1984; Peters *et al.*, 1984; Sept, 1984, 1986, 1990; Vincent, 1984, 1985; Peters 1987; O'Brien & Peters 1991). Studies of sub Saharan plant part toxicity, nutritional value, and availability also suggest the highest quality of plant food resources available to early hominids may have been C_3 nuts, mature fleshy fruits, flowers, dry fruits, and beans (Peters & O'Brien, 1984; Peters, 1987; O'Brien & Peters, 1991; see also Stahl, 1984). Because procurement costs are high, deeply-buried underground storage organs (USOs) (C_3 plant parts) are ranked below these above-ground plant foods (O'Brien & Peters, 1991). In semi-arid habitats, however, seasonal exploitation of USOs appears to be cost effective in terms of calories (Vincent 1984, 1985).

Woody C_3 fruit-bearing plants are abundant today in arid and semi-arid riverine and woodland settings in East Africa and the Western Rift Valley (Sept, 1984, 1986, 1990). Woodland C_3 nut-bearing species have a discontinuous regional distribution in the Sudano-Zambezian phytogeographic zone and are comparatively rare north of the Tanzanian miombo woodlands (Sept, 1984; Peters, 1987; O'Brien & Peters, 1991). Fruit-bearing C_3 plants also occur frequently in woody grasslands and bushlands or grassy woodlands, but are very rare in open C_4 grasslands (Sept, 1990). Open grassland in the South African uplands (>1200 m) has few edible plant species; woodland nuts and other edible C_3 plant foods are abundant at lower altitude in riverine valley settings (Peters & Maguire, 1981; O'Brien & Peters, 1991). In East Africa, USOs are scattered in diffuse patches in dry grasslands, and are rare in arid areas (Vincent, 1984, 1985). USOs are most abundant in well-drained, small sandy channels in semi-arid bush and woodland habitats (Vincent 1984, 1985). Near-surface edible bulbs and corms also occur in well-drained soils near channel margins (Sept, 1984).

HOMINID HABITAT PREFERENCES

At one riverine locality in southern Kenya, the availability of fruits (C_3 parts) may have been sufficient to support hominids year-round in a semi-arid setting (Sept, 1984, 1986). The abundance of plant foods on the sandy, basement rock-derived soils in this locality supports evidence for the combined influence of soil parent material and rainfall on the dominance of grass or woody vegetation in the tropical savanna mosaic in Africa (Clark, 1967; Bell, 1982; Sept, 1984; Ambrose, 1986). In general, low nutrient soils (leached of nutrients in well-drained areas with high rainfall, or derived from basement rock regardless of yearly rainfall amount) support a greater diversity of edible C_3 plant species compared to high nutrient soils on volcanic parent material. In high nutrient soil areas with annual rainfall between $\sim$400–1400 mm, C_4 grassland with a high animal biomass, but a very low ratio of plants edible by hominids, is promoted over C_3 woody taxa with edible parts. Areas in Africa with rainfall less than 400 or greater than 1400 mm are usually covered by treeless to sparsely wooded grassland and desert steppe or forest, respectively (Bell, 1982).

Studies of the distribution, abundance, and predictability of animal resources in East Africa are limited in number and geographic diversity. Abandoned medium-sized, adult herbivore lion kills were regularly encountered in riparian woodlands during the dry season in the migratory Serengeti ecosystem (Blumenschine, 1986, 1987). Given that modern scavengers, spotted hyenas in particular, were rarely observed venturing into densely wooded areas (*contra* observations by Bunn *et al.*, 1988; O'Connell *et al.*, 1988), "hammerstone-wielding hominid denizens" may have occupied a vacant riparian woodland passive scavenging "niche" (Blumenschine, 1987:394). A similar lack of bone consumers may have effected regular scavenging by early hominids of carcasses abandoned by extinct sabertooth cats in densely wooded areas, possibly providing a predictable food resource regardless of season (Blumenschine, 1987; Marean, 1989). Natural mortality peaks in today's Serengeti reduce competition for carcasses in open C_4 grasslands during the late dry season (Blumenschine, 1986, 1987), but such fat-depleted carcasses may have been too energetically costly for hominid protein metabolism (Speth, 1987, 1989). Modern cheetahs leave the digestive tracts of their prey intact, and opportunistic hominids may have scavenged from cheetah kills near woodland-savanna ecotones (Bunn & Ezzo, 1993). It also has been suggested that observed routine abandonment by modern territorial leopards of smaller-sized carcasses cached in favored feeding trees may have provided observant, tree-climbing hominids with a low risk, fairly predictable woodland protein resource (Cavallo & Blumenschine, 1989).

As noted by others (e.g., Bunn & Ezzo, 1993; Plummer & Bishop, 1994; Potts, 1994), faunal resource availability models based on modern analogs like the Serengeti may not have wholesale applicability to the Plio-Pleistocene. Such criticisms illustrate the need for additional field research in a variety of habitats. Archaeological evidence (i.e. cutmarks) also suggests that hominids may have had earlier access to meatier carcasses by aggressive rather than passive scavenging, and/or by hunting small mammals (see e.g., Isaac & Crader, 1981; Bunn & Kroll, 1986; Potts, 1988; Bunn & Ezzo, 1993; Oliver, 1994). In addition, extensive overlap in Bed I sites at Olduvai (e.g., Potts, 1984, 1988) for example, questions whether bone-crushers were uncommon at all Plio-Pleistocene sites (also see Oliver, 1994). The rarity of sabertooths and their large prey in fossil assemblages, plus the lack of evidence for closed lake margin habitats at Olduvai, leads some (e.g., Bunn & Ezzo, 1993) to question the sabertooth model. This study suggests, however, that there may have been an association between relatively closed habitats on the paleolake margin at Olduvai and large sabertooth-size prey (e.g., buffalo, *Elephas recki*, giraffids) recovered from FLK *Zinjanthropus*, HWK-E level 1, and the 1989 basal Bed II trenches (Leakey, 1971; Potts, 1988; Blumenschine & Masao, 1991). Elemental analyses

(Sr/Ca ratios) of the Omo-Shungura fossil assemblage support a closed habitat preference by sabertooths, suggesting *Homotherium* may have fed primarily on browsing species (Sillen, 1992:497 on Sillen, 1986).

As an alternative source of protein, Plio-Pleistocene hominids may have eaten fish stranded during the dry season in riverine settings or trapped in shallow pools during spawning migrations in the wet season (Stewart, 1994). Modern fishers in Africa benefit from this predictable seasonal pattern, and similar harvests are documented archaeologically in late Pleistocene sites over much of the continent.

In sum, modern ecological studies of the distribution and abundance of food resources in sub Saharan Africa indicate woodlands, particularly in riverine or riparian settings, may have offered hominids the best opportunity for access to a varied, dependable range of edible plant parts, as well as dry season aquatic resources and scavengeable, abandoned carcasses (including sabertooth kills). In contrast to more wooded areas, open C_4 savanna grasslands may have been a less successful setting for opportunistic hominid foragers. Edible plant foods, including USOs are scattered in diffuse patches in more open habitats, and although numerous carcasses have been observed in open savanna, competition for carcasses among lions, hyenas, and other scavengers is more intense (e.g., Blumenschine, 1987; Behrensmeyer, 1993), and any remains are also subject to rapid decay (Behrensmeyer, 1978b; Behrensmeyer *et al.*, 1979; Behrensmeyer & Boaz, 1980). As noted, dry season ungulate carcass gluts in open C_4 grasslands may be too fat-depleted for hominid consumption (Speth, 1987, 1989). Further, foraging may have been riskier away from water and/or the arboreal refuge offered by woodland habitats (e.g., Isaac, 1984; Sept, 1984; Blumenschine, 1986).

Implications for early hominid habitat preferences

Behavioral hypotheses

The foregoing ecologically based models of site location and the ecological relationships indicated by modern studies of plant and animal resource abundance and distribution can be integrated into behavioral hypotheses of hominid foraging and land-use patterns in analogous Plio-Pleistocene habitats. For example, if early hominids preferred woodland habitats, as suggested by the modern distribution and abundance of exploitable plant and/or animal resources, we can expect the number and density of archaeological sites to vary with woody vegetation biomass. As a byproduct of recurrent use, one would expect the highest density of archaeological materials and sites in more wooded areas. Conversely, sites created in open grasslands should be less numerous and have a more diffuse scatter of artifacts. Finally, within regions and local areas where archaeological traces are abundant, one would expect surfaces across the paleolandscape with scant or no evidence of hominid activity to be in non-woodland settings.

These behavioral hypotheses can be restated in terms of C_3 versus C_4 floral biomass. If early hominids preferentially exploited more closed (woody C_3) microhabitats with trees in the savanna mosaic, we would expect the highest density of archaeological materials and sites to correlate with more negative C_3 carbon isotopic values. If hominids preferred gallery forests or riparian woodlands with a relatively closed canopy, then we would expect the highest densities to correlate with C_3 isotopic values for this type of microhabitat. Less numerous and more diffuse artifact scatters would then correspond to less negative, wooded grassland and open grassland C_4 carbon isotopic ratios. If, however, research fails to demonstrate a correlation with either C_3 or C_4 isotopic values, then the distribution of sites over the Plio-Pleistocene landscape may be less constrained by vegetation structure than by other variables, such as fresh water or stone sources.

Paleosol carbon isotopic evidence

The stable carbon isotopic ratios from the basal Bed II paleosol at Olduvai Gorge indicate that the 1 km^2 area in the eastern paleolake margin supported a relatively closed C_3 riparian forest to grassy woodland ~1·74 Myr ago. Ten of the 16 basal Bed II trenches containing archaeological traces were sampled for isotopic analysis. There is no correlation between artifact or bone densities and the slight variation in pedogenic carbonate and organic carbon isotopic ratios with the small study area. However, only a small proportion of available paleohabitats has been analysed (less than one-fifth of the lateral extent of the paleosol). Continuing excavation across the basal Bed II paleolandscape combined with stable carbon isotopic analysis of the paleosol may yet reveal an association between artifact density and microhabitat use.

There is, nevertheless, a positive relationship between flaked stone and hammerstone-modified bone densities in the 16 trenches. Blumenschine & Masao (1991) suggest that hominids may have preferentially discarded stone at animal carcass concentrations across the paleolandscape, possibly related to marrow extraction. Modern resource studies to date suggest the best opportunity for predictable access to scavengeable carcasses by early hominids may have occurred in woodlands, possibly riparian woodlands (Blumenschine, 1986, 1987; Cavallo & Blumenschine, 1989; also see Marean, 1989). Riverine and woodland settings may also provide a predictable, perhaps year-round source of above-ground plant foods, particularly fruits (Sept, 1984, 1986, 1990).

The density of artifact discard within the 16 basal Bed II trenches was apparently not related to distance from either the paleolake or the nearest quartz outcrop (Blumenschine & Masao, 1991). Although these stone discard and paleogeographic conclusions are based on an admittedly small area and sample size (Blumenschine & Masao, 1991), the evidence from this preliminary research for intensive use of this riparian to grassy woodland habitat suggests food resources rather than water or stone sources were the primary determinants of habitat exploitation or preference by the Oldowan hominids.

Stable carbon isotopic analyses suggest a resemblance between the eastern lake margin floral mosaic in the basal Bed II study area and the middle Bed I FLK *Zinjanthropus* site. The *Zinjanthropus* level has a higher concentration of archaeological traces (stone and bone) than any other excavated Plio-Pleistocene assemblage at Olduvai (Leakey, 1971; Bunn, 1982; Potts, 1988; Blumenschine & Masao, 1991). The site may have been a "favored locality" to which hominids transported carcass parts for consumption (Bunn & Kroll, 1986:442; also see Potts, 1988; Bunn, 1991; Bunn & Ezzo, 1993), or may represent a core/refuge area (Oliver, 1994). Although sampling is obviously limited, the carbon isotopic evidence for a riparian to grassy woodland on the *Zinjanthropus* paleosol provides further support for a preference for woody C_3 habitats by Oldowan toolmakers.

In addition to the stable carbon isotopic analyses presented here, geochemical evidence from diachronic studies of East African hominid locality paleosols and carbonates (Cerling & Hay, 1986; Cerling et al., 1988; Cerling, 1992; Kingston, 1992) tend to support the suggestion that Plio-Pleistocene hominids exploited more closed C_3 woodland habitats. Summarizing research at Baringo, Laetoli, Olduvai, and Turkana, Cerling (1992) argues that although C_4 plants are present in East Africa by 9·4 Myr, they comprised <50% of the floral biomass in hominid sites until $\sim 1\cdot7$ Myr. At that time forests and grassy woodlands were briefly replaced by wooded grasslands and open grasslands with 60–80% C_4 biomass (including the middle Bed II Lemuta Member at Olduvai). Cerling (1992) found that open grasslands with >90% C_4 biomass were not consistently present at East African hominid sites until ~ 1 Myr. In the 15 Myr Tugen Hills (Baringo basin, Kenya) sequence, where rainfall is high, there is no carbon isotopic evidence for open C_4 grasslands at any time (Kingston, 1992).

Rainfall
Stable oxygen isotopic values from carbonates reflect meteoric rainfall and temperature, and suggest the modern ecosystems at Olduvai and Turkana are hotter and drier than in the past (Cerling *et al.*, 1977; Cerling & Hay, 1986; Cerling *et al.*, 1988; see above). Fossil evidence (i.e., pollen, mollusks) also suggests rainfall in East African Plio-Pleistocene hominid localities was greater than at present (e.g., Hay, 1976; Bonnefille & Riollet, 1980; Bonnefille, 1984*a*; Bonnefille & Vincens, 1985). The presence of *Brachystegia* in the macrofossil assemblage from the Omo-Shungura Formation (Dechamps & Maes, 1985), a tree that is now exotic to the area and characteristic of miombo woodland, likewise argues for greater rainfall in the past (Feibel *et al.*, 1991). Rainfall in *Brachystegia* miombo averages 500–1200 mm; rainfall in the Turkana basin today is less than 300–200 mm.

As noted, there is a general positive correlation over Africa between rainfall and plant biomass (Clark, 1967; Bell, 1982) which may be influenced by soil nutrient status (Bell, 1982). Such influence may be fairly dramatic, as evidenced by the contrast between the Serengeti short grasslands on rich volcanic ash bounded by *Brachystegia* miombo woodlands on the Tanzanian granitic shield (Bell, 1982). However, grasslands such as the Serengeti are unusual and comprise relatively little of the modern East African landscape (Pratt *et al.*, 1966; White, 1983). In contrast, woodlands cover approximately one-third of East Africa, while wooded grasslands are the most abundant vegetation type today (Kingdon, 1974). Higher rainfall and/or soils with low nutrient status, as well as topographic features, may have promoted more woody C_3 vegetation in the East African savanna mosaic in the past.

Discussion

The actualistic, ecologic, and isotopic evidence presented here suggests that dense concentrations of the archaeological traces of hominid activity during the Plio-Pleistocene in East Africa may coincide with the abundance and diversity of plant foods available in wooded C_3 plant communities on well-drained or low nutrient soils, and/or in C_3 riparian woodlands rich with fruits, nuts, beans, and berries. Plant foods were an essential and presumably major ingredient in the diet of early hominids. Groves of trees, possibly growing near water sources, may have also supplied hominids with shady spots, vantage points, sleeping platforms, and refuges against predation. In addition, scavengeable carcasses from sabertooth, leopard, and cheetah kills may have been available in woodlands year-round. Carcasses from lion kills and aquatic resources may have been more accessible during the dry season in riparian settings.

Although some actualistic (e.g., Bunn *et al.*, 1988; O'Connell *et al.*, 1988) and archaeological evidence (e.g., Potts, 1988; Bunn & Ezzo, 1993; Plummer & Bishop, 1994; Oliver, 1994) question different aspects of the availability of animal carcasses (with meat and/or marrow) and/or safety of early hominids in the likely presence of other scavengers in woodland and/or riparian woodland settings, the modern ecological distribution of edible plant foods supports the isotopic evidence that suggests a hominid preference for woody habitats. In addition, some East African Plio-Pleistocene sites have evidence of trees (e.g., FxJj50 tree rootcasts, FxJj1 leaf impressions, Bunn *et al.*, 1980) or anomalous features that may have been caused by roosting raptors (FLK *Zinjanthropus* microfauna, Klein, 1986), tree root growth (DK stone circle, Potts, 1984), or tree fall (FLK *Zinjanthropus* irregular hollow, Leakey, 1971). Given the constraints against living in the open African savanna, such as susceptibility to predation, heat stress (Wheeler, 1993), and rarity of suitable plant foods, Plio-Pleistocene hominids may have remained close to more wooded areas. Like modern human forager groups (e.g., Hadza, Woodburn, 1968; Bunn *et al.*, 1988; O'Connell *et al.*, 1988; Okiek, Ambrose, 1986; Marshall 1993), early hominids may have exploited food resources from both relatively wooded and more open habitats. But, within such a foraging pattern in a known home range area, the evidence presented here suggests Plio-Pleistocene hominids in East Africa may have preferred more woody habitats, namely riparian forest and/or grassy woodland.

The anatomical traits of contemporary "robust" australopithecines and at least one species of early *Homo* (cf. Susman & Stern, 1982; Susman, 1988; Wood, 1992) support a more arboreal or tree-centered dependence. Unfortunately, the specific identity of the Plio-Pleistocene tool maker(s) and/or user(s) remains unknown—although it is commonly assumed that members of the genus *Homo* made the stone tools. Recent arguments, however, for "robust" australopithecine tool use (Susman, 1988, 1991) and synchronic species diversity within the genus *Homo* (e.g., Wood, 1992) make it more difficult to use the ecological context of archaeological traces as evidence for the habitat preference of only one hominid species. Likewise, the first appearance of both "robust" australopithecines and early *Homo* is contemporaneous with the earliest evidence for stone tools ~ 2.5 Myr ago (e.g., Wood, 1993; Kibunjia, 1994).

Plio-Pleistocene archaeological sites may have been preferentially located in or near more wooded areas, such as riparian or grassy woodlands, as suggested by the carbon isotopic evidence. Until, however, the behavioral hypotheses presented here can be tested with paleosol isotopic evidence from predominantly C_4 floral microhabitats (grasslands and wooded grasslands), an actual preference by early hominids for more woody C_3 habitats (grassy woodland, riparian forest, or closed canopy forest) cannot be adequately demonstrated.

Although the isotopic and rainfall evidence from East African hominid sites questions whether C_4 grasses were a major component (>50%) of the vegetation mosaic during the Plio-Pleistocene, C_4 grasslands were likely to have been present nearby in this volcanically active area. With higher annual rainfall (~ 400–1400 mm), grasslands are promoted on high nutrient soils derived from volcanic ashfall (Bell, 1982). The high proportion of grazing species transported by hominids to archaeological sites (e.g., HWK-E levels 1-2, FLK-N levels 1-6, Kappelman, 1984; Plummer & Bishop, 1994; FxJj50, Bunn *et al.*, 1980) tends to support this hypothesis. The isotopic evidence for a synchronous expansion of C_4 ecosystems in Pakistan and North America between 7–5 Myr (Cerling *et al.*, 1993) suggests atmospheric CO_2 levels would not have prohibited the presence of C_4 grasslands in East Africa. In addition, isotopic studies of fossilized tooth enamel demonstrate that obligate grazers were consuming C_4 grasses by ~ 3 Myr at Makapansgat in South Africa (Lee-Thorp, 1989) and by ~ 2 Myr in the Omo basin (Ericson *et al.*, 1981). Future isotopic analyses from continuing excavations in the basal Bed II paleosol, or from other landscape archaeology projects, may thus uncover C_4-dominant microhabitats that are needed to adequately test the behavioral hypotheses outlined above.

Summary

To understand early hominid habitat preference and to generate information on the microhabitat context of hominid behavior and resource use during the Plio-Pleistocene, we need to examine the synchronic variation of hominid activity preserved in archaeological sites in a diversity of floral microhabitats across paleolandscapes. Stable isotopic analysis of paleosols is a powerful tool for floral microhabitat and climatic reconstruction. The stable carbon isotopic composition of paleosol organic matter and pedogenic carbonate can be used to estimate the original proportion of tropical grasses (C_4) to woody (C_3) vegetation growing on the Plio-Pleistocene land surface in alkaline or freshwater ecosystems in East Africa. Annual temperature and moisture may be estimated from carbonate oxygen isotopic ratios.

Paleosol carbon isotopic evidence from landscape archaeological excavation of a basal Bed II paleosol (equivalent to HWK-E level 1) indicates a 1 km^2 portion of the eastern paleolake margin near the HWK and FLK localities at Olduvai Gorge supported a riparian forest to grassy woodland $\sim 1 \cdot 74$ Myr ago. Archaeological traces recovered from 16 trenches suggest hominids may have preferentially discarded stone at carcass concentrations related to marrow extraction. Isotopic evidence from the middle Bed I FLK *Zinjanthropus* paleosol indicates the dense artifact concentration was formed under a C_3 floral biomass resembling that in basal Bed II.

The isotopic data thus question whether hominids were predominantly associated with open C_4 Serengeti-type grasslands or wooded grasslands, and instead support actualistic and ecological studies that suggest hominids preferentially inhabited woody C_3 habitats, perhaps C_3 riparian woodland in particular, because of an abundance of plant foods, scavenging opportunities, trees for shade, predator refuge, and sleeping platforms. Such settings may have also offered access to water, aquatic resources, and primary and/or secondary stone sources. In contrast, open C_4 grasslands and wooded grasslands may have been a riskier and less successful setting for opportunistic scavenging, plant food gathering, or sleeping.

The behavioral correlation suggested here between C_3 woodlands and hominid habitat preference requires testing with paleosol carbon isotopic evidence from more open C_4 floral microhabitats. It is suggested that continued excavation across the Olduvai paleolandscape may yet reveal an association between artifact density and microhabit use. If no correlation were demonstrated with isotopic values, the Plio-Pleistocene hominids may have been less constrained by vegetation structure than by other variables, such as fresh water or stone sources.

The application of geochemical techniques to recover the carbon isotopic signals of soil organic matter and pedogenic carbonate, preserved *in situ*, within buried, fossilized soils at open-air archaeological sites is a viable avenue for future paleoenvironmental reconstruction of site-specific Plio-Pleistocene floral microhabitats. By combining geochemical analyses with excavation of the synchronic evidence of hominid activity across paleolandscapes, we can begin to make accurate statements about the floral microhabitat context of early hominid habitat preferences. In conjunction with paleoenvironmental information obtained by more traditional methods, we can then proceed toward a better understanding of the dynamics of cultural behavior and land-use patterns during the Plio-Pleistocene.

Acknowledgements

This manuscript has benefited from comments by S. H. Ambrose, P. A. Garber, R. L. Hay, J. S. Oliver (who also suggested adding the hominid to Figure 1), and three reviewers, including R. J. Klein and R. Potts. I am indebted to R. J. Blumenschine, R. L. Hay and F. T. Masao for supplying me with the samples from Olduvai Gorge. Dissertation fieldwork in

Kenya was funded by the Boise Fund, Leakey Foundation, National Science Foundation (BNS 9014160) and University of Illinois, with permission granted by the Office of the President and Kenya Wildlife Service, and greatly facilitated by the National Museums of Kenya and East African Herbarium. Laboratory research in Illinois was supported by the Wenner-Gren Foundation and Sigma Xi Research Society. Isotopic analyes were performed in the Department of Anthropology's Stable Isotope Laboratory at Urbana, while funded by National Science Foundation grants to S. H. Ambrose, who helped guide this research project.

References

Ambrose, S. H. (1986). Hunter-gatherer adaptations to non-marginal environments: an ecological and archaeological assessment of the Dorobo model. *Sprache und Geschichte in Afrika* **7**(2), 11–42.

Ambrose, S. H. & Sikes, N. E. (1991). Soil organic carbon isotopic evidence for vegetation change in the Kenya Rift Valley. *Science* **253**, 1402–1405.

Behrensmeyer, A. K. (1975a). Taphonomy and paleoecology in the hominid fossil record. *Yrbk phys. Anthrop.* **19**, 36–50.

Behrensmeyer, A. K. (1975b). The taphonomy and paleoecology of the Plio-Pleistocene vertebrate assemblages east of Lake Rudolf, Kenya. *Bull. Mus. Comp. Zool.* **146**, 473–578.

Behrensmeyer, A. K. (1976). Fossil assemblages in relation to sedimentary environments in the East Rudolf succession. In (Y. Coppens, F. C. Howell, G. L. Isaac & R. E. F. Leakey, Eds) *Earliest Man and Environments in the Lake Rudolf Basin*, pp. 383–401. Chicago: University of Chicago Press.

Behrensmeyer, A. K. (1978a). The habitat of Plio-Pleistocene hominids in East Africa: Taphonomic and micro-stratigraphic evidence. In (C. Jolly, Ed.) *Early Hominids of Africa*, pp. 165–189. London: Duckworth.

Behrensmeyer, A. K. (1978b). Taphonomic and ecologic information from bone weathering. *Paleobiology* **4**, 150–162.

Behrensmeyer, A. K. (1982). Geological context of human evolution *An. Rev. Earth planet. Sci.* **10**, 39–60.

Behrensmeyer, A. K. (1985). Taphonomy and the paleoecologic reconstruction of hominid habitats in the Koobi Fora Formation. In *L'Environnement des Hominidés au Plio-Pléistocène*, pp. 309–323. Paris: Masson.

Behrensmeyer, A. K. (1993). The bones of Amboseli. *Nat. Geog. Res. Expl.* **9**(4), 402–421.

Behrensmeyer, A. K. & Cooke, H. B. S. (1985). Paleoenvironments, stratigraphy, and taphonomy in the African Pliocene and early Pleistocene. In (E. Delson, Ed.) *Ancestors: The Hard Evidence*, pp. 60–62. New York: Alan R. Liss.

Behrensmeyer, A. K. & Dechant-Boaz, D. E. (1980). The recent bones of Amboseli National Park, Kenya, in relation to East African paleoecology. In (A. K. Behrensmeyer & A. P. Hill, Eds) *Fossils in the Making: Vertebrate Taphonomy and Paleoecology*, pp. 72–92. Chicago: University of Chicago Press.

Behrensmeyer, A. K. Western, D. & Dechant-Boaz, D. E. (1979). New perspectives in vertebrate paleoecology from a recent bone assemblage. *Paleobiology* **5**, 12–21.

Bell, R. H. V. (1982). The effect of soil nutrient availability on community structure in African ecosystems. In (B. J. Huntley & B. H. Walker, Eds) *Ecology of Tropical Savannas*, pp. 193–216. Berlin: Springer-Verlag.

Binford, L. R. (1984). *Faunal Remains from Klasies River Mouth*. Orlando: Academic Press.

Björkman, O. & Berry, J. (1973). High-efficiency photosynthesis. *Sci. Am.* **229**(4), 80–93.

Blumenschine, R. J. (1986). *Early Hominid Scavenging Opportunities: Implications of Carcass Availability in the Serengeti and Ngorongoro Ecosystems*. Oxford: British Archaeological Reports International Series 283.

Blumenschine, R. J. (1987). Characteristics of an early hominid scavenging niche. *Curr. Anthrop.* **28**, 383–407.

Blumenschine, R. J. (1991). Hominid carnivory and foraging strategies, and the socio-economic function of early archaeological sites. *Phil. Trans. R. Soc. Lond.* **B334**, 211–221.

Blumenschine, R. J., Cavallo, J. A. & Capaldo, S. D. (1994). Competition for carcasses and early hominid behavioral ecology: A case study and a conceptual framework. *J. hum. Evol.* **27**, 197–213.

Blumenschine, R. J. & Masao, F. T. (1991). Living sites at Olduvai Gorge, Tanzania? Preliminary landscape archaeology results in the basal Bed II lake margin zone. *J. hum. Evol.* **21**, 451–462.

Boaz, N. T. (1977). Paleoecology of early Hominidae in Africa. *Kroeber Anthrop. Soc. Pap.* **50**, 37–62.

Bonnefille, R. (1984a). Cenozoic vegetation and environments of early hominids in East Africa. In (R. O. Whyte, Ed.) *The Evolution of the East Asian Environment*, pp. 579–612. Hong Kong: Centre of Asian Studies, University of Hong Kong.

Bonnefille, R. (1984b). Palynological research at Olduvai Gorge. Research Reports No. 17, pp. 227–243. National Geographic Society.

Bonnefille, R. & Riollet, G. (1980). Palynologie, végétation et climats de Bed I et Bed II à Olduvai, Tanzanie. In (R. E. Leakey & B. A. Ogot, Eds) *Proc. 8th Pan-African Congress of Prehistory & Quaternary Studies*, pp. 123–127. Nairobi: The International Louis Leakey Memorial Institute for African Prehistory.

Bonnefille, R. & Vincens A. (1985). Apport de la palynologie à l'environnement des Hominidés d'Afrique orientale. In *L'Environnement des Hominidés au Plio-Pléistocène*, pp. 237–278. Paris: Masson.

Boulière, F. (Ed.) (1983). *Ecosystems of the World 13: Tropical Savannas*. Amsterdam: Elsevier Scientific.

Brain, C. K. (1981). The evolution of man in Africa: Was it a consequence of Cainozoic cooling? *Ann. Geol. Soc. S. Afr.* **84**, 1–19.

Brown, F. H. & Feibel, C. S. (1988). "Robust" hominids and Plio-Pleistocene paleogeography of the Turkana Basin, Kenya and Ethiopia. In (F. Grine, Ed.) *Evolutionary History of the "Robust" Australopithecines*, pp. 325–341. New York: Aldine de Gruyter.

Brown F. H. & Feibel, C. S. (1991). Stratigraphy, depositional environments and palaeogeography of the Koobi Fora Formation. In (J. M. Harris, Ed.) *Koobi Fora Research Project, Vol. 3*, pp. 1–30. Oxford: Clarendon Press.

Bunn, H. T. (1982). Meat-Eating and Human Evolution: Studies on the Diet and Subsistence Patterns of Plio-Pleistocene Hominids in East Africa. Ph.D. Dissertation, University of California, Berkeley.

Bunn, H. T. (1991). A taphonomic perspective on the archaeology of human origins. *Ann. Rev. Anthrop.* **20**, 433–467.

Bunn, H. T., Bartram, L. E. & Kroll, E. M. (1988). Variability in bone assemblage formation from Hadza hunting, scavenging, and carcass processing. *J. Anthrop. Archaeol.* **7**, 412–457.

Bunn, H. T. & Ezzo, J. A. (1993). Hunting and scavenging by Plio-Pleistocene hominids: Nutritional constraints, archaeological patterns, and behavioral implications. *J. Archaeol. Sci.* **20**, 365–398.

Bunn, H. T., Harris, J. W. K., Isaac, G., Kaufulu, Z., Kroll, E., Schick, K., Toth, N. & Behrensmeyer, A. K. (1980). FxJj50: An early Pleistocene site in northern Kenya. *World Archaeol.* **12**, 109–139.

Bunn, H. T. & Kroll, E. M. (1986). Systematic butchery by Plio/Pleistocene hominids at Olduvai Gorge, Tanzania. *Curr. Anthrop.* **27**, 431–452.

Cavallo, J. A. & Blumenschine, R. J. (1989). Tree-stored leopard kills: expanding the hominid scavenging niche. *J. hum. Evol.* **18**, 393–399.

Cerling, T. E. (1984). The stable isotopic composition of modern soil carbonate and its relationship to climate. *Earth planet. Sci. Lett.* **71**, 229–240.

Cerling, T. E. (1992). Development of grasslands and savannas in East Africa during the Neogene. *Palaeogeogr. Palaeoclimat. Palaeoecol.* **97**, 241–247.

Cerling, T. E., Bowman, J. R. & O'Neil, J. R. (1988). An isotopic study of a fluvial-lacustrine sequence: the Plio-Pleistocene Koobi Fora sequence, East Africa. *Palaeogeogr. Palaeoclimat. Palaeoecol.* **63**, 335–356.

Cerling, T. E. & Hay, R. L. (1986). An isotopic study of paleosol carbonates from Olduvai Gorge. *Quatern. Res.* **25**, 63–78.

Cerling, T. E., Hay, R. L. & O'Neil, J. R. (1977). Isotopic evidence for dramatic climatic changes in East Africa during the Pleistocene. *Nature* **267**, 137–138.

Cerling, T. E., Kappelman, J., Quade, J., Ambrose, S. H., Sikes, N. E. & Andrews, P. (1992). Reply to comment on the paleoenvironment of *Kenyapithecus* at Fort Ternan. *J. hum. Evol.* **23**, 371–377.

Cerling, T. E., Quade, J., Ambrose, S. H. & Sikes, N. E. (1991). Miocene fossil soils, grasses and carbon isotopes from Fort Ternan (Kenya): grassland or woodland? *J. hum. Evol.* **21**, 295–306.

Cerling, T. E., Quade, J., Wang, Y. & Bowman, J. R. (1989). Carbon isotopes in soils and paleosols as ecology and paleoecology indicators. *Nature* **341**, 138–139.

Cerling, T. E., Wang, Y. & Quade, J. (1993). Expansion of C_4 ecosystems as an indicator of global ecological change in the late Miocene. *Nature* **361**, 344–345.

Clark, J. D. (1967). *Atlas of African Prehistory*. Chicago: University of Chicago Press.

Clark, J. D. (1987). Transitions: *Homo erectus* and the Acheulian: the Ethiopian sites of Gadeb and the Middle Awash. *J. hum. Evol.* **16**, 809–826.

Coppens, Y. (1980). The differences between *Australopithecus* and *Homo*; preliminary conclusions from the Omo Research Expedition's studies. In (L. K. Königsson, Ed.) *Current Argument on Early Man*, pp. 207–225. Oxford: Pergamon Press.

Darwin, C. (1871). *The Descent of Man, and Selection in Relation to Sex*. London: John Murray.

Dechamps, R. & Maes, F. (1985). Essai de reconstitution des climats et des végétations de la basse valée de l'Omo au Plio-Pléistocène à l'aide des bois fossiles. In *L'Environnement des Hominidés au Plio-Pléistocène*, pp. 175–222. Paris: Masson.

Deines, P. (1980). The isotopic composition of reduced organic carbon. In (P. Fritz & J. C. Fontes, Eds) *Handbook of Environmental Isotope Geochemistry. Vol. 1: The Terrestrial Environment*, pp. 329–406. Amsterdam: Elsevier.

Dzurec, R. S., Boutton, T. W., Caldwell, M. M. & Smith, B. N. (1985). Carbon isotope ratios of soil organic matter and their use in assessing community composition changes in Curlew Valley, Utah. *Oecologia* **66**, 17–24.

Ehleringer, J. R. & Cooper, T. A. (1988). Correlation between carbon isotope ratio and microhabitat in desert plants. *Oecologia* **76**, 562–566.

Ehleringer, J. R., Sage, R. F., Flanagan, L. B. & Pearcy, R. W. (1991). Climate change and the evolution of C_4 photosynthesis. *Trends Ecol. Evol.* **6**, 95–99.

Ericson, J. H., Sullivan, C. H. & Boaz, N. T. (1981). Diets of Pliocene mammals from Omo, Ethiopia, deduced from carbon isotope ratios in tooth apatite. *Palaeogeogr. Palaeoclimat. Palaeoecol.* **36**, 69–73.

Feibel, C. S., Harris, J. M. & Brown, F. H. (1991). Palaeoenvironmental context for the late Neogene of the Turkana Basin. In (J. M. Harris, Ed.) *Koobi Fora Research Project, Vol. 3*, pp. 321–370. Oxford: Clarendon Press.

Foley, R. (Ed.) (1984). *Hominid Evolution and Community Ecology: Prehistoric Human Adaptation in Biological Perspective*. London: Academic Press.

Foley, R. (1987). *Another Unique Species: Patterns in Human Evolutionary Ecology*. Essex: Longman Scientific & Technical.

Grine, F. E. (1981). Trophic differences between "gracile" and " robust" australopithecines: a scanning electron microscope analysis of occlusal events. *S. Afr. J. Sci.* **77**, 203–230.

Grine, F. E. (1988). Evolutionary history of the "robust" australopithecines: A summary and historical perspective. In (F. E. Grine, Ed.) *Evolutionary History of the "Robust" Australopithecines*, pp. 509–520. New York: Aldine de Gruyter.

Harris, D. R. (1980). Tropical savanna environments: Definition, distribution, diversity and development. In (D. R. Harris, Ed.) *Human Ecology in Savanna Environments*, pp. 1–27. London: Academic Press.

Harris, J. M. (1987). Summary. In (M. D. Leakey & J. M. Harris, Eds) *Laetoli: A Pliocene Site in Northern Tanzania*, pp. 524–531. Oxford: Clarendon Press.

Hay, R. L. (1976). *Geology of the Olduvai Gorge*. Berkeley: University of California Press.

Hay, R. L. (1989a). Holocene carbonatite-nephelinite tephra deposits of Oldoinyo Lengai, Tanzania. *J. Volcan. Geotherm. Res.* **37**, 77–91.

Hay, R. L. (1989b). Stratigraphy and paleoenvironment of lowermost Bed II in a part of Olduvai Gorge. Research Reports. National Geographic Society.

Hill, A. & Ward, S. (1988). Origin of the Hominidae: The record of African large hominoid evolution between 14 My and 4 My. *Yrbk phys. Anthrop.* **3**, 49–83.

Isaac, G. L. (1966). New evidence from Olorgesailie relating to the character of Acheulean occupation sites. In (L. D. Cuscoy, Ed.) *Actas del V Congreso Panafricano de Prehistoria y de Estudio del Cuaternario*, Vol. 2, pp. 135–145. Santa Cruz de Tenerife: Museo Arqueologico.

Isaac, G. L. (1972). Comparative studies of Pleistocene site locations in East Africa. In (P. J. Ucko & G. W. Dimbleby, Eds) *Man, Settlement and Urbanism*, pp. 165–176. London: Duckworth.

Isaac, G. L. (1976). The activities of early African hominids. In (G. L. Isaac & E. R. McCown, Eds) *Human Origins: Louis Leakey and the East African Evidence*, pp. 483–514. Menlo Park: W. A. Benjamin.

Isaac, G. L. (1978a). The archaeological evidence for the activities of early African hominids. In (C. Jolly, Ed.) *Early Hominids of Africa*, pp. 219–254. London: G. Duckworth & Co.

Isaac, G. L. (1978b). Food sharing and human evolution: archaeological evidence from the Plio-Pleistocene of East Africa. *J. Anthrop. Res.* **34**, 311–325.

Isaac, G. L. (1978c). The food-sharing behavior of protohuman hominids. *Sci. Am.* **238**(4), 90–108.

Isaac, G. L. (1981). Stone Age visiting cards: approaches to the study of early land use patterns. In (I. Hodder, G. Isaac & N. Hammond, Eds) *Patterns of the Past*, pp. 131–155. Cambridge: Cambridge University Press.

Isaac, G. L. (1983a). Aspects of human evolution. In (D. S. Bendall, Ed.) *Evolution from Molecules to Men*, pp. 509–543. Cambridge: Cambridge University Press.

Isaac, G. L. (1983b). Bones in contention: Competing explanations for the juxtaposition of early Pleistocene artifacts and faunal remains. In (J. Clutton-Brock & C. Grigson, Eds) *Animals and Archaeology. 1. Hunters and their Prey*, pp. 3–19. Oxford: British Archaeological Reports International Series 163.

Isaac, G. L. (1984). The archaeology of human origins: studies of the Lower Pleistocene in East Africa 1971–1981. *Adv. World Archaeol.* **3**, 1–87.

Isaac, G. L. & Crader, D. C. (1981). To what extent were early hominids carnivorous? An archaeological perspective. In (R. S. O. Harding & G. Teleki, Eds) *Omnivorous Primates: Gathering and Hunting in Human Evolution*, pp. 37–103. New York: Columbia University Press.

Isaac, G. L., Harris, J. W. K. & Marshall, F. (1980). A method for determining the characteristics of artifacts between sites in the Upper Member of the Koobi Fora Formation, East Lake Turkana. In (R. E. Leakey & B. A. Ogot, Eds) *Proc. 8th Pan-African Congress of Prehistory & Quaternary Studies*, pp. 19–22. Nairobi: The International Louis Leakey Memorial Institute for African Prehistory.

Jolly, C. (1970). The seed-eaters: a new model of hominid differentiation based on a baboon analogy. *Man* **5**, 1–26.

Kappelman, J. (1984). Plio-Pleistocene environments of Bed I and lower Bed II, Olduvai Gorge, Tanzania. *Palaeogeogr. Palaeoclimat. Palaeoecol.* **48**, 171–196.

Kappelman, J. (1993). Review of Miocene paleosols and ape habitats of Pakistan and Kenya. *Am. J. phys. Anthrop.* **92**, 117–123.

Kibunjia, M. (1994). Pliocene archaeological occurrences in the Lake Turkana Basin. A review of patterns and gaps in the record. *J. hum. Evol.* **27**, 159–171.

Kingdon, J. (1974). *East African Mammals: An Atlas of Evolution in Africa. Vol. 1.* Chicago: University of Chicago Press.

Kingston, J. D. (1992). Stable Isotopic Evidence for Hominid Paleoenvironments in East Africa. Ph.D. Dissertation, Harvard University.

Klein, R. G. (1986). Comment on H. T. Bunn and E. M. Kroll, Systematic butchery by Plio/Pleistocene hominids at Olduvai Gorge, Tanzania. *Curr. Anthrop.* **27**, 446–447.

Kroll, E. M. & Isaac, G. L. (1984). Configurations of artifacts and bones at early Pleistocene sites in East Africa. In (H. J. Hietala & P. A. Larson, Eds) *Intrasite Spatial Analysis in Archaeology*, pp. 4–31. Cambridge: Cambridge University Press.

Leakey, M. D. (1971). *Olduvai Gorge, Vol. 3: Excavations in Beds I and II, 1960–1963*. Cambridge: Cambridge University Press.

Lee-Thorp, J. A. (1989). Stable Carbon Isotopes in Deep Time: The Diets of Fossil Fauna and Hominids. Ph.D. Dissertation, University of Cape Town.

Marean, C. W. (1989). Sabertooth cats and their relevance for early hominid diet and evolution. *J. hum. Evol.* **18,** 559–582.

Marshall, F. (1993). Food sharing and the faunal record. In (J. Hudson, Ed.) *From Bones to Behavior: Ethnoarchaeological and Experimental Contributions to the Interpretations of Faunal Remains*, pp. 228–246. Carbondale: Southern Illinois University.

Martin, A., Mariotti, A., Balesdent, J., Lavelle, P. & Vauttoux, R. (1990). Estimate of organic matter turnover rate in a savanna soil by ^{13}C natural abundance measurements. *Soil Biol. Biochem.* **22,** 517–523.

Nadelhoffer, K. J. & Fry, B. (1988). Controls of natural nitrogen-15 and carbon-13 abundances in forest soil organic matter. *Soil Sci. Soc. Am. J.* **52,** 1633–1640.

O'Brien, E. M. & Peters, C. R. (1991). Ecobotanical contexts for African hominids. In (J. D. Clark, Ed.) *Cultural Beginnings: Approaches to Understanding Early Hominid Life-Ways in the African Savanna*, pp. 1–15. Bonn: Dr R. Habelt GMBH.

O'Connell, J. F., Hawkes, K. & Blurton-Jones, N. (1988). Hadza hunting, butchering, and bone transport and their archaeological implications. *J. Anthrop. Res.* **44,** 113–161.

Oliver, J. S. (1994). Socioecological implications of early hominid faunal assemblages. *J. hum. Evol.* **27,** 267–294.

Peters, C. R. (1987). Nut-like oil seeds: food for monkeys, chimpanzees, humans, and probably ape-men. *Am. J. phys. Anthrop.* **73,** 333–363.

Peters, C. R. & Maguire, B. (1981). Wild plant foods of the Makapansgat area: a modern ecosystems analogue for *Australopithecus africanus* adaptations. *J. hum. Evol.* **10,** 565–583.

Peters, C. R. & O'Brien, E. M. (1981). The early hominid plant-food niche: insights from an analysis of plant exploitation by *Homo, Pan*, and *Papio* in eastern and southern Africa. *Curr. Anthrop.* **22,** 127–140.

Peters, C. R. & O'Brien, E. M. (1982). On early hominid plant food niches: Reply. *Curr. Anthrop.* **23,** 214–218.

Peters, C. R. & O'Brien, E. M. (1984). On hominid diet before fire. *Curr. Anthrop.* **25,** 358–360.

Peters, C. R., O'Brien, E. M. & Box, E. O. (1984). Plant types and seasonality of wild plant foods, Tanzania to southwestern Africa: Resources for models of the natural environment. *J. hum. Evol.* **13,** 397–414.

Plummer, T. & Bishop, L. C. (1994). Hominid paleoecology as indicated by artiodactyl remains from sites at Olduvai Gorge, Tanzania. *J. hum. Evol.* **27,** 47–75.

Potts, R. (1984). Home bases and early hominids. *Am. Sci.* **72,** 338–347.

Potts, R. (1988). *Early Hominid Activities at Olduvai*. New York: Aldine.

Potts, R. (1989). Olorgesailie: New excavations and findings in Early and Middle Pleistocene contexts, southern Kenya rift valley. *J. hum. Evol.* **18,** 477–484.

Potts, R. (1991). Why the Oldowan? Plio-Pleistocene toolmaking and the transport of resources. *J. Anthrop. Res.* **47**(2), 153–176.

Potts, R. (1994). Variables versus models of early Pleistocene hominid land use. *J. hum. Evol.* **27,** 7–24.

Pratt, D. J., Greenway, P. J. & Gwynne, M. D. (1966). A classification of East African rangeland, with an appendix on terminology. *J. Appl. Ecol.* **3,** 369–382.

Pratt, D. J. & Gwynne, M. D. (1977). *Rangeland Management and Ecology in East Africa*. Huntington, NY: Krieger.

Schick, K. D. (1987). Modeling the formation of stone age artifact concentrations. *J. hum. Evol.* **16,** 789–807.

Schick, K. D. (1991). On making behavioral inferences for early archaeological sites. In (J. D. Clark, Ed.) *Cultural Beginnings: Approaches to Understanding Early Hominid Life-Ways in the African Savanna*, pp. 79–107. Bonn: Dr R. Habelt GMBH.

Schwartz, D., Mariotti, A., Landranchi, R. & Guillet, B. (1986). ^{13}C/^{12}C ratios of soil organic matter as indicators of vegetation changes in the Congo. *Geoderma* **39,** 97–103.

Sept, J. M. (1984). Plants and Early Hominins in East Africa: A Study of Vegetation in Situations Comparable to Early Archaeological Site Locations. Ph.D. Dissertation, University of California, Berkeley.

Sept, J. M. (1986). Plant foods and early hominids at site FxJj50, Koobi Fora, Kenya. *J. hum. Evol.* **15,** 751–770.

Sept, J. M. (1990). Vegetation studies in the Semliki Valley, Zaire, as a guide to paleoanthropological research. In (N. T. Boaz, Ed.) *Evolution of Environments and Hominidae in the African Western Rift Valley*, pp. 95–121. Martinsville: Virginia Mus. Nat. Hist. Memoir No. 1.

Sept, J. M. (1992*a*). Archaeological evidence and ecological perspectives for reconstructing early hominid subsistence behavior. In (M. B. Schiffer, Ed.) *Advances in Archaeological Method and Theory*, Vol. 4, pp. 1–56. Tucson: University of Arizona Press.

Sept, J. M. (1992*b*). Was there no place like home? A new perspective on early hominid archaeological sites from the mapping of chimpanzee nests. *Curr. Anthrop.* **33,** 187–207.

Shipman, P. & Harris, J. M. (1988). Habitat preference and paleoecology of *Australopithecus boisei* in eastern Africa. In (F. Grine, Ed.) *Evolutionary History of the "Robust" Australopithecines*, pp. 343–381. New York: Aldine de Gruyter.

Sikes, N. E. (in prep.). Pleistocene Hominid Habitat Preferences in East Africa: Stable Isotopic Evidence from Paleosols. Ph.D. Dissertation, University of Illinois-Urbana.

Sikes, N. E., Hay, R. L., Blumenschine, R. J., Ambrose, S. H. & Masao, F. T. (in prep.). Paleosol carbon isotopic evidence for woodland habitat use by hominids in basal Bed II, Olduvai Gorge, Tanzania.

Sillen, A. (1986). Biogenic and diagenetic Sr/Ca in Plio-Pleistocene fossils in the Omo Shungura Formation. *Paleobiology* **12**, 322–323.

Sillen, A. (1992). Strontium-calcium ratios (Sr/Ca) of *Australopithecus robustus* and associated fauna from Swartkrans. *J. hum. Evol.* **23**, 495–516.

Smith, B. N. (1972). Natural abundance of the stable isotopes of carbon in biological systems. *BioScience* **22**, 226–231.

Smith, B. N. & Epstein, S. (1971). Two categories of $^{13}C/^{12}C$ ratios for higher plants. *Plant Physiology* **47**, 380–384.

Speth, J. D. (1987). Early hominid subsistence strategies in seasonal habitats. *J. Archaeol. Sci.* **14**, 13–29.

Speth, J. D. (1989). Early hominid hunting and scavenging: the role of meat as an energy source. *J. hum. Evol.* **18**, 329–343.

Stahl, A. B. (1984). Hominid dietary selection before fire. *Curr. Anthrop.* **25**, 151–168.

Stern, N. (1993). The structure of the lower Pleistocene archaeological record: A case study from the Koobi Fora Formation. *Curr. Anthrop.* **34**(3), 201–225.

Stewart, K. (1994). Early hominid utilization of fish resources and implications for seasonality and behavior. *J. hum. Evol.* **27**, 229–245.

Susman, R. L. (1988). New postcranial remains from Swartkrans and their bearing on the functional morphology and behavior of *Paranthropus robustus*. In (F. Grine, Ed.) *Evolutionary History of the "Robust" Australopithecines*, pp. 149–172. New York: Aldine de Gruyter.

Susman, R. L. (1991). Who made the Oldowan tools? Fossil evidence for tool behavior in Plio-Pleistocene hominids. *J. Anthrop. Res.* **47**, 129–151.

Susman, R. L. & Stern, J. T. (1982). Functional morphology of *Homo habilis*. *Science* **217**, 931–934.

Tieszen, L., Senyimba, M., Imbamba, S. & Troughton, J. (1979). The distribution of C_3 and C_4 grasses along an altitudinal and moisture gradient in Kenya. *Oecologia* **37**, 337–350.

Toth, N. (1987). Behavioral inferences from early stone artifact assemblages: An experimental model. *J. hum. Evol.* **16**, 763–787.

Toth, N. & Schick, K. D. (1986). The first million years: the archaeology of protohuman culture. In (M. B. Schiffer, Ed.) *Advances in Archaeological Method and Theory*, Vol. 8, pp. 1–96. Orlando: Academic Press.

Vincent, A. S. (1984). Plant foods in savanna environments: a preliminary report of tubers eaten by the Hadza of northern Tanzania. *World Archaeol.* **17**, 131–148.

Vincent, A. S. (1985). Wild Tubers as a Harvestable Resource in the East African Savannas: Ecological and Ethnographic Studies. Ph.D. Dissertation, University of California, Berkeley.

Vogel, J. C. (1978). Recycling of carbon in a forest environment. *Oecologia Plantarum* **13**, 89–94.

Vrba, E. S. (1985). Ecological and adaptive changes associated with early hominid evolution. In (E. Delson, Ed.) *Ancestors: The Hard Evidence*, pp. 63–71. New York: Alan R. Liss.

Vrba, E. S. (1988). Late Pliocene climatic events and hominid evolution. In (F. Grine, Ed.) *Evolutionary History of the "Robust" Australopithecines*, pp. 405–426. New York: Aldine de Gruyter.

Walter, R. C., Manega, P. C., Hay, R. L., Drake, R. E. & Curtis, G. H. (1991). Laser-fusion $^{40}Ar/^{39}Ar$ dating of Bed I, Olduvai Gorge, Tanzania. *Nature* **354**, 145–149.

Wesselman, H. B. (1984). *The Omo Micromammals: Systematics and Paleoecology of Early Man Sites from Ethiopia*. Basel: S. Karger.

Wheeler, P. E. (1993). The influence of stature and body form on hominid energy and water budgets; a comparison of *Australopithecus* and early *Homo* physiques. *J. hum. Evol.* **24**, 13–28.

White, F. (1983). *The Vegetation of Africa*. La Chaux-de-Fonds: UNESCO.

White, T. D. (1988). The comparative biology of "robust" Australopithecus: Clues from context. In (F. Grine, Ed.) *Evolutionary History of the "Robust" Australopithecines*, pp. 449–483. New York: Aldine de Gruyter.

Wood, B. (1992). Origin and evolution of the genus *Homo*. *Nature* **355**, 783–790.

Wood, B. (1993). Rift on the record. *Nature* **365**, 789–790.

Woodburn, J. (1968). An introduction to Hadza ecology. In (R. B. Lee & I. DeVore, Eds) *Man the Hunter*, pp. 49–55. Chicago: Aldine Publishing Company.

Thomas W. Plummer
Department of Anthropology, National Museum of Natural History, Smithsonian Institution, Washington, DC 20560, U.S.A.

Laura C. Bishop
Department of Anthropology, Yale University, P.O. Box 208277, New Haven CT 06520, U.S.A.

Received 25 September 1993
Revision received 9 March 1994
and accepted 12 March 1994

Keywords: Environment, ecology, Plio-Pleistocene, Olduvai Gorge, hominid, bovid.

Hominid paleoecology at Olduvai Gorge, Tanzania as indicated by antelope remains

Bed I Olduvai Gorge, Tanzania has provided abundant fauna in both paleontological and archeological contexts. These have been used to reconstruct the general paleoecological setting of the sites as well as provide more specific inferences on hominid habitat use. Previous paleoecological studies utilizing the Olduvai fauna have been taxon-based, substituting the habitat preferences of modern taxa for those of their extinct relatives. Here we investigate the relationship between bovid metapodial functional anatomy and habitat preference using taxon-free discriminant function analyses. We develop discriminant function models linking metapodial morphology to three broadly-defined habitat categories (open, intermediate, closed) using modern bovids of known ecology. The models developed for complete, proximal and distal metapodials are then applied to metapodials from four Bed I archeological localities: DK I, FLK NN I, FLK I and FLK N I. Results support the presence of the drying trend previously noted from middle to upper Bed I. They contradict taxon-based studies of the Olduvai bovids, suggesting a higher proportion of intermediate and closed habitats. The results of this study do not support theories of hominid foraging based on the exploitation of a single habitat type. The bovids from each locality exhibit a range of morphologies, suggesting that hominids at Olduvai were utilizing habitats ranging from open to closed, perhaps the full range available in the lake margin zone.

Journal of Human Evolution (1994) **27,** 47–75

Introduction

Bed I Olduvai Gorge, Tanzania has yielded well-dated hominid fossils and abundant evidence of their activities (Table 1). It provides one of the best records of Plio-Pleistocene hominid biological and cultural evolution, as well as the environmental context in which they occurred. The ecological setting for hominid activities has been reconstructed using uniformitarian principles, with the habitat preferences of modern taxa being substituted for those of their extinct relatives. This procedure is a direct projection of present conditions to the past (actualism) and it can not identify unique adaptations or behaviors in fossil taxa. Here we examine the relationship between functional anatomy and ecological preference in a taxon-free, multivariate analysis of modern antelope (Artiodactyla: Bovidae) metapodials. The Bovidae are well suited for this approach, as they are taxonomically diverse and exhibit a wide range of habitat specificities (Scott, 1985). Metapodial functional anatomy correlates well with habitat preference in modern bovids. We apply this relationship to fossil metapodials from archeological localities at Olduvai Gorge. Since hominids were involved in the formation of these assemblages (Potts, 1988), the habitat preferences of the bovids provide an indirect assessment of hominid paleoecology.

Antelope metapodials were chosen for this study because: (1) they have specific morphologies which can be linked to locomotor adaptation (Gentry, 1970; Scott, 1979, 1985); (2) they are among the most durable of the bovid long bones (Brain, 1981); and (3) they are well-represented at Olduvai sites (Potts, 1988). As Kappelman (1986, 1988, 1991) has documented for the bovid femur, habitat-specific metapodial morphologies reflect the degree to which cursoriality is used as a predator avoidance strategy, and are mainly related to

Table 1   **Stratigraphic position and dating of Olduvai Gorge, Bed I sites in this study (after Hay 1976, Walter *et al.*, 1991)**

		1·749 ± 0·007 Ma
TUFF IF		
	FLK N I L/1–6	
		1·750 ± 0·020 Ma
TUFF IE		
		FLK I L/13
		FLK I L/15
		1·764 ± 0·014 Ma
TUFF ID		
		1·761 ± 0·028 Ma
TUFF IC		
	FLK NN I L/1	FLK I L/22
	FLK NN I L/2	
	FLK NN I L/3	
		1·859 ± 0·007 Ma to 1·798 ± 0·004 Ma
TUFF IB		
	DK I L/1–3	
		1·976 ± 0·015 Ma
TUFF IA		

DK I is bracketed by Tuffs IA and IB, dated at 1·976 ± 0·015 Ma and 1·859 ± 0·007 Ma respectively (Walter *et al.*, 1991). The FLK sites are bracketed between Tuffs IB and IF, and so were deposited in an interval of less than 50,000 years.

differences in joint stabilization, shaft shape and lever arm length (Plummer & Bishop, in prep.).

Previous research on Bed I paleoecology

The Bed I localities have been the focus of intense geological and paleoenvironmental research. Sedimentological study has suggested that the climate during Bed I times was semi-arid, but generally wetter than at present day Olduvai (approximately 566 mm of rainfall per annum) (Hay, 1976). The rarity of soil carbonates is an indication that rainfall was relatively high and conditions moist during Bed I deposition (Cerling & Hay, 1986). Lithological and sedimentary evidence suggests that paleolake Olduvai was at its maximum extent prior to the deposition of Tuff IB, during the brief interval in which DK I was formed. Intermediate lake levels prevailed between the deposition of Tuff IB and Tuff ID, corresponding to the deposition of FLK NN I and FLK I level 22. The lake was at its lowest level during the rest of the deposition of FLK I. Finally, the lake expanded again during FLK N I deposition, just prior to Tuff IF.

Fossil pollen analysis has provided evidence of environmental fluctuations during Bed I times (Bonnefille, 1984). A sediment sample from FLK NN I contained a high proportion of pollen from Afro-montane arboreal taxa, indicating a greater regional presence of montane forest than at present. Below Tuff ID, more steppe-like conditions near the lake were suggested by a sample of lacustrine deposits from RHC, which exhibited a reduction in forest pollen and an increase in *Acacia* and *Commiphora*. Arid steppe conditions were suggested by the increase of

Sudano-Zambezian pollen and decrease of forest pollen in samples from FLK I and FLK N I, below Tuff IF. These samples suggest that environmental conditions towards the top of Bed I were more arid than they are today. Further evidence for a marked decrease in rainfall near the top of Bed I comes from $\delta^{18}O$ values in pedogenic and groundwater carbonates (Cerling et al., 1977).

Analyses of both micromammalian and macromammalian fauna have largely corroborated the interpretation of the pollen record. Changes in insectivore representation, particularly in the families Erinaceidae, Soricidae and Macroscelididae, suggest that the climate became more arid between FLK NN I and FLK I times (Butler & Greenwood, 1976). Analysis of the habitat affinities of murid rodents led Jaeger (1976) to suggest that wetter, more closed environments were succeeded by drier, more open ones in the middle of Bed I, between FLK NN I (level 2) and FLK I (level 22). FLK N I levels demonstrated a higher proportion of open habitat murids than earlier sites (Jaeger, 1976; Kappelman, 1984).

The Olduvai bovids

Bovidae (antelopes) have been the focus of many Bed I paleoenvironmental studies (Gentry & Gentry, 1978a,b; Kappelman, 1984; Potts, 1988; Shipman & Harris, 1988) because they are the most common macromammalian family at all Olduvai sites (Leakey, 1971; Potts, 1988), and because they are a speciose group inhabiting a broad range of habitats. Most studies have used some form of the Antilopine-Alcelaphine criterion (AAC) developed by Vrba (1980). Vrba found that the tribes Antilopini and Alcelaphini always made up a high proportion (greater than 60%) of the total bovid sample in modern game parks and reserves dominated by open habitats. Areas with greater bush and tree coverage invariably had a much lower (less than 40%) representation of these tribes. These findings were used by Vrba (1980) to evaluate the paleoenvironmental setting of early Pleistocene South African sites.

The AAC was subsequently applied to Bed I assemblages by Potts (1982, 1988) and Kappelman (1984). Kappelman (1984) also used the combined proportion representation of bovids from the tribes Reduncini, Tragelaphini and Hippotragini to indicate closed and/or moist habitats. Potts' calculations were based on his minimum number of individual estimates (MNI); Kappelman used the taxonomic attributions and number of identifiable specimen (NISP) counts provided by Gentry & Gentry (1978a,b). A comparison of MNI and NISP calculations of the AAC shows they produce similar patterns (Figure 1). The DK I levels had moderate to high AAC values, the FLK NN I levels had low values, the FLK I levels had moderate to high values and the FLK N I level values were all high. Their results suggested that during DK I deposition the near-lake environments were mixed but contained significant areas of grassland, that conditions became moister and more closed habitat dominated during FLK NN I deposition, and that there was a trend of increasing aridity through FLK I deposition into FLK N I. Deposition of the latter locality was thought to have taken place during open, arid conditions. Paleocommunity analysis has served to further emphasize the high taxonomic (and by extension habitat) diversity of seven Bed I levels (Potts, 1988).

In an expansion of the AAC, Shipman & Harris (1988) used the proportional representation of three pairs of bovid tribes to elucidate habitat availability in Plio-Pleistocene ecosystems. In particular, the percentage representation of Antilopini plus Alcelaphini was taken to be indicative of arid, open habitats, representation of Bovini plus Reduncini was thought to be suggestive of wetter, closed habitats, and Tragelaphini plus Aepycerotini representation was indicative of closed, dry habitats. Relative proportions of tribes at the Olduvai Bed I sites

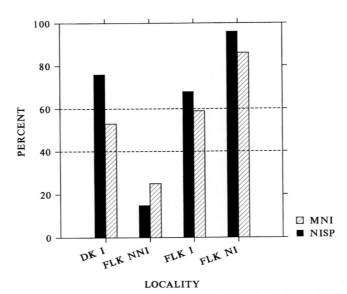

LOCALITY

Figure 1. AAC for Bed I localities calculated using Number of Identifiable Specimens (NISP) and Minimum Numbers of Individuals (MNI). Data from Gentry & Gentry (1978a,b), Kappelman (1984) and Potts (1988). According to Vrba (1980), an AAC of less than 40% is likely to occur in closed habitats, while an AAC of greater than 60% is likely to represent open habitats.

demonstrated a wide range of habitat types between open/arid and closed/wet. In particular, the authors concluded that all of the FLK N I levels clustered with the open/dry habitats, FLK NN I levels 2 and 3 were most likely closed/wet, and that FLK I level 22 and DK I were between the two extremes (Shipman & Harris, 1988).

In summary, past geological and paleontological research at Olduvai has suggested that Bed I deposition occurred under a semi-arid climate generally wetter than occurs at Olduvai today (Hay, 1976; Cerling & Hay, 1986). Faunal (Butler & Greenwood, 1976; Jaeger, 1976; Gentry & Gentry, 1978a,b; Kappelman, 1984; Shipman & Harris, 1988; Potts, 1988) and pollen (Bonnefille, 1984) analyses have suggested that a savanna-mosaic containing tracts of grassland, wooded grassland, woodland and bushland existed around the lake. Waterside vegetation zones including swamps, reedy bogs and riparian woodlands were also present (Potts, 1988). Both faunal and pollen analyses have noted a drying trend from middle to upper Bed I, with the top of Bed I deposited under conditions more arid than today.

Materials and methods

This study concentrates on bovids living in three broadly defined habitat categories: open (grasslands), intermediate (bushland, woodland, swamp, ecotone) and closed (continuous tree canopy, including forest). These characterizations are based on the frameworks of Scott (1979, 1985) and Kappelman (1986, 1988, 1991), and represent a partitioning of the continuum from habitats lacking trees and bushes to those with a continuous tree canopy. Cliff, mountain and high mountain plateau dwelling taxa were excluded. Bovids weighing more than 250 kg (genera *Taurotragus* and *Syncerus*) were excluded to minimize differences in shape and size scaling which occur at the largest body weights (Scott, 1979).

Table 2   **Modern antelope species used in this study, with habitat preferences from Scott (1979). Taxonomy above the species level after Gentry (1992)**

Subfamily **Tribe** *Species*	Habitat type
Bovinae	
Tragelaphini	
Tragelaphus spekei	Intermediate
Tragelaphus scriptus	Closed
Tragelaphus imberbis	Intermediate
Tragelaphus strepsiceros	Intermediate
Tragelaphus euryceros	Closed
Cephalophini	
Cephalophus natalensis	Closed
Cephalophus leucogaster	Closed
Cephalophus sylvicultor	Closed
Cephalophus monticola	Closed
Cephalophus nigrifrons	Closed
Cephalophus dorsalis	Closed
Sylvicapra grimmia	Intermediate
Antilopinae	
Neotragini	
Ourebia ourebi	Intermediate
Raphicerus campestris	Intermediate
Madoqua kirki	Intermediate
Nesotragus moschatus	Closed
Antilopini	
Litocranius walleri	Intermediate
Gazella thomsoni	Open
Gazella granti	Open
Antidorcas marsupialis	Open
Hippotraginae	
Reduncini	
Kobus ellipsiprymnus	Intermediate
Kobus kob	Open
Kobus megaceros	Intermediate
Redunca arundinum	Intermediate
Redunca fulvorufula	Intermediate
Redunca redunca	Intermediate
Hippotragini	
Addax nasomasculatus	Open
Oryx gazella	Open
Oryx leucoryx	Open
Hippotragus equinus	Intermediate
Hippotragus niger	Intermediate
Alcelaphinae	
Aepycerotini	
Aepyceros melampus	Intermediate
Alcelaphini	
Damaliscus lunatus	Open
Damaliscus dorcas	Open
Alcelaphus buselaphus	Open
Connochaetes gnou	Open
Connochaetes taurinus	Open

Table 3   **Dimensionless ratios used in the discriminant function models**

Metacarpal and Metatarsal
  PROXI = PAP/PML
  LENI1 = L/MAXLENG
  LENI3 = 10*(MML/L)
  PDI1 = PML/DML
  PDI3 = (PAP*PML/DAP*DML)
  MIDI2 = (PAP*PML)/(MAP*MML)
  MIDI3 = (DAP*DML)/(MAP*MML)
  MIDI4 = PML/MML
  MIDI5 = DML/MML
  DISI = DAP/DML
  TROI1 = TAPSMALL/TAPLARGE
  TROI2 = TMLMIN/TMLMAX
  TROI3 = TAPLARGE/DAP
  TROI4 = TAPSMALL/DAP
  TROI5 = TMLMIN/DML
  TROI6 = TMLMAX/DML

Metacarpal only
  MAGI = MGAP/MGML
  MAGI2 = MGML/PML

Metatarsal only
  PRONGI = PRONGAP/PRONGML

We measured and analysed a sample of 306 metacarpals and 301 metatarsals of 37 extant African antelope species from the collections of the American Museum of Natural History (New York, NY) and the National Museum of Natural History (Washington, DC) (Table 2). The number of individuals measured per species ranged from 3 to 19 (mean of 8). Measurements were taken on adult animals of both sexes, and wild-shot animals were preferentially measured. Less than 10% of the sample were zoo specimens, many of which had been wild-caught. Fourteen measurements were taken on both the metacarpal and metatarsal (Figure 2). These were used to generate dimensionless ratios reflecting shape and relative proportions of particular morphological features related to habitat preference (Table 3). Summary statistics for each metapodial ratio, by habitat group, are provided in Tables 4 and 5.

In order to insure that morphology rather than body weight was determining habitat group membership, we investigated the relationship between our ratios and body size. Femoral length is strongly correlated with body weight for antelope of the size range investigated here (Scott, 1985). The femoral length of each antelope in our study sample was measured and regressed against metapodial ratio values to test the degree of size-dependence. The correlation coefficient and coefficient of determination between each ratio and femoral length were calculated in the MGLH module of SYSTAT (Wilkinson, 1989) and are presented in Table 6. The amount of variance in most ratios explained by femoral length is low (maximum $R^2 = 0.306$). The observed distribution of each ratio was indistinguishable from a normal and significant correlations were not found between ratio means and variances, so ratios were not log transformed.

Discriminant function analysis was used to test whether metapodial morphology could distinguish among bovids from different habitats (Kappelman, 1986, 1988, 1991; Van Valkenburgh, 1987; Soulonias & Dawson-Saunders, 1988). Discriminant function analysis is a

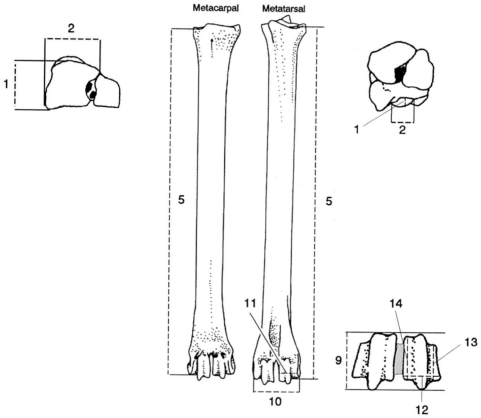

Figure 2. Measurements taken on right metapodials and abbreviations.
Metacarpal
  1. Magnum facet A-P   MGAP
  2. Magnum facet M-L   MGML
Metatarsal
  1. Posterior naviculocuboid facet A-P   PRONGAP
  2. Posterior naviculocuboid facet M-L   PRONGML
Metacarpal and Metatarsal
  3. Proximal articulation A-P (not figured)   PAP
  4. Proximal articulation M-L (not figured)   PML
  5. Functional length   L
  6. Maximum length (not figured)   MAXLENG
  7. Midshaft A-P (not figured)   MAP
  8. Midshaft M-L (not figured)   MML
  9. Distal articulation A-P   DAP
  10. Distal articulation M-L   DML
  11. Medial trochlea minimum M-L   TMLMIN
  12. Medial trochlea maximum M-L   TMLMAX
  13. Medial trochlea minimum A-P   TAPSMALL
  14. Medial trochlea maximum A-P (off ridge)   TAPLARGE

form of classification algorithm which classifies cases into previously determined, naturally occurring groups (James, 1985). Feature selection is an important part of discriminant function analysis, as it is desirable to limit the number of variables (in this case metapodial ratios) in any

Table 4   **Metacarpal ratio statistics of extant bovids by habitat group**

Variable	n	Minimum	Maximum	Mean	Std dev.
LEN13					
Open	118	0·620	1·250	0·901	0·165
Intermediate	121	0·490	1·220	0·841	0·164
Closed	67	0·770	1·610	1·099	0·217
PROXI					
Open	118	0·550	0·750	0·647	0·040
Intermediate	121	0·580	0·760	0·678	0·042
Closed	67	0·610	0·750	0·673	0·033
MAGI					
Open	118	0·800	1·060	0·923	0·059
Intermediate	121	0·820	1·320	0·996	0·084
Closed	67	0·940	1·320	1·083	0·079
MAGI2					
Open	118	0·580	0·770	0·679	0·037
Intermediate	121	0·540	0·800	0·658	0·050
Closed	67	0·550	0·690	0·615	0·035
MIDI4					
Open	118	1·530	2·240	1·748	0·120
Intermediate	121	1·460	1·930	1·658	0·095
Closed	67	1·380	1·910	1·600	0·123
DISI					
Open	118	0·550	0·840	0·682	0·077
Intermediate	121	0·580	0·900	0·732	0·067
Closed	67	0·590	0·810	0·688	0·052
PDI1					
Open	118	0·850	1·110	0·966	0·053
Intermediate	121	0·890	1·170	1·020	0·061
Closed	67	0·890	1·250	1·027	0·067
PDI3					
Open	118	0·770	1·000	0·887	0·054
Intermediate	121	0·780	1·190	0·967	0·084
Closed	67	0·860	1·390	1·033	0·103
TROI2					
Open	118	0·750	0·980	0·872	0·047
Intermediate	121	0·740	0·980	0·866	0·048
Closed	67	0·760	0·960	0·853	0·052
TROI3					
Open	118	0·840	0·940	0·894	0·018
Intermediate	121	0·830	0·950	0·896	0·025
Closed	67	0·880	0·960	0·922	0·020
TROI4					
Open	118	0·700	0·850	0·780	0·028
Intermediate	121	0·680	0·870	0·769	0·034
Closed	67	0·710	0·830	0·775	0·029
TROI5					
Open	118	0·350	0·460	0·402	0·021
Intermediate	121	0·340	0·440	0·394	0·021
Closed	67	0·340	0·430	0·384	0·022

particular model. The STEPDISC stepwise discriminant analysis procedure in PC-SAS was used as an aid in selecting the subset of ratios providing the best discrimination between habitat categories (SAS Institute, Inc., 1985). This procedure also limited intercorrelation between variables within each model.

Table 5   **Metatarsal ratio statistics of extant bovids by habitat group**

Variable	n	Minimum	Maximum	Mean	Std dev.
LENI1					
Open	118	0·920	0·970	0·950	0·011
Intermediate	118	0·920	0·970	0·950	0·010
Closed	65	0·920	0·970	0·946	0·011
LEN13					
Open	118	0·540	1·130	0·807	0·148
Intermediate	118	0·570	1·170	0·815	0·146
Closed	65	0·620	1·440	0·997	0·210
PROXI					
Open	118	0·910	1·140	0·018	0·049
Intermediate	118	0·880	1·150	1·019	0·052
Closed	65	0·880	1·060	0·970	0·047
PRONGI					
Open	118	0·370	0·860	0·589	0·111
Intermediate	118	0·270	0·730	0·490	0·100
Closed	65	0·220	0·540	0·364	0·066
MIDI2					
Open	118	2·030	3·550	2·525	0·266
Intermediate	118	1·720	2·970	2·378	0·238
Closed	65	1·880	2·850	2·279	0·226
MIDI3					
Open	118	1·900	3·020	2·260	0·198
Intermediate	118	1·590	2·850	2·076	0·235
Closed	65	1·420	2·190	1·815	0·169
MIDI5					
Open	118	1·760	2·300	1·954	0·111
Intermediate	118	1·470	2·140	1·735	0·141
Closed	65	1·330	1·750	1·571	0·093
DISI					
Open	118	0·570	0·790	0·687	0·054
Intermediate	118	0·590	0·860	0·734	0·059
Closed	65	0·630	0·810	0·718	0·044
PDI3					
Open	118	0·980	1·280	1·227	0·056
Intermediate	118	0·910	1·380	1·150	0·082
Closed	65	1·060	1·500	1·258	0·100
TROI1					
Open	118	0·780	0·930	0·851	0·031
Intermediate	118	0·760	0·930	0·849	0·041
Closed	65	0·740	0·930	0·830	0·040
TROI2					
Open	118	0·720	0·950	0·839	0·041
Intermediate	118	0·720	0·940	0·831	0·046
Closed	65	0·680	0·930	0·816	0·046
TROI3					
Open	118	0·820	0·920	0·871	0·019
Intermediate	118	0·810	0·960	0·878	0·028
Closed	65	0·820	0·960	0·904	0·029
TROI4					
Open	118	0·670	0·810	0·741	0·031
Intermediate	118	0·640	0·850	0·745	0·046
Closed	65	0·680	0·820	0·749	0·030
TROI6					
Open	118	0·440	0·480	0·457	0·011
Intermediate	118	0·430	0·520	0·460	0·014
Closed	65	0·420	0·470	0·447	0·011

Table 6   The correlation coefficient (R) and the adjusted coefficient of determination (R^2) between femoral length and metapodial ratios of extant bovids

Metacarpal ratio	R	R^2	P	Metatarsal ratio	R	R^2	P
MAGI	0·162	0·023	0·006	PRONGI	0·022	0·000	0·714
MAGI2	0·085	0·004	0·154				
LENI1	0·248	0·058	<0·001		0·333	0·108	<0·001
LENI3	0·532	0·281	<0·001		0·555	0·306	<0·001
PROXI	0·426	0·179	<0·001		0·117	0·028	0·003
MIDI1	0·040	0·000	0·502		0·109	0·008	0·070
MIDI2	0·219	0·045	<0·001		0·131	0·014	0·029
MIDI3	0·105	0·007	0·078		0·019	0·000	0·746
MIDI4	0·015	0·000	0·799		0·104	0·007	0·084
MIDI5	0·271	0·070	<0·001		0·116	0·010	0·053
MIDI6	0·316	0·097	<0·001		0·124	0·012	0·039
MIDI7	0·191	0·033	0·001		0·287	0·079	<0·001
DISI	0·392	0·151	<0·001		0·465	0·213	<0·001
TROI1	0·266	0·068	<0·001		0·365	0·130	<0·001
TROI2	0·257	0·063	<0·001		0·065	0·001	0·276
TROI3	0·050	0·000	0·405		0·133	0·014	0·026
TROI4	0·280	0·075	<0·001		0·399	0·156	<0·001
TROI5	0·026	0·000	0·663		0·195	0·035	0·001
TROI6	0·451	0·200	<0·001		0·269	0·069	<0·001
PDI1	0·433	0·185	<0·001		0·329	0·105	<0·001
PDI2	0·213	0·042	<0·001		0·207	0·039	<0·001
PDI3	0·447	0·197	<0·001		0·166	0·024	0·005

Using the forward selection option, a stepwise discriminant analysis begins with no variables. The variable that contributes most to the discriminatory power of the model, as measured by Wilks' *lambda*, is entered at each step. Additional ratios were entered according to the significance level of an F test from an analysis of covariance, where the variables already chosen act as covariates and the variable under consideration is the dependent variable. The selection process was completed when none of the remaining variables met the default entry criterion (significance level of 0·15). In most cases it was possible to further reduce the number of ratios per model by eliminating those with squared partial correlations with the habitat variable of 3% or less.

The DISCRIM procedure in PC-SAS developed a discriminant model to classify each modern metapodial into one of the three habitat groups (SAS Institute, Inc., 1985). The discriminant function is determined by a measure of the generalized squared (Mahalanobis) distance and can be based on either the pooled covariance matrix (linear discriminant function) or the individual within-group covariance matrices (quadratic discriminant function). The "pool=test" option of the DISCRIM procedure provides a likelihood ratio test of the homogeneity of the within-group covariance matrices to determine which of the two methods is most applicable. If significant differences do not exist, the pooled covariance matrix should be used to compute linear discriminant functions to classify observations (Morrison, 1976). If significant differences exist, quadratic discriminant functions should be computed from the within-group covariance matrices (James, 1985; Reyment, 1991). Canonical coefficients generated for each analysis indicate the relative contributions of different variables to the separation of the three habitat groups.

Table 7   **Olduvai metapodial sample by portion and site**

	Complete	Proximal	Distal	Total
DKI				
MC	2	5	7	14
MT	5	5	5	15
FLK NN I				
MC	11	5	3	19
MT	1	2	0	3
FLK I				
MC	3	16	9	28
MT	3	8	13	24
FLK N I				
MC	33	46	32	111
MT	21	43	41	105
Total	79	130	110	319

MC=metacarpal, MT=metatarsal.

Each observation is placed in the habitat category from which it has the smallest generalized squared distance. The success of the generated model is indicated by multivariate statistics testing the significance of differences between group means, as well as by how well the discriminant function classifies specimens of known habitat. As bones from archeological sites are rarely complete, we used this method to create three separate discriminant function models representing preserved morphology (complete, proximal, distal) using different combinations of ratios for both the metacarpal and the metatarsal. The classification criteria developed can be applied to fossil specimens to provide a taxon-free indication of the habitat classification of extinct antelopes.

Metapodials were analysed from all levels of DK I, FLK NN I, FLK I and FLK N I. These sites were deposited in lake margin sediments on the eastern border of paleolake Olduvai (Hay, 1976). We applied the discriminant function models generated with the modern antelope metapodials to a total of 319 Bed I fossil metapodials housed in the National Museums of Kenya, Nairobi (Table 7). Data for all levels were pooled by locality to maximize sample size. This is justifiable given that antelope taxonomic representation varies only slightly between levels at each locality (Gentry & Gentry, 1978a,b). Most levels yielded artefacts and were accumulated largely through hominid activity (Potts, 1988; Bunn, 1986). The only major exception to this generalization is FLK NN I level 2, which was probably a carnivore accumulation (Potts, 1982, 1988; Bunn, 1986). The proportion representation of the bovid taxa from level 2 is very similar to that of the other FLK NN I levels (Kappelman, 1984; Shipman, 1986; Potts, 1988). Thus, the combination of these levels does not introduce any obvious bias to the investigation of FLK NN I paleoecology. Future analysis by level is planned for the larger samples.

Results

Discriminant function analysis of modern metapodials

The results of the six discriminant function models created to classify complete and fragmentary metapodials into habitat preference categories are presented here. All of

Table 8   **Results of the complete metacarpal discriminant function analysis**

A. Pooled within-class standardized canonical coefficients

	Function 1	Function 2
MAGI	0·57272	0·49126
LENI3	0·75867	0·93425
PDI1	0·50588	− 0·02059
MIDI4	− 0·33687	0·71100
PDI3	0·39197	− 0·29693
TROI3	0·32187	0·33036
TROI4	− 0·29837	0·16772
MAGI2	0·18530	0·63221
% Variance	89	11
Significance	$P=0.0001$	$P=0.0001$

B. Classification results from quadratic discriminant function (total correct=84%)

	Open	Intermediate	Closed	Total
Open	105	13	0	118
	89%	11%	0%	100%
Intermediate	21	92	8	121
	17%	76%	7%	100%
Closed	0	6	61	67
	0%	9%	91%	100%

C. Multivariate statistics testing hypothesis that class means are equal

	$S=2$	$M=2.5$	$n=147$		
Statistic	Value	F	Num DF	Den DF	Pr>F
Wilks' Lambda	0·2424	38·1491	16	592	0·0001
Pillai's Trace	0·9088	30·9165	16	594	0·0001
Hotelling-Lawley Trace	2·5016	46·1234	16	590	0·0001
Roy's Greatest Root	2·2208	82·4477	8	297	0·0001

the likelihood ratio tests of the within-group covariance matrices calculated very small probabilities ($P=0.0001$ to $P=0.0005$), indicating significant differences between the within-group covariance matrices. Consequently, quadratic discriminant functions were used (James, 1985). For all models, multivariate analysis of variance demonstrated highly significant differences ($P=0.0001$) between class means for the 3 habitat categories (Tables 8–13).

The complete metacarpal model calculated two discriminant functions using eight variables (Table 8). The two functions accounted for 89% and 11% of the variance, respectively. The variable coefficients suggest the degree to which each variable contributes to separation of groups on that function. MAGI, PDI1, PDI3 and TROI4 had their greatest values on Function 1, while the LENI3, MIDI4, TROI3 and MAGI2 coefficients were greater on Function 2. This model correctly classified 84% of the modern antelope sample.

Table 9   Results of the proximal metacarpal discriminant function analysis

A. Pooled within-class standardized canonical coefficients

	Function 1	Function 2
PROXI	− 0·06819	1·48795
MAGI	0·98234	− 1·00981
MAGI2	− 0·06744	− 0·74115
% Variance	89	11
Significance	P=0·0001	P=0·0001

B. Classification results from quadratic discriminant function (total correct=60%)

	Open	Intermediate	Closed	Total
Open	90 76%	19 16%	9 8%	118 100%
Intermediate	38 31%	47 39%	36 30%	121 100%
Closed	10 15%	9 13%	48 72%	67 100%

C. Multivariate statistics testing hypothesis that class means are equal

Statistic	S=2 Value	M=0 F	n=149·5 Num DF	Den DF	Pr>F
Wilks' Lambda	0·5523	34·6689	6	602	0·0001
Pillai's Trace	0·4791	31·7089	6	604	0·0001
Hotelling-Lawley Trace	0·7536	37·6796	6	600	0·0001
Roy's Greatest Root	0·6685	67·2964	3	302	0·0001

Classification success varied with habitat category, with 89% of the open country bovids, 76% of the intermediate habitat bovids and 91% of the closed habitat bovids being correctly classified.

The proximal metacarpal model used three variables (Table 9). The two functions accounted for 89% and 11% of the variance, respectively. MAGI had a high positive coefficient on Function 1, while PROXI and MAGI2 had high positive and negative coefficients on Function 2, respectively. This model correctly classified 60% of the total modern antelope sample, with 76% of the open country bovids, 39% of the intermediate habitat bovids and 72% of the closed habitat bovids assigned to the correct habitat type.

Five ratios were used in the distal metacarpal model (Table 10). The two discriminant functions accounted for 76% and 24% of the variance, respectively. All variables except DISI had their highest coefficients on Function 1. Seventy-four percent of the open country bovids, 55% of the intermediate habitat bovids and 81% of the closed habitat bovids were correctly classified, which when combined represents an overall success rate of 68%.

The complete metatarsal model included ten variables (Table 11). Eighty-four percent and 16% of the variance were accounted for by the 1st and 2nd discriminant functions,

Table 10   Results of the distal metacarpal discriminant function analysis

A. Pooled within-class standardized canonical coefficients

	Function 1	Function 2
DISI	0·01923	− 0·92748
TROI2	− 0·66853	0·09006
TROI3	− 0·98163	0·21834
TROI4	0·52772	0·39962
TROI5	1·11570	0·22446
% Variance	76	24
Significance	P=0·0001	P=0·0001

B. Classification results from quadratic discriminant function (total correct=68%)

	Open	Intermediate	Closed	Total
Open	87	22	9	118
	74%	19%	7%	100%
Intermediate	32	66	23	121
	26%	55%	19%	100%
Closed	5	8	54	67
	7%	12%	81%	100%

C. Multivariate statistics testing hypothesis that class means are equal

	S=2	M=1	n=148·5		
Statistic	Value	F	Num DF	Den DF	Pr>F
Wilks' Lambda	0·5388	21·6706	10	598	0·0001
Pillai's Trace	0·5165	20·8874	10	600	0·0001
Hotelling-Lawley Trace	0·7536	22·4570	10	596	0·0001
Roy's Greatest Root	0·5755	34·5286	5	300	0·0001

respectively. The highest coefficients for MIDI3, PRONGI and TROI1 were on Function 1; the other variables had higher coefficients on Function 2. This model correctly classified 89% of the modern antelope sample. As in the metacarpal models, classification success varied with habitat. Ninety-six percent of the open country bovids, 78% of the intermediate habitat bovids and 95% of the closed habitat bovids were correctly classified.

For the proximal metatarsal, the best discrimination was achieved using a two variable model (Table 12). The two discriminant functions accounted for 94% and 6% of the variance, respectively. The largest coefficient for PRONGI was on Function 1, while that for PROXI was on Function 2. This model correctly classified 64% of the open country bovids, 44% of the intermediate habitat bovids and 89% of the closed habitat bovids. The overall success rate for the model was 62%.

Finally, the distal metatarsal model used five variables (Table 13). The two discriminant functions accounted for 74% and 26% of the variance, respectively. All of the ratios except DISI had their highest coefficients with Function 1. Seventy-nine percent of the open country

Table 11　Results of the complete metatarsal discriminant function analysis

A. Pooled within-class standardized canonical coefficients

	Function 1	Function 2
MIDI5	1·05201	−1·35780
MIDI3	−2·03729	−0·92787
TROI6	0·18057	0·47056
PROXI	0·12405	0·26273
PDI3	−1·26977	−1·78415
PRONGI	0·33523	−0·08134
TROI1	0·35429	0·21716
MIDI2	1·65946	2·17476
LENI3	−0·22396	−0·57410
LENI1	−0·04653	−0·42809
% Variance	84	16
Significance	$P=0·0001$	$P=0·0001$

B. Classification results from quadratic discriminant function (total correct=89%)

	Open	Intermediate	Closed	Total
Open	113	5	0	118
	96%	4%	0%	100%
Intermediate	17	92	9	118
	14%	78%	8%	100%
Closed	0	3	62	65
	0%	5%	95%	100%

C. Multivariate statistics testing hypothesis that class means are equal

	$S=2$	$M=3·5$	$n=143·5$		
Statistic	Value	F	Num DF	Den DF	Pr>F
Wilks' Lambda	0·1711	40·9759	20	578	0·0001
Pillai's Trace	1·0876	34·5679	20	580	0·0001
Hotelling-Lawley Trace	3·3340	48·0089	20	576	0·0001
Roy's Greatest Root	2·7925	80·9819	10	290	0·0001

bovids, 59% of the intermediate habitat bovids and 75% of the closed habitat bovids were correctly classified, for an overall success rate of 70%.

In summary, the model which best predicted bovid habitat preference was that for the complete metatarsal, with an overall correct classification rate of 89%. The complete metacarpal model also yielded excellent results, with 84% of the cases correctly classified. The overall success rates of the partial metapodial models were lower, around 60% for the proximal ends and 70% for the distal ends.

Classification errors

Discriminant function analysis inevitably results in some classification errors, when a bovid specimen is incorrectly assigned to a habitat preference group. Error occurrences can be

Table 12   Results of the proximal metatarsal discriminant function analysis

A. Pooled within-class standardized canonical coefficients

	Function 1	Function 2
PRONGI	0·95300	− 0·40911
PROXI	0·14185	1·02735
% Variance	94	6
Significance	$P=0·0001$	$P=0·0001$

B. Classification results from quadratic discriminant function (total correct=62%)

	Open	Intermediate	Closed	Total
Open	76	35	7	118
	64%	30%	6%	100%
Intermediate	37	52	29	118
	31%	44%	25%	100%
Closed	1	6	58	65
	2%	9%	89%	100%

C. Multivariate statistics testing hypothesis that class means are equal

	$S=2$	$M=-0·5$	$n=147·5$		
Statistic	Value	F	Num DF	Den DF	Pr>F
Wilks' Lambda	0·5402	53·5462	4	594	0·0001
Pillai's Trace	0·4814	47·2309	4	596	0·0001
Hotelling-Lawley Trace	0·8112	60·0316	4	592	0·0001
Roy's Greatest Root	0·7586	113·0291	2	298	0·0001

divided into three groups. The first group consists of species in which one half or less of the specimens are misclassified. These misclassifications probably represent normal morphological variation within a species and are not considered further here. The second group includes taxa in which over one half of the specimens are misclassified in a single discriminant function model (Table 14). The last group includes taxa in which over one half of the specimens are misclassified in several discriminant function models (Table 15). Following the methodology of Kappelman (1991), we present possible behavioral and ecological explanations for these errors below.

The misclassifications in Table 14 are from the proximal and distal metapodial models and include taxa from all three habitat categories. These taxa only missed in one of six models, so providing a post hoc explanation for each misclassification may be overinterpreting our results. Discriminant function analysis can not be expected to work perfectly at all times and errors may be more attributable to the limits of the technique than to systematic ecological differences between species. One definite pattern does emerge; open habitat antelopes from the tribe Alcelaphini (*A. buselaphus*, *D. lunatus*, *C. taurinus* and *C. gnou*) were misclassified only by the proximal metatarsal model. Scott (1979) suggested that there are two speed-adapted morphologies among open country antelopes, one emphasizing the forelimb over the hindlimb

Table 13 Results of the distal metatarsal discriminant function analysis

A. Pooled within-class standardized canonical coefficients

	Function 1	Function 2
DISI	0·16912	0·89621
TROI3	0·86687	0·13362
TROI6	− 0·57248	0·45683
TROI2	− 0·33619	0·03683
TROI4	− 0·27559	− 0·00085
% Variance	74	26
Significance	$P=0{\cdot}0001$	$P=0{\cdot}0001$

B. Classification results from quadratic discriminant function (total correct=70%)

	Open	Intermediate	Closed	Total
Open	93 79%	18 15%	7 6%	118 100%
Intermediate	38 32%	70 59%	10 8%	118 100%
Closed	7 11%	9 14%	49 75%	65 100%

C. Multivariate statistics testing hypothesis that class means are equal

	$S=2$	$M=1$	$n=146$		
Statistic	Value	F	Num DF	Den DF	Pr>F
Wilks' Lambda	0·5643	19·4728	10	588	0·0001
Pillai's Trace	0·4857	18·9255	10	590	0·0001
Hotelling-Lawley Trace	0·6833	20·0208	10	586	0·0001
Roy's Greatest Root	0·5090	30·0323	5	295	0·0001

(the Alcelaphini) and the other in which fore- and hindlimbs are of approximately equal importance (the Antilopini). The misclassification of alcelaphine, but not open habitat antilopine, metatarsals may partially reflect these locomotor differences.

Of the 37 species included in this study, 11 had greater than one half of their individuals misclassified by two or more of the discriminant function models (Table 15). The bulk of these misclassifications were by the partial metapodial models. The complete metapodial models were extremely successful, with only two taxa (*Tragelaphus spekei*, *Hippotragus niger*) being mostly misclassified by the metacarpal model, and one taxa (*Sylvicapra grimmia*) by the metatarsal model. The taxa in Table 15 are discussed below.

The distal metacarpals and metatarsals of the springbok, *Antidorcas marsupialis*, were frequently reclassified from open to intermediate habitats. Springbok exhibit a unique behavior; when startled they "pronk", or jump up to 2 m in the air with their limbs in a stiff-legged pose (Dorst & Dandelot, 1970; Estes, 1991). This may affect the morphology of their distal metapodials, and lead to their misclassification.

Table 14   **Summary table of taxa for which more than one-half of the individuals were misclassified in a single discriminant function model**

Species	Reclassified/total[1]		Original habitat	Most likely reclassification
Alcelaphus buselaphus	7/13	TP	Open	Intermediate
Damaliscus lunatus	6/10	TP	Open	Intermediate
Connochaetes taurinus	6/9	TP	Open	Intermediate
Connochaetes gnou	2/3	TP	Open	Intermediate
Gazella thomsoni	13/19	CD	Open	Intermediate
Tragelaphus strepsiceros	6/6	CP	Intermediate	Closed
Tragelaphus imberbis	2/3	CP	Intermediate	Open, Closed
Madoqua kirki	5/9	CD	Intermediate	Closed
Raphicerus campestris	5/6	CP	Intermediate	Open
Redunca fulvorufula	5/9	CP	Intermediate	Open
Cephalophus leucogaster	6/10	CP	Closed	Open
Cephalophus sylvicultor	5/9	CP	Closed	Open
Nesotragus moschatus	3/4	TD	Closed	Intermediate

[1]By Discriminant function model: CC—Complete metacarpal, CP—Proximal metacarpal, CD—Distal metacarpal, TC—Complete metatarsal, TP—Proximal metatarsal, TD—Distal metatarsal.

The proximal metapodials of *Kobus kob*, another open country antelope, were frequently reclassified into the intermediate habitat category. The kob was also reclassified from open to intermediate habitat in the discriminant function analysis of the antelope femur (Kappelman, 1991). Kob are generally found in green, well-watered pastures (Estes, 1991). They are the swiftest members of their tribe, but they are less cursorial than antilopines and alcelaphines and sometimes seek cover when pursued. This suggests that they may depend on cover more than other open habitat antelopes.

The swamp-dwelling sitatunga (*Tragelaphus spekei*) is reclassified from intermediate to closed habitats by all of the metacarpal models and the proximal metatarsal model. Discriminant function analysis of the antelope femur led to the same reclassification (Kappelman, 1991). Like many forest antelopes, the sitatunga is primarily solitary and uses a series of connected paths weaving through thick cover (Kappelman, 1991; Leuthold, 1977). Locomotor patterns required for its swamp microhabitat may closely approximate those needed to move in a forest substrate.

The grey duiker, *Sylvicapra grimmia*, was misclassified from intermediate to closed habitat by both the proximal metacarpal and the complete metatarsal models. Its femora were also reclassified in the same way (Kappelman, 1991). It is classified by Kappelman (1988 after Scott, 1979, 1985) as a "run-to-cover" form which often lies in tall grass to avoid predators (Leuthold, 1977). This more sedentary predator avoidance strategy may be reflected in its reclassification to closed habitats. Alternatively, this misclassification may be linked to its phylogeny, as the other members of its tribe (Cephalophini) are forest inhabitants (Scott, 1979; Estes, 1991).

Aepyceros melampus, *Hippotragus niger* and *H. equinus* are all intermediate habitat forms which were reclassified by two or more models into the open habitat category. This same reclassification occurred in the discriminant function analysis of their femora (Kappelman,

Table 15

Summary of taxa for which more than one-half of the individuals studied were incorrectly classified by more than one discriminant function model

Species	Reclassified/total[1]		Original habitat	Most likely reclassification
Antidorcas marsupialis	5/7	CD	Open	Intermediate
	6/8	TD		
Kobus kob	6/10	CP	Open	Intermediate
	9/10	TP		
Tragelaphus spekei	4/7	CC	Intermediate	Closed
	6/7	CP		
	7/7	CD		
	7/7	TP		
Sylvicapra grimmia	6/10	CP	Intermediate	Closed
	5/7	TC		
Aepyceros melampus	7/10	CP	Intermediate	Open
	5/8	TP		
Hippotragus niger	8/11	CC	Intermediate	Open
	10/11	CP		
	8/11	CD		
	8/11	TP		
	9/11	TD		
Hippotragus equinus	4/4	CP	Intermediate	Open
	3/4	CD		
	3/4	TP		
	4/4	TD		
Redunca redunca	5/9	TP	Intermediate	Open
	6/9	TD		
Kobus megaceros	5/6	CD	Intermediate	Closed
	6/7	TD		Open
Kobus ellipsiprymnus	5/7	CP	Intermediate	Closed
	6/7	CD		Open, Closed
	5/6	TP		Open
	5/6	TD		Open
Tragelaphus scriptus	6/11	CD	Closed	Intermediate
	4/7	TD		

[1]By Discriminant function model. Abbreviations as in Table 14.

1991). All three of these taxa are edge specialists or ecotonal, feeding in the open and running to heavy cover when threatened (Scott, 1979; Estes, 1991). They are probably subject to selection pressures for cursorial locomotion during the time they spend in the open. Additionally, *A. melampus* is known as a leaper, which may result in adaptations similar to those for cursoriality (Scott, 1979).

Redunca redunca, the bohor reedbuck, is a grazing antelope found close to water. It lies-out in long grass or reed beds during the heat of the day, but at night emerges from thick vegetation to graze (Kingdon, 1982). Where it overlaps with the southern reedbuck (*R. arundinum*), the bohor is found in more open settings. The reclassification of its proximal and distal metatarsal from intermediate to open may reflect this more frequent exploitation of open microhabitats.

Kobus megaceros is a little-studied reduncine which lives in papyrus swamps. The distal metacarpal model placed it in closed habitat, while the distal metatarsals were classified as open. In her study of antelope limb proportions, Scott (1979) suggested that the forelimbs and hindlimbs of swamp dwelling antelopes (*T. spekei*, *K. leche* and *K. megaceros*) were modified in different directions. The forelimb has been modified in the direction of force, while the hindlimb shows speed or bounding adaptations. The forelimb modifications may be related to the powerful swimming abilities of these taxa, while bounding may be a useful strategy on mucky ground. The contradictory results of the two distal metapodial models may reflect these different limb modifications. A somewhat similar situation exists with our *T. spekei* results; all three metacarpal models were suggestive of closed habitat, while two of the three metatarsal models suggested intermediate habitat.

Kobus ellipsiprymnus, the waterbuck, is one of the most water-dependent of all antelopes (Estes, 1991). It is a grazer most commonly found in woodland clearings close to water, running to heavy cover when threatened (Dorst & Dandelot, 1970). The contradiction between its metacarpal (generally from intermediate to closed) and metatarsal (from intermediate to open) misclassifications is reminiscent of our *K. megaceros* results. The waterbuck is known to enter water to escape predators and occasionally to feed (Kingdon, 1982; Estes, 1991) and its misclassifications may partially reflect this.

The only closed habitat species misclassified in more than one model (the distal metapodial models) is *Tragelaphus scriptus*, the bushbuck. Its distal metapodials were reclassified into the intermediate habitat category. Bushbuck live in a wide range of habitats (e.g., forest, forest-fringe, forest-clearings) but are essentially dependent on thick cover (Dorst & Dandelot, 1970; Kingdon, 1982; Estes, 1991). The reclassification of its distal metapodials may reflect selection pressures while in ecotonal settings.

Analysis of Olduvai antelope metapodials
The reclassifications discussed above generally agree with more detailed examinations of each species' habitat characteristics. This suggests that metapodial morphology can provide a means for distinguishing among antelopes from different habitats. The classification criteria developed were applied to complete, proximal and distal Olduvai metapodials to infer the habitat preferences of the antelopes collected by Bed I hominids.

For both metacarpal and metatarsal, the overall success rates of the complete (84% and 89%), proximal (60% and 62%) and distal (68% and 70%) models were similar (Tables 8–13). In the following figures we combine the classification results of the Olduvai metacarpals and metatarsals, by model. Systematic error is potentially introduced by combining the results of the proximal metacarpal and metatarsal models. The proximal metatarsal model correctly classified fewer modern open habitat antelopes (64% versus 76%) and more closed habitat antelopes (89% versus 72%) than the proximal metacarpal model. However, we feel the benefits of increased sample size and greater clarity of presentation outweigh this concern.

In Figures 3–6 the results of the discriminant function analyses are presented from the oldest to youngest locality. Habitat assignments are presented by fossil portion, and results from all models are totaled in the final column. Approximately 60% of the complete metapodials from DK I are classified as intermediate, but samples of distal and proximal fragments contribute larger proportions of bones with the open habitat morphology to the total (Figure 3). FLK NN I is striking in its high representation of bovids with the closed habitat morphology, in similar proportions for both complete bones and the small number of distal fragments (Figure 4). The small sample of FLK I complete metapodials is dominated by intermediate and closed habitat morphologies, but the distal and proximal samples include a greater proportion of the open habitat morphology (Figure 5). The largest sample of metapodials was from FLK N I (Figure

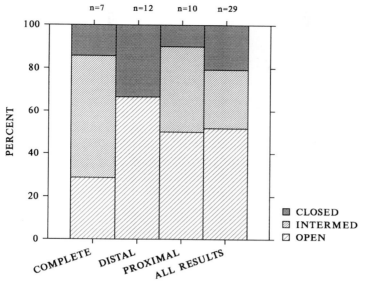

Figure 3. Results of discriminant function analyses of DK I metapodials, by metapodial portion.

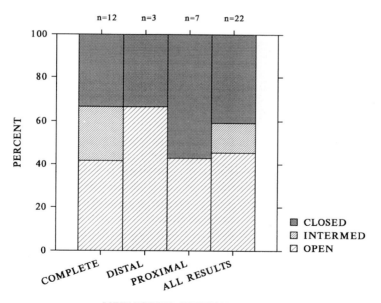

Figure 4. Results of discriminant function analyses of FLK NN I metapodials, by metapodial portion.

6). Proportionately more of the complete fossils were classified as having open habitat morphologies, but all models classify some bones into the intermediate and closed habitat categories.

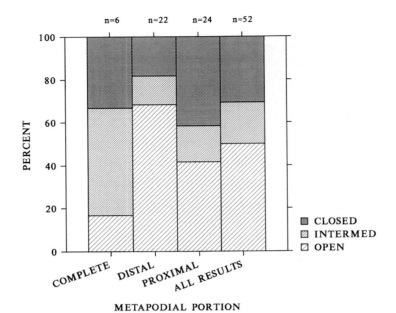

Figure 5. Results of discriminant function analyses of FLK I metapodials, by metapodial portion.

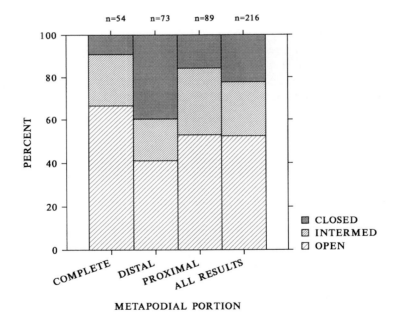

Figure 6. Results of discriminant function analyses of FLK N I metapodials, by metapodial portion.

The merged classifications from all of the metapodial models (Figure 7) suggest that the proportion of open habitat bovids in the assemblages of DK I, FLK NN I and FLK I was relatively constant, with most fluctuation occurring in the proportion of specimens with

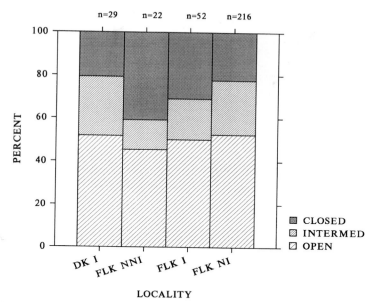

Figure 7. Summary of all results for Bed I localities in this study, presented in chronological order from left to right. Note the relatively high proportions of closed and intermediate habitat bovids at all sites.

intermediate and closed habitat morphologies. The FLK NN I and FLK N I assemblages, respectively, represent the extremes in closed and open habitat representation. All localities show higher proportions of closed and intermediate habitat groups than would be expected from AAC or other taxon-based paleohabitat reconstructions. Further, no locality has a proportion of open-habitat morphologies greater than 52%.

Discussion

Interpretable results relating bovid remains to past environments can be derived without relying on taxonomy. The results presented here are in broad agreement with other faunal and isotopic reconstructions of Olduvai paleoenvironments (e.g., Kappelman, 1984; Potts, 1988; Cerling et al., 1977; Cerling & Hay, 1986), particularly the proposed increase in aridity from middle to upper Bed I. There is a marked reduction in the proportion of metapodials with closed habitat morphologies and an increase in those with open habitat morphologies from FLK NN I to FLK N I (Figure 7). Although some localities have small metapodial samples and classification results are pooled from different discriminant function models, the concurrence between our results and those of previous studies suggest that the totals in Figure 7 accurately reflect past environmental conditions at Olduvai.

Bovid morphology, taxonomy and habitat preference
Our results differ markedly from taxon-based analyses of the bovid samples. Three of the assemblages studied here (DK I, FLK I and FLK N I) have AAC values near to or above 60 percent whether calculated using number of identifiable specimens (NISP) or minimum number of individuals (MNI) (Figure 1). This suggests paleoenvironments that were

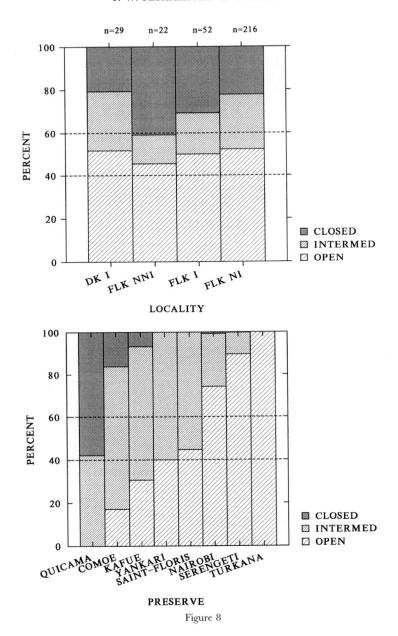

Figure 8

open-habitat dominated. However, in our study the percentages of fossils with open habitat morphologies is approximately 50% for each of these localities. Even the complete metapodials from FLK N I, a relatively large sample ($n=54$) classified by models with high success rates, had a much lower proportion of open morphologies (67%, Figure 6) than would be expected from the AAC (86–96%, Figure 1). What could explain this discrepancy between AAC calculations and the discriminant function results?

In this study, complete metacarpals of the most common Bed I antilopine, *Antidorcas recki*, are consistently classified as having an intermediate habitat morphology. This contradicts the usual assumption of open-habitat preference for this tribe. *Antidorcas recki* is not an isolated case: bovids from several other tribes have also been classified by this method into groups other than those predicted by their taxonomy. Two interpretations are suggested by these results. The first possibility is that the observed locomotor adaptations of fossil taxa are truly reflective of contemporary paleoenvironment. Although the Olduvai antelopes exhibit the range of modern morphological adaptations, it does not therefore follow that the modern relationship of taxonomy to habitat preference extends to the Bed I sample. Paleoecologists would have to uncouple taxonomy from assumptions about antelope habitat preference for time periods as recent as the Plio-Pleistocene. The presence of a high proportion of *A. recki* even during FLK N I deposition, commonly considered to be a period of great aridity, may instead indicate the presence of significant patches of intermediate habitats in the lake margin zone.

The alternative possibility, less convincing in our view, is that there was an evolutionary lag between habitat usage and morphological change. Thus Olduvai antilopines may have been exploiting open habitats even though they did not possess all of the modern morphological correlates of that habitat preference. If this were the case, the discriminant function assignments would underestimate the proportions of the bovid sample living in open country, and thus would be lower than the AAC. However, our results still show the same general trend noted with the AAC calculations; the maximum amount of closed habitat was during FLK NN I deposition, while the most open habitat was present during FLK N I deposition. Further functional anatomical and microwear studies of the fauna as well as isotopic and pollen analyses of the relevant stratigraphic sequences should provide support for one or the other alternative in the future.

Hominid habitat specificity

Our results have important implications for interpreting the paleoenvironmental context of hominid foraging. On the basis of actualistic studies in the Serengeti, Blumenschine (1986, 1987) has argued that Plio-Pleistocene hominids could have filled a dry season

Figure 8. Habitat preference representation from this study (top) compared with a subset of the antelope census data from modern game parks summarized in Shipman & Harris (1988) (bottom). Modern genera *Alcelaphus*, *Connochaetes*, *Damaliscus*, *Antidorcas*, *Gazella*, and *Oryx* were included in the open habitat group. *Aepyceros*, *Kobus*, *Redunca*, *Hippotragus*, *Ourebia*, *Raphicerus*, and *Sylvicapra* were included in the intermediate habitat group. *Cephalophus* was included in the closed habitat group. *Tragelaphus* census data were included in the intermediate and/or closed habitat groups, depending on the habitat preference of the species in question. The proportion of closed habitat for Quiçama may be overrepresented, as *Tragelaphus* species attributions were not available, and all counts for this genus were included in the closed habitat category. Census data from the following representative wildlife preserves is included in this figure:
Quiçama, Angola. Zambezian phytochorion—100% woodland.
Comoe, Ivory Coast. Sudanian phytochorion—73% woodland, Guinea-Congoland/Sudanian phytochorion—17% forest/grassland.
Kafue National Park, Zambia. Zambezian phytochorion—2% forest, 89% woodland, 9% grassland.
Yankari, Nigeria. Sudanian phytochorion—100% woodland.
Saint-Floris, Central African Republic. Sudanian phytochorion—100% woodland.
Nairobi National Park, Kenya. Somalia-Masai phytochorion—100% bushland/grassland.
Serengeti Ecological Unit, Tanzania. Somalia-Masai phytochorion—73% bushland, 19% grassland, 8% bushland/grassland.
Lake Turkana Game Reserve, Kenya. Somalia-Masai phytochorion—73% bushland, 27% bushland/grassland.

scavenging niche in a riparian woodland setting. According to the model, hyenas avoid riparian woodlands, which would thus be more resource-rich and safer for scavenging hominids. Hunting and scavenging in open habitats is thought to be too risky for hominids which, in this interpretation, depended upon trees for refuge from predators. In an expansion of this idea, Cavallo & Blumenschine (1989) noted that tree-stored leopard kills could have been scavenged at little risk by hominids. Marean (1989) suggested that large scale scavenging opportunities were potentially available from sabertooth cat kills in closed habitats (woodland to forest).

The metapodials from each of the sampled localities span the range of morphologies in modern antelopes from open, intermediate and closed habitats. This result suggests two alternatives: either antelopes with a wide range of habitat preferences were available to hominids in a preferred habitat zone (e.g., riparian woodlands), or hominids were ranging through and collecting faunal remains from a variety of habitats. Analysis of modern bovid census and mortality data reveals that antelopes do, in general, die where they habitually live (Shipman & Harris, 1988, using data from Behrensmeyer & Dechant Boaz, 1980). In a modern setting, it would be unusual to find bovids with such diverse habitat preferences as documented here naturally accumulating in one habitat zone.

During the dry season in the Serengeti, open country antelopes travel through riparian woodlands to reach permanent water sources (Blumenschine, 1986). If the modern Serengeti provides an accurate analogy for the ecosystem dynamics around paleolake Olduvai, then hominids could have scavenged and/or hunted open country bovids without leaving a riparian woodland setting. However, the best dry season scavenging opportunities in the Serengeti are provided by only a few species of migratory ungulates. The Bed I Olduvai site faunas are very diverse, are not dominated by a few taxa, and include species which were almost certainly resident (browsing ungulates, suids, primates) (Potts, 1988). None of the Bed I ungulate taxa are known to have been migratory, and there is no evidence for seasonal ungulate migrations on the scale of those of the modern Serengeti during the East African Plio-Pleistocene.

The Bed I assemblages thought to have been primarily collected by carnivores shed additional light on this issue. The available data suggest that lions, leopards, cheetahs, hyenas and wild dogs kill and scavenge antelopes with different habitat preferences in proportions reflecting local relative abundance (Kruuk & Turner, 1967; Pienaar, 1969; Kruuk, 1972; Schaller, 1972; Hill, 1983, 1989). Two predominantly carnivore accumulated assemblages are known from Bed I: FLK NN I level 2 from Middle Bed I and Long K from Upper Bed I (Potts, 1988). Antelope taxa at FLK NN I level 2 and Long K are present in similar proportions to those found at the hominid assemblages with which they were roughly coeval (FLK NN I level 3 and the FLK N I levels, respectively) (Gentry & Gentry, 1978a,b; Kappelman, 1984; Shipman, 1986; Potts, 1988). The FLK NN I level 2 and Long K assemblages are similar to those of spotted hyenas (*Crocuta crocuta*) in their patterns of collection and modification and occur in the same lake margin facies as the hominid accumulated levels. The co-occurrence of hyena-like and hominid activity in one environmental setting conflicts with a basic premise of the riparian woodlands scavenging model (Blumenschine, 1986, 1987). The similarity in antelope composition between the hominid and carnivore formed accumulations is a further contrast, as presumably the carnivores were not hunting/scavenging solely in the riparian woodlands. It seems possible that the relative proportions of bovids with different habitat preferences in the Bed I carnivore assemblages reflect their availability in the paleocommunity.

These results suggest that hominid foraging was more generalized and exhibited more habitat diversity than is suggested by Blumenschine's (1986, 1987) riparian woodlands scavenging model [see Bunn & Ezzo (1993) for additional criticism]. Whether obtained through hunting or scavenging, the antelope remains collected by hominids provide an indirect measure of the habitats they were ranging through. The bovid metapodials at each of the sampled localities span a range of morphologies, strongly suggesting that hominids were collecting antelope remains from a variety of habitats, very possibly the full range available to them. This is true even during the environmental extremes represented by FLK NN I and FLK N I. The high taxonomic diversity of Bed I sites primarily accumulated by hominids led Potts (1982, 1984, 1988) to conclude the hominids transported faunal remains from taxa living in more than one habitat, exploiting both open, grassy plains and a more densely vegetated zone. His work has suggested that hominids ranged through and collected faunal resources from a habitat mosaic existing in and around the ancient lake margin. The results presented here support this conclusion.

Habitat abundance

Our results suggest striking differences between the overall habitat composition of the Olduvai lake margin zone and the proportion representation of habitats in modern African ecosystems (Figure 8). Shipman & Harris (1988) noted that the proportional representation of bovid tribes in modern African game parks reflects well the habitat classifications of those parks based on vegetation categories. None of the East African game parks whose census data are summarized by Shipman & Harris (1988) has bovid taxon counts which reflect the proportions of habitat preferences present at Olduvai Bed I localities. Specifically, the lake margin setting of the Bed I localities contained a higher proportion of closed habitats than modern open-habitat dominated ecosystems. Moreover, there may have been significant differences in the way habitats were structured during the Plio-Pleistocene (Cerling, 1992). Analogues for Olduvai should be chosen from settings with greater proportions of intermediate and closed habitats (e.g., Tappen, 1990). Modern grassland-dominated ecosystems, such as the Serengeti, are inappropriate analogues for Olduvai paleoenvironments during Bed I times. Moreover, the extent of environmental variation through Bed I argues against the use of a single modern analogue for the entire sequence. Data permitting, analogies should be carried out by level, as in Shipman & Harris (1988).

Conclusions

Several points can be drawn from this study. Taxon-free analysis of antelope metapodials can provide meaningful paleoenvironmental information. More taxon-free analyses should be carried out in the future. The presence of metapodials with a wide range of morphologies is consistent with the great taxonomic diversity at the Olduvai sites. The proposed increase in aridity from middle to upper Bed I noted by previous paleoenvironmental studies was also noted here. More importantly, our results suggest the presence of a higher proportion of intermediate and closed habitats in the paleolake margin zone during Bed I times than has been proposed in the past. These results also suggest that the paleoenvironmental setting at Olduvai was considerably different from that of modern, open-habitat dominated East African ecosystems. The evidence indicates that hominids were foraging in a variety of habitats, possibly the full range available to them around the paleolake.

Acknowledgements

We would like to thank the Government of the United Republic of Tanzania for permission to study the Olduvai materials. The staffs of the National Museums of Kenya, the American Museum of Natural History, the National Museum of Natural History and the Natural History Museum, London, provided invaluable assistance. Support by the Human Origins Program of the Smithsonian Institution (T.W.P.) is gratefully acknowledged. This paper was greatly improved through discussions with Rick Potts, Andrew Hill, Stanley Ambrose, Alan Gentry and Lee-Ann Hayek and constructive reviews from Kay Behrensmeyer, John Kappelman and Pat Shipman. We are grateful to Jennifer Clark for drawing the metapodial figures. Finally, thanks to Jim Oliver, Nancy Sikes and Kathy Stewart for including us in this symposium.

References

Behrensmeyer, A. K. & Dechant Boaz, D. (1980). The recent bones of Amboseli Park, Kenya in relation to East African paleoecology. In (A. K. Behrensmeyer & A. P. Hill, Eds) *Fossils in the Making*, pp. 72–92. Chicago: University of Chicago Press.

Blumenschine, R. J. (1986). *Early Hominid Scavenging Opportunities: Implications of Carcass Availability in the Serengeti and Ngorongoro Ecosystems*. Oxford: B.A.R. International Series 283.

Blumenschine, R. J. (1987). Characteristics of an early hominid scavenging niche. *Curr. Anthrop.* **28,** 383–407.

Bonnefille, R. (1984). Palynological research at Olduvai Gorge. *Nat. Geog. Soc. Res. Rep.* **17,** 227–243.

Bonnefille, R., Lobreau, D. & Riollet, G. (1982). Pollen fossile de Ximenia (Olacaceae) dans le Pliestocene Inferieure d'Olduvai en Tanzanie: implications paleoecologiques. *J. Biogeog.* **9,** 469–486.

Brain, C. K. (1981). *The Hunters or the Hunted?* Chicago: University of Chicago Press.

Bunn, H. T. (1986). Patterns of skeletal representation and hominid subsistence activities at Olduvai Gorge, Tanzania, and Koobi Fora, Kenya. *J. hum. Evol.* **15,** 673–690.

Bunn, H. T. & Ezzo, J. A. (1993). Hunting and scavenging by Plio-Pleistocene hominids: nutritional constraints, archaeological patterns and behavioural implications. *J. Archaeol. Sci.* **20,** 365–398.

Butler, P. M. & Greenwood, M. (1976). Elephant shrews (Macroscelididae) from Olduvai and Makapansgat. In (R. J. G. Savage & S. C. Coryndon, Eds) *Fossil Vertebrates of Africa* **4,** 1–56. London: Academic Press.

Cavallo, J. & Blumenschine, R. (1989). Tree-stored leopard kills: expanding the hominid scavenging niche. *J. hum. Evol.* **18,** 393–399.

Cerling, T. E. (1992). Development of grasslands and savannas in East Africa during the Neogene. *Palaeogeogr., Palaeoclimat., Palaeoecol.* **97,** 241–247.

Cerling, T. E., Hay, R. L. & O'Neil, J. R. (1977). Isotopic evidence for dramatic climatic changes in East Africa during the Pleistocene. *Nature* **267,** 137–138.

Cerling, T. E. & Hay, R. L. (1986). An isotopic study of paleosol carbonates from Olduvai Gorge. *Quat. Res.* **25,** 63–78.

Dorst, J. & Dandelot, P. (1970). *A Field Guide to the Larger Mammals of Africa.* London: Collins.

Estes, R. D. (1991). *The Behavior Guide to African Mammals.* Los Angeles: The University of California Press.

Gentry, A. W. (1970). The Bovidae (Mammalia) of the Fort Ternan fossil fauna. In (L. S. B. Leakey & R. J. G. Savage, Eds) *Fossil Vertebrates of Africa* **2,** 243–323. London: Academic Press.

Gentry, A. W. (1992). The subfamilies and tribes of the family Bovidae. *Mammal Rev.* **22,** 1–32.

Gentry, A. W. & Gentry, A. (1978a). Fossil Bovidae (Mammalia) of Olduvai Gorge, Tanzania. Part I. *Bull. Br. Mus. (Nat. Hist.) Geol. series* **29,** 289–446.

Gentry, A. W. & Gentry, A. (1978b). Fossil Bovidae (Mammalia) of Olduvai Gorge, Tanzania. Part II. *Bull. Br. Mus. (Nat. Hist.) Geol. series* **30,** 1–83.

Hay, R. L. (1976). *Geology of the Olduvai Gorge.* Berkeley: University of California Press.

Hill, A. (1983). Hyenas and early hominids. In (J. Clutton-Brock & C. Grigson, Eds) *Animals and Archaeology, Volume 1: Hunters and their Prey.* Oxford: B.A.R. International Series **163,** 87–92.

Hill, A. (1989). Bone modification by modern spotted hyenas. In (R. Bonnichsen & M. Sorg, Eds) *Bone Modification.* Center for the Study of the First Americans. Orono: University of Maine, pp. 169–178.

Jaeger, J. J. (1976). Les rongeurs (Mammalia, Rodentia) du Pleistocene inferieur d'Olduvai Bed I (Tanzanie) Iiere partie: les Murides. In (R. J. G. Savage & S. C. Coryndon, Eds) *Fossil Vertebrates of Africa* **4,** pp. 57–120. London: Academic Press.

James, M. (1985). *Classification Algorithms.* New York: John Wiley & Sons.

Kappelman, J. (1984). Plio-Pleistocene environments of Bed I and lower Bed II, Olduvai Gorge, Tanzania. *Palaeogeogr., Palaeoclimat., Palaeoecol.* **48,** 171–196.

Kappelman, J. (1986). The Paleoecology and Chronology of the Middle Miocene Hominoids from the Chinji Formation of Pakistan. Ph.D. Dissertation, Harvard University, Cambridge, Massachusetts.

Kappelman, J. (1988). Morphology and locomotor adaptations of the bovid femur in relation to habitat. *J. Morph.* **198,** 119–130.

Kappelman, J. (1991). The paleoenvironment of *Kenyapithecus* at Fort Ternan. *J. hum. Evol.* **20,** 95–129.

Kingdon, J. (1982). *East African Mammals. An Atlas of Evolution in Africa.* III. Parts C and D (Bovids). London: Academic Press.

Kruuk, H. (1972). *The Spotted Hyena.* Chicago: University of Chicago Press.

Kruuk, H. & Turner, M. (1976). Comparative notes on predation by lion, leopard, cheetah and wild dog in the Serengeti area, East Africa. *Mammalia* **31,** 1–27.

Leakey, M. D. (1971). *Olduvai Gorge. 3. Excavations in Beds I and II, 1960–1963.* Cambridge: Cambridge University Press.

Leuthold, W. (1977). *African Ungulates: A Comparative Review of their Ethology and Behavioral Ecology.* New York: Springer-Verlag.

Marean, C. W. (1989). Sabertooth cats and their relevance for early hominid diet and evolution. *J. hum. Evol.* **18,** 559–582.

Morrison, D. F. (1976). *Multivariate Statistical Methods.* New York: McGraw-Hill.

Pienaar, U. de V. (1969). Predator-prey relationships amongst the larger mammals of the Kruger National Park. *Koedoe* **12,** 108–187.

Potts, R. (1982). Lower Pleistocene site formation and hominid activities at Olduvai Gorge, Tanzania. Ph.D. Dissertation, Harvard University. Cambridge, Massachusetts.

Potts, R. (1984). Home bases and early hominids. *Am. Sci.* **72,** 338–347.

Potts, R. (1988). *Early Hominid Activities at Olduvai Gorge.* Hawthorne, NY: Aldine de Gruyter.

Reyment, R. A. (1991). *Multidimensional Palaeobiology.* New York: Pergamon Press.

SAS Institute, Inc. (1985). *SAS User's Guide: Statistics, Version 5 Edition.* Cary, NC: SAS Institute, Inc.

Schaller, G. B. (1972). *The Serengeti Lion.* Chicago: University of Chicago Press.

Scott, K. M. (1979). Adaptation and allometry in bovid postcranial proportions. Ph.D. Dissertation. Yale University. New Haven, Connecticut.

Scott, K. M. (1985). Allometric trends and locomotor adaptations in the Bovidae. *Bull. Am. Mus. Nat. Hist.* **197,** 197–288.

Shipman, P. (1986). Studies of hominid-faunal interactions at Olduvai Gorge. *J. hum. Evol.* **15,** 691–706.

Shipman, P. & Harris, J. M. (1988). Habitat preference and paleoecology of *Australopithecus boisei* in Eastern Africa. In (F. E. Grine, Ed.) *Evolutionary History of the "Robust" Australopithecines,* pp. 343–381. Hawthorne, NY: Aldine de Gruyter.

Solounias, N. & Dawson-Saunders, B. (1988). Dietary adaptations and paleoecology of the late Miocene ruminants from Pikermi and Samos in Greece. *Palaeogeogr., Palaeoclimat., Palaeoecol.* **65,** 149–172.

Tappen, M. J. (1990). Present day bone deposition in Parc des Virunga, Zaire. (Abstract). 6th meeting of the International Council on Archaeozoology. Washington, D.C.: Smithsonian Institution.

Van Valkenburgh, B. (1987). Skeletal indicators of locomotor behavior in living and extinct carnivores. *J. Vert. Paleont.* **7,** 162–182.

Vrba, E. S. (1980). The significance of bovid remains as an indicator of environment and predation patterns. In (A. K. Behrensmeyer & A. P. Hill, Eds) *Fossils in the Making,* pp. 247–272. Chicago: University of Chicago Press.

Walter, R. C., Manega, P. C., Hay, R. L., Drake, R. E. & Curtis, G. H. (1991). Laser-fusion ^{40}Ar/^{39}Ar dating of Bed I, Olduvai Gorge, Tanzania. *Nature* **354,** 145–149.

Wilkinson, L. (1989). *SYSTAT: The System for Statistics.* Evanston, IL: SYSTAT, Inc.

Henry M. McHenry
Department of Anthropology, University of California, Davis, CA 95616, U.S.A.

Received 24 June 1993
Revision received 16 December 1993 and accepted 17 December 1993

Keywords: Australopithecus, Homo habilis, Homo erectus, body weight, encephalization, sexual dimorphism, life history

Behavioral ecological implications of early hominid body size

Opinions vary but one interpretation of the vastly expanded hominid postcranial sample from 4 to 1·3 M.Y. hominids leads to the following working hypotheses: (1) The average female of the earliest known species of Hominidae, *Australopithecus afarensis,* weighed about 29 kg and the male, 45 kg. (2) Average female body weight remains between 29 and 34 kg in all species of hominids before the appearance of *Homo erectus* at 1·7 M.Y. (3) Average male body weight ranged between 40 and 52 kg in these pre-*erectus* species. (4) The origin of *H. erectus* marked a dramatic increase in body size especially in the female. (5) The brain size increase from *A. afarensis* to *A. africanus* to the "robust" australopithecines does not appear to be an artifact of body size increase but reflects progressive encephalization. (6) The expansion of absolute brain size with the appearance of *Homo* is beyond what would be expected from body size increase alone. These working hypotheses have implications for how members of early hominid species behaved to enhance their chances of survival and reproduction within the constraints and requirements of their environments. For example, the relatively high level of body size sexual dimorphism in the earliest species implies a polygynous mating system and a ranging pattern in which females foraged in smaller territories than the males. Although one might expect from analogy with *Pan* to have social groups consisting of closely related males and less closely related females, the high level of sexual dimorphism is not expected. Perhaps the substantial body size increase and reduction in sexual dimorphism apparent in *Homo* by 1·7 M.Y. is related to a significant expansion in ranging area. The energetic requirement of the expanded brain may imply altered feeding strategies in both the "robust" australopithecine and *Homo* lineages.

Journal of Human Evolution (1994) **27,** 77–87

Introduction

Behavioral ecology is the study of what animals do to enhance their chances of survival and reproductive success within the constraints and requirements of their environment (Krebs & Davies, 1981). It is useful to refer to what animals do in terms of problem-solving strategies to optimize reproductive success, although such language must be understood as a kind of shorthand (Foley, 1987). Obviously animals are not making decisions about social organization, ranging pattern, and predator avoidance as a conscious strategy. Terms such as strategy, decision-making, and optimality are meant to be analogous to how natural selection leads to adaptation.

The interaction between the environment and the behavior of an animal is complex. Because of this complexity, research strategies in human behavioral ecology employ many simplifying assumptions and borrow methods from a variety of disciplines, including economics, psychology, political science and engineering (Smith, E. A., 1992).

Complexity multiplies when approaching *pre*-human behavioral ecology. Environments must be reconstructed and the intrinsic qualities of the animal differ from modern humans. The subject of this paper, body size in early hominids, is that intrinsic quality had important consequences for the behavioral ecology of early hominids. Body size variation in mammals is related to numerous life history variables (e.g., neonatal brain size, neonatal body weight, gestation length, weaning age, age at maturity, age at first breeding, interbirth interval, and life span (Harvey & Clutton-Brock, 1985)). It is also related to metabolic cost of locomotion,

Table 1 Body weight (kg)[1]

	Geological age (Ma)	Male	Female	Species
A. afarensis	4·0–2·9	44·6 ± 18·5	29·3 ± 15·7	37·0 ± 17·1
A. africanus	3·0–2·4	40·8 ± 17·3	30·2 ± 19·5	35·5 ± 18·4
A. robustus	1·8–1·6	40·2 ± 15·8	31·9 ± 21·5	36·1 ± 18·7
A. boisei	2·0–1·3	48·6 ± 34·6	34·0 ± 13·7	44·3 ± 24·2
H. habilis	2·4–1·6	51·6 ± 22·6	31·5 ± 22·5	41·6 ± 22·6
African H. erectus	1·7–0·7	63·0[2]	52·3[3]	57·7

[1]Body weight estimates from McHenry (1992) except where indicated. 95% confidence intervals are the average confidence intervals of all individual predictions that go into deriving each body weight estimate.
[2]Based on KNM-ER 736, 1808, and WT 15 000 from Ruff & Walker (1993).
[3]Based on KNM-ER 737, O.H. 28 and 34 from Ruff & Walker (1993).

population density, home range size, social organization (especially the relative sizes of the two sexes), diet, cognitive skills (relative brain-size) and much more (Blumenberg, 1981, 1983, 1984, 1985; Calder, 1984; Damuth, 1981a,b; Damuth & MacFadden, 1990; Foley, 1987, 1992; Martin, R. D., 1983, 1984).

Body size in early hominids

Table 1 presents body weight estimates for male and female of early hominid species based on McHenry (1992) with the addition of the African *H. erectus* from Ruff & Walker (1993). McHenry (1992) found that hindlimb joint size gave the most reasonable predictions of body weight and that formulae based on a comparative sample of humans were more reasonable than those based on all species of Hominoidea. The confidence intervals are calculated from the standard errors of estimate given in McHenry (1992) and represent the average of those derived from all the individual fossils that went into calculating each weight estimate.

Clearly these weight estimates are only rough approximations. Their 95% confidence intervals give some picture of the breadth of potential error, but in many cases these statistically derived ranges underestimate the probable error. One source of error is the fact that the small fossil samples probably are not exactly representative of the true species averages. Another is that body proportions in early hominids are unlike any living hominoids. Yet another source of error arises from the use of log-transformed data. For example, the weight estimates based on femoral head size underestimate the true weight by 0·2% to 0·65% according to various estimators of bias reviewed in Smith, R. J. (1992, 1993). A further complication comes from sorting specimens into male and female categories. For early *Australopithecus* the available sample of postcrania falls into two distinct size morphs so the assumption is that the large ones are male and the small ones are female. This assumption becomes more doubtful when applied to species that are apparently less dimorphic such as African *H. erectus*. In the case of African *H. erectus*, three of the specimens have associated pelvic bones from which sex can be more reliably assessed. Of these, KNM-WT 15 000 has many features indicating maleness and O.H. 28 has a broad and evenly curved sciatic notch like a modern female (Ruff & Walker, 1993). Another specimen, KNM-ER 1808, has a broad sciatic notch and reduced brow-ridges of a female, but the sciatic notch has the asymmetry of a male (McHenry, 1991b; Ruff & Walker, 1993). McHenry (1991b) concludes that it is best considered a male because of its asymmetrical sciatic notch and its enormous size.

One reassuring aspect of the weights presented in Table 1 is the fact that they are all thoroughly checked by comparisons of all hominid postcranial specimens with equivalent parts of humans and apes derived from individuals of known body weight at death. By this method the size of all postcranial specimens are matched with standard specimens of different size categories. McHenry (1991*c*) explains this method in detail.

Taxonomic difficulties

As Tuttle (1988:397) pointed out "Students, teachers, and paleophiles, beware! A new era of taxonomic splitting is upon us." Table 1 takes a lumper's approach. Splitting can be done almost at every point. Specimens from Hadar lumped here as *A. afarensis* are divided into two species by some (e.g., Coppens, 1986; Olson, 1981; Olson, 1985; Senut, 1980; Senut & Tardieu, 1985). The Sterkfontein sample, *A. africanus*, may contain a mix of two species (e.g., Clarke, 1988; Kimbel & White, 1988). Two species of South African "robust" australo-pithecines is championed by many (e.g., Grine, 1985; Howell, 1978). Specimens assigned to *H. habilis* may well belong to two species: Wood (1992) reviews the evidence and recommends that the small specimens (e.g., all those from Olduvai plus KNM-ER 1813) be retained in *H. habilis* and that large specimens (e.g., KNM-ER 1470, 1590, and 1481) be placed in a separate species, *H. rudolfensis*. Even *H. erectus* has not survived fragmentation; early *H. erectus* (only in Africa) is now referred to *H. ergaster* by some (e.g., Groves, 1989; Wood, 1992), but later *H. erectus* from both Africa (e.g., O.H. 9) and Asia (the Java and Peking people) remain in that taxon (Wood, 1992).

The two weight estimates in Table 1 that are most affected by these taxonomic difficulties are *A. afarensis* and *H. habilis*. The case against dividing the Hadar sample into two species is very strong (e.g., Kimbel *et al.*, 1985; White, 1985). The degree of body size dimorphism is not necessarily excessive (McHenry, 1991*a*; but see Hartwig-Scherer & Martin, R.D., 1991, for a counter-view). There is still strong evidence that specimens attributed to *H. habilis* should be regarded as a single species (Tobias, 1991), but the appeal of dividing the sample into two is strong (e.g., Chamberlain & Wood, 1987; Groves, 1989; Leakey, R. E. F. & Leakey, M. G., 1978; Lieberman *et al.*, 1988; Rightmire, 1993; Stringer, 1986; Wood, 1991; Wood, 1992; Wood *et al.*, 1991; but see Miller, 1991).

Table 2 divides the *Homo* sample into four species following Wood (1992). The first species is *H. habilis sensu stricto* which contains all of the Olduvai material (including the O.H. 8 foot and the O.H. 35 tibia) and the smaller bodied East Turkana fossils (including the KNM-ER 3735 partial skeleton). McHenry (1992) uses O.H. 8 and O.H. 35 to estimate the body weight of the female *H. habilis* (31·5 kg) since these are among the smaller morph of specimens referred to that species. The body weight of the male of Wood's (1992) *H. habilis sensu stricto* might be about 37·0 kg. This estimate derives from KNM-ER 3735 which is approximately the same size as KNM-ER 1503 and 1822 (Leakey, R. E. F. *et al.*, 1989; McHenry, 1992). The second species, *H. rudolfensis*, contains the large hip and thigh fossils, KNM-ER 3228 and 1481a, which provide the male estimate in Table 2 (59·6 kg). The female body weight of *H. rudolfensis* is estimated from KNM-ER 813 and 1472 (50·8 kg). *Homo ergaster* contains specimens previously assigned to early African members of *H. erectus*. Table 2 gives the estimate for the male of this species as 63 kg based on Ruff & Walker's (1993) value for KNM-ER 737. Wood (1992) keeps the later African *H. erectus* specimens in that species and refers them to *H. erectus sensu stricto*. Two femora provide the basis of the estimate of the female *H. erectus sensu stricto* (O.H. 28 and 34) with estimated weights of 54

Table 2   **Early *Homo*[1] body weight (kg)**

	Male	Female	Species
H. habilis sensu stricto	37·0[2]	31·5[3]	34·3
H. rudolfensis	59·6[4]	50·8[5]	55·2
H. ergaster	63·0[6]	52·0[7]	57·5
H. erectus sensu stricto (African)	63·0[8]	52·5[9]	57·8

[1]As defined by (Wood, 1992*b*).
[2]Based on KNM-ER 3735 which is about the same size as KNM-ER 1503 (McHenry, 1992*a*).
[3]Based on O.H. 8 and 35 (McHenry, 1992*a*).
[4]Based on KNM-ER 1481 and 3228 (McHenry, 1992*a*).
[5]Based on KNM-ER 813 and 1472 (McHenry, 1992*a*).
[6]Based on KNM-ER 736, 1808 and KNM-WT 15 000 (Ruff & Walker, 1993).
[7]Based on KNM-ER 737 (Ruff & Walker, 1993).
[8]Based on O.H. 28 and 34 (Ruff & Walker, 1993).

and 51 kg as reported in Ruff & Walker (1993). Since there are no positively identified male postcrania of this species, the value for early African *H. erectus* (=*H. ergaster*) is used.

General significance of body weights

The general pattern revealed by Table 1 is clear. The average female of the earliest known species of Hominidae, *Australopithecus afarensis*, weighed about 29 kg and the male, 45 kg. Average female body weight remains between 29 and 34 kg in all species of hominids before the appearance of *Homo erectus* at 1·7 M.Y. Average male body weight ranged between 40 and 52 kg in these pre-*erectus* species. The origin of *H. erectus* marked a dramatic increase in body size especially in the female. Female body weight remains small in all species of hominid until the emergence of *H. erectus* and male body weight increases slightly through time with a big jump to *H. erectus*. This pattern appears to describe the entire postcranial sample, but one must always keep in mind the broad confidence intervals and the many assumptions that go into deriving these estimates.

Sexual dimorphism

By these estimates of body weight, the degree of sexual dimorphism in early hominid species is more than that seen in modern humans but much less than of *Gorilla* and *Pongo*. The earliest species of hominid, *A. afarensis*, has a level of body size dimorphism only slightly above modern *Pan*. McHenry (1991*a*) discusses the implications of the degree of dimorphism to the reconstruction of social structure in *A. afarensis*. Although dimorphism appears slightly larger than in modern chimps, the fact that it is well below that of modern gorillas and orangs does not rule out a social structure with kin-related males forming alliances and mutually defending territory occupied by several females (Foley & Lee, 1989). This level of sexual dimorphism in body size appears to be incompatible with what is known about primate monogamy (contra Lovejoy, 1981).

The degree of body size sexual dimorphism is high in the estimates of male and female *H. habilis sensu lato*, but still well below that seen in species of *Gorilla* and *Pongo*. If the *H. habilis sensu lato* sample is broken into two species by the scheme presented in Wood (1992), then the dimorphism drops to a level much like that seen in modern humans. It appears that the degree of sexual dimorphism significantly reduces with the origin of *H. erectus* (*sensu lato*). Females increase in size significantly (by as much as 20 kg, perhaps). All conclusions are vulnerable to change pending new discoveries, but current evidence appears to show a dramatic change in body size sexual dimorphism with the origin of *H. erectus* at 1·7 M.Y. In the period immediately preceding the appearance of *H. erectus* (1·9 to 1·8 M.Y.), there are many tiny individuals represented by postcrania such as the O.H. 62 *H. habilis* partial skeleton. But by 1·7 M.Y. the East African hominid postcranial sample contains only large-bodied adult individuals. The only small individuals that are present are either immature (e.g., the KNM-ER 1463 femur) or are attributed to *A. boisei* by depositional association in Area 6A of Ileret (e.g., KNM-ER 1823, 1824, and 1825). All the other postcranial specimens from 1·7 on are relatively large-bodied and some (e.g., KNM-ER 739 and 5428) are enormous, matching modern humans of body sizes between 72 and 86 kg.

Relative brain size

Table 3 presents estimates of relative brain size. Brain size in the fossils is derived from endocranial volume by formula number 4 in Aiello & Dunbar (1993). The first estimate (EQ 1) is the ratio of observed endocranial volume and brain size predicted from female body weight, 0·045 (female body weight)$^{0.86}$, based on a large sample of primate species (Harvey & Clutton-Brock, 1985). The predicting equation uses the exponent 0·86 which is higher than the 0·76 exponent derived for mammals by Martin, R. D. (1981) using major axis technique, but below the 0·92 found for primates by Harvey & Pagel (1991) using the structural relations model with $\lambda = 0.20$. One effect of using female body weight instead of species body weight is relatively higher EQ's in species with high levels of body size sexual dimorphism. One justification for using female body weight is the apparent fact that brain growth is closely linked to maternal metabolism (Martin, 1981, 1983). The second estimate of relative brain size (EQ 2) is the ratio of actual endocranial volume divided by 0·0589 (body weight)$^{0.76}$. This formula derives from Martin, R. D. (1981). The values can be regarded as roughly the number of times larger the actual brain size is relative to the average mammal at the same body weight. Finally, EQ 3 is actual endocranial volume divided by 0·48 (body weight)$^{0.6}$ which derives from the major axis formula relating brain and body size in Old World monkeys and apes (Martin, R. D., 1983).

Table 3 divides species of early *Homo* in several different ways. *Homo habilis sensu lato* and *H. erectus sensu lato* follow Tobias (1991) and Howell (1978). Several authors have divided the specimens referred to early *Homo* differently, however. The attribution in Table 3 of *H. habilis sensu stricto*, *H. rudolfensis*, *H. ergaster* and *H. erectus sensu stricto* follow Wood (1992). The most striking consequence of Wood's classification is the diminutive relative brain size of *H. rudolfensis*. According to Wood (1992) this species contains KNM-ER 1470 with its 752 cc endocranial volume. This is the largest brained specimen in the record before 1·7 M.Y., but because of the large body size, its relative brain-size is much closer to *A. robustus* than to any other species of *Homo*. The contemporary species of *Homo*, according to Wood's (1991, 1992) classification is *H. habilis sensu stricto*. The body size of this species is small (31·5 kg for the female, 37·0 kg for the male) and consequently its relative endocranial volume is much larger than *H. rudolfensis*. The facial architecture of *H. habilis sensu stricto* is more similar to early *Homo erectus* (=*H. ergaster*). Perhaps these shared traits between

Table 3 Encephalization

	Female body weight (kg)	Species body weight (kg)	ECV (cc)	Brain (cc)[1]	EQ1[2]	EQ2[3]	EQ3[4]
A. afarensis	29·3	37·0	404[5]	384	1·23	2·20	1·45
A. africanus	30·2	35·5	442[6]	420	1·31	2·48	1·62
A. boisei	34·0	41·3	515[7]	488	1·37	2·57	1·72
A. robustus	31·9	36·1	530[8]	502	1·49	2·93	1·92
H. habilis sensu lato	31·5	41·6	632[9]	597	1·79	3·13	2·10
H. habilis sensu stricto	31·5	34·3[10]	612[11]	579	1·74	3·51	2·29
H. rudolfensis	50·8[12]	55·2[13]	752[14]	709	1·41	2·99	2·11
African *H. erectus sensu lato*	52·3	57·7	871[15]	820	1·59	3·35	2·38
H. ergaster	52·3[16]	57·5[17]	854[18]	804	1·56	3·29	2·33
African *H. erectus sensu stricto*	52·5[19]	57·8[20]	897[21]	844	1·63	3·44	2·44
H. sapiens[22]	40·1	44·0		1250	3·05	6·28	4·26
P. troglodytes[22]	31·1	36·4		410	1·25	2·38	1·57

[1]Brain size predicted by equation (4) in Aiello & Clutton-Brock (1993).

[2]EQ1 based on Harvey & Clutton-Brock (1985) predicted brain size of 0.045 (female body weight).$^{0.86}$

[3]EQ2 based on Martin's (1981b) predicted brain size of 0.0589 (body weight)$^{0.76}$ for mammals.

[4]EQ3 based on Martin's (1983) predicted brain size of 0.48 (body weight)$^{0.6}$ based on individual genera of Old World monkeys and apes.

[5]From Falk (1987).

[6]From Holloway (1983).

[7]Based on Holloway (1983) when ER 406 is 520, ER 407 is 510, ER 732 is 500 and O.H. 5 is 530.

[8]Based on SK 1585 reported in Holloway (1983).

[9]From Tobias (1991) where O.H. 7 is 674, O.H. 13 is 673, O.H. 16 is 638, O.H. 24 is 594, ER 1470 is 752, ER 1805 is 582 and ER 1813 is 510.

[10]Midpoint between female for *H. habilis sensu lato* (31·5 kg) and 37·0 kg which is the estimated weight for the male as represented by the KNM-ER 3735 partial skeleton.

[11]The same as *H. habilis sensu lato* minus 752 cc for KNM-ER 1470.

[12]Based on KNM-ER 1472 and 813.

[13]Midpoint between female estimate (50·8 kg) and male (59·6 kg, the average value for KNM-ER 1481 and 3228).

[14]The value of KNM-ER 1470 given in Holloway (1983).

[15]Based on O.H. 9 (1067 cc), 12 (727 cc), KNM-ER 3733 (848 cc), KNM-ER 3883 (804 cc) reported in Holloway (1983) and KNM-WT 15 000 (909 cc) reported in Begun & Walker (1993).

[16]Based on KNM-ER 737, O.H. 28 and O.H. 34 given in Ruff & Walker (1993).

[17]Midpoint between female (52·3 kg, footnote 16) and male (62·7 kg, KNM-ER 736, 1808, and the projected adult weight of KNM-WT 15 000 given in Ruff & Walker, 1993).

[18]Based on KNM-ER 3733 and 3883 in Holloway (1983) and KNM-WT 15 000 in Begun & Walker (1993).

[19]Based on O.H. 28 and 34 reported in Ruff & Walker (1993).

[20]Midpoint between female (footnote 19) and an assumed male body weight equal to that of the male of *H. ergaster* (63·0 kg, footnote 17).

[21]Based on O.H. 9 (1067 cc) and O.H. 12 (727) reported by Holloway (1983).

[22]From Harvey & Clutton-Brock (1985).

H. habilis sensu stricto and early *Homo erectus* ($=H.$ *ergaster*) are homologous and therefore indicate that the two species form a sister group relative to *H. rudolfensis* and the australopithecines. The postcranial morphology of *H. rudolfensis*, on the other hand, appears to share many derived traits with early *H. erectus* ($=H.$ *ergaster*) that are not present in *H. habilis sensu stricto*.

However the species of early *Homo* are divided, a few general points are apparent. The brain size increase from *A. afarensis* to *A. africanus* to the "robust" australopithecines does not appear to be an artifact of body size increase but reflects progressive encephalization. It also appears that the expansion of absolute brain size with the appearance of *Homo* is beyond what would be expected from body size increase alone. Given the energetic costs of brain-size increase

(Foley & Lee, 1991; Martin, R. D., 1983; Parker, 1990) this remarkable change in relative brain size implies that a major alteration in subsistence occurred.

Home range area

There is a predictable relationship between body size and home range area in mammalian species (McNab, 1963; Harestad & Bunnell, 1979; Martin, R. A., 1981; LaBarbabera, 1989; Milton & May, 1975). Harestad & Bunnell (1979) note that for large bodied mammalian omnivores the relationship between home range (H in hectares) and body weight (W) in grams is

$$H = 0.05 W^{0.92}$$

with a high correlation ($r = 0.95$). Using the estimated body weights from Tables 1 and 2 this equation predicts a home range for *A. afarensis* of 941 ha and one for early African *H. erectus* of 1416 ha. These areas would be about 200 times larger if the equation for carnivores is applied.

Life history parameters

Table 4 presents estimates of life history variables predicted from body weights using the formula in Harvey & Clutton-Brock (1985). Because of the small body size of females predicted for all pre-*erectus* species of hominid, these species appear to have chimp-like parameters. Relative to modern humans, these pre-*erectus* species have particularly small neonatal brain and body sizes. The origin of *H. erectus* marks a dramatic change in all parameters because of the apparent increase in female body size. This increase is most striking in neonatal brain and body size. If *H. habilis sensu lato* is partitioned into two species as Wood (1991, 1992) and others suggest, then this change may have occurred by 1·9 M.Y., but by 1·7 M.Y. small-bodied hominids disappear from the East African fossil record. This means that despite any problems in attributing postcranial specimens to species from 1·7 M.Y. on, one can be relatively certain that sexual dimorphism is substantially reduced.

Compared to other mammalian species, *H. sapiens* has an unusual relationship between life history parameters and female body weight. Actual neonatal brain and body weight far exceed the mammalian norm. The predicted size of female early *H. erectus* yields an estimate for these parameters which is a jump ahead of earlier hominid species, but may be an underestimation of the true values because the actual values are not known. The actual values for modern *H. sapiens* are conspicuously different from the predicted mammalian norm in many other life history parameters as well, but as with neonatal size, one is left with no empirical evidence of when these changes took place. Age of weaning and interbirth interval are delayed in modern *H. sapiens* and *P. troglodytes* from that predicted from mammalian females. Predicted age at maturity and first breeding in chimps is like other mammals, but strikingly different in modern humans. This uniquely human characteristic of delayed maturation may have not been fully developed in early *H. erectus*. The evidence is KNM-WT 15 000. Smith (1993) notes the discrepancy between dental age of this specimen (11 years on a modern human scale), skeletal age (13–13·5 years) and stature (15 years). Perhaps the best guide to the first appearances of uniquely human life history parameters (i.e., neonatal brain size, neonatal body weight, gestation length, age at maturity, and age at first breeding) are provided by the fossil evidence for adult brain size which occurred between 500–200 000 years ago. In this sense, the fossils

Table 4 Life history parameters[1]

	Neonatal brain size (g)	Neonatal body wt (kg)	Gestation length (mo)[1]	Weaning age (mo)[1]	Age at maturity (yr)[1]	Age at first breeding (yr)[1]	Interbirth interval (mo)[1]	Lifespan (yr)[1]
A. afarensis	162	1·7	7·6	28·7	9·3	11·0	43·6	42·2
A. africanus	166	1·8	7·6	29·2	9·4	11·2	44·1	42·6
A. robustus	175	1·9	7·6	30·1	9·7	11·4	45·0	43·3
A. boisei	185	2·0	7·7	31·2	10·0	11·8	46·1	44·1
H. habilis	173	1·8	7·6	29·8	9·7	11·4	44·8	43·1
African *H. erectus*	270	2·9	8·2	39·6	12·5	14·2	54·0	50·0
P. troglodytes predicted	171	1·8	7·6	29·6	9·6	11·3	44·6	43·0
P. troglodytes actual	128	1·8	7·6	48·7	9·8	11·5	60·8	44·5
H. sapiens predicted	214	2·3	7·9	34·2	10·9	12·6	49·0	46·3
H. sapiens actual	384	3·3	8·9	24·0	16·5	19·3	48·0	70·0

[1]Formulae from Harvey & Clutton-Brock (1985).

attributed to *H. erectus*, especially the KNM-WT 15 000 partial skeleton, document the transition (Walker, 1993).

Conclusions

The dance between precision and uncertainty enlivens paleoanthropology. Precise dates of postcranial elements provide a baseline. Taxonomic attribution of those elements and body-size reconstructions remain debatable. Present evidence reveals a pattern, but new discoveries can always disrupt the pattern. Present evidence appears to show that hominid species remained small and sexually dimorphic in body size relative to modern humans until 1·7 M.Y. with the appearance of *H. erectus*. Perhaps the pattern changed by 1·9 M.Y. with two species formerly lumped into *H. habilis*, but definitely by 1·7 M.Y. female body size increased dramatically. This was a fundamental change that related to changes in brain size, ranging behavior, and a host of life history parameters.

Acknowledgements

The author thanks R. E. and M. G. Leakey and the staff of the National Museums of Kenya, M. D. Leakey, F. C. Howell, D. C. Johanson and the staff of the Cleveland Museum of Natural History and the Institute of Human Origins, Tadessa Terfa, Mammo Tessema, Berhane Asfaw and the staff of the National Museum of Ethiopia, C. K. Brain and the staff of the Transvaal Museum, P. V. Tobias and the staff of the Department of Anatomy and Human Biology, University of Witwatersrand, for permission to study the original fossil material in their charge and for numerous kindnesses. The author also thanks the late L. Barton, D. R. Howlett, C. Powell-Cotton and staff of the Powell-Cotton Museum; M. Rutzmoser and staff of the Museum of Comparative Zoology, Harvard University; R. Thorington and the staff of the Division of Mammology, Smithsonian Institution; D. J. Ortner and the staff of the Department of Anthropology, Smithsonian Institution; D. F. E. T. van den Audenaerde and M. Lovette and the staff of the Musee d'Afrique Centrale, Tervuren; R. D. Martin and the staff of the Anthropologische Institut, Zurich; W. W. Howells and the staff of the Peabody Museum, Harvard University; C. Edelstamm and the staff of the Natur Historiska Rismuseet, Stockholm; R. L. Susman and W. L. Jungers for many kindnesses and for permission to study the comparative material in their charge; L. J. McHenry, C. B. Ruff, and R. J. Smith for their insightful comments on early drafts; J. Martini for her help in preparing this paper; and J. S. Oliver, N. Sikes, and K. Stewart for inviting me to participate in this symposium. Partial funding was provided by the Committee of Research, University of California, Davis.

References

Aiello, L. C. & Dunbar, R. (1993). Neocortex size, group size, and the evolution of language. *Curr. Anthrop.* **34,** 184–193.

Begun, D. & Walker, A. (1993). The endocast. In (A. Walker & R. Leakey, Eds) *The Nariokotome Homo erectus Skeleton,* 326–358. Cambridge: Harvard University Press.

Blumenberg, B. (1981). Observations on the paleoecology, population structure and body weight of some Tertiary hominoids. *J. hum. Evol.* **10,** 543–564.

Blumenberg, B. (1983). The evolution of the advanced hominid brain. *Curr. Anthrop.* **24,** 589–623.

Blumenberg, B. (1984). Allometry and evolution of Tertiary hominoids. *J. hum. Evol.* **13,** 613–676.

Blumenberg, B. (1985). Population characteristics of extinct hominid endocranial volume. *Am. J. phys. Anthrop.* **68,** 269–280.

Calder, W. A. (1984). *Size, Function, and Life History*. Cambridge: Harvard University Press.

Chamberlain, A. T. & Wood, B. A. (1987). Early hominid phylogeny. *J. hum. Evol.* **16**, 119–133.

Clarke, R. J. (1988). A new *Australopithecus* cranium from Sterkfontein and its bearing on the ancestry of *Paranthropus*. In (F. E. Grine, Ed.) *Evolutionary History of the "Robust" Australopithecines*, 285–292. New York: Aldine de Gruyter.

Coppens, Y. (1986). Evolution de l'homme. *La Vie des Sciences, Comptes Rendus* **3**, 227–223.

Damuth, J. (1981*a*). Population density and body size in mammals. *Nature* **290**, 699–700.

Damuth, J. (1981*b*). Home range, home range overlap, and species energy use among herbivorous animals. *Biol. J. Linn. Soc.* **15**, 185–193.

Damuth, J. & MacFadden, B. J. (1990). *Body Size in Mammalian Paleobiology: Estimation and Biological Implications*. Cambridge: Cambridge University Press.

Foley, R. A. (1987). *Another Unique Species: Patterns in Human Evolutionary Ecology*. Harlow: Longman Scientific & Technical.

Foley, R. A. (1992). Evolutionary ecology of fossil hominids. In (E. A. Smith & B. Winterhalder, Ed.) *Evolutionary Ecology and Human Behaviour*, 131–164. New York: Aldine de Gruyter.

Foley, R. F. & Lee, P. C. (1989). Finite social space, evolutionary pathways, and reconstructing hominid behavior. *Science* **243**, 901–906.

Foley, R. A. & Lee, P. C. (1991). Ecology and energetics of encephalization in hominid evolution. *Phil. Trans. R. Soc. Lond. (series B)* **334**, 223–232.

Grine, F. E. (1985). Australopithecine evolution: the deciduous dental evidence. In (E. Delson, Ed.) *Ancestors: The Hard Evidence*, 153–167. New York: Alan R. Liss, Inc.

Groves, C. P. (1989). *A Theory of Human and Primate Evolution*. Oxford: Clarendon Press.

Harestad, A. S. & Bunnell, F. L. (1979). Home range and body weight—a reevaluation. *Ecology* **60**, 389–402.

Hartwig-Scherer, S. & Martin, R. D. (1991). Was "Lucy" more human than her "child"? *J. hum. Evol.* **21**, 439–449.

Harvey, P. H. & Clutton-Brock, T. H. (1985). Life history variation in primates. *Evolution* **39**, 559–581.

Harvey, P. H. & Pagel, M. D. (1991). *The Comparative Method in Evolutionary Biology*. Oxford: Oxford University Press.

Howell, F. C. (1978). Hominidae. In (V. J. Maglio & H. B. S. Cooke, Ed.) *The Evolution of African Mammals*, 154–248. Cambridge: Harvard University Press.

Kimbel, W. H. & White, T. D. (1988). Variation, sexual dimorphism and the taxonomy of *Australopithecus*. In (F. E. Grine, Ed.) *Evolutionary History of the "Robust" Australopithecines*, 175–192. New York: Aldine de Gruyter.

Kimbel, W. H., White, T. D. & Johanson, D. C. (1985). Craniodental morphology of the hominids from Hadar and Laetoli: Evidence of "*Paranthropus*" and *Homo* in the Mid-Pliocene of Eastern Africa. In (Eric Delson, Ed.) *Ancestors: The Hard Evidence*, 120–137. New York: Alan R. Liss, Inc.

Krebs, J. R. & Davies, N. B. (1981). *An Introduction to Behavioural Ecology*. Sunderlan, Massachusetts: Sinauer Associates, Inc.

LaBarbabera, M. (1989). Analyzing body size as a factor in ecology and evolution. *Ann. Rev. Ecol. Syst.* **20**, 97–117.

Leakey, R. E. F. & Leakey, M. G. (1978). *The Fossil Hominids and an Introduction to their Context, 1967–1974. Koobi Fora Research Project. 1*. Oxford: Clarendon Press.

Leakey, R. E. F., Walker, A., Ward, C. V. & Grausz, H. M. (1989). A partial skeleton of a gracile Hominid from the Upper Burgi Member of the Koobi Fora Formation, East Lake Turkana, Kenya. In (Giacomo Giacobini, Ed.) *Hominidae: Proceedings of the 2nd International Congress of Human Paleontology Turin, September 28–October 3, 1987*, 167–174. Milan: Jaka.

Lieberman, D. E., Pilbeam, D. R. & Wood, B. A. (1988). A probabilistic approach to the problem of sexual dimorphism in *Homo habilis*: A comparison of KNM-ER 1470 and KNM-ER 1813. *J. hum. Evol.* **17**, 503–512.

Lovejoy, C. O. (1981). The origin of man. *Science* **211**, 341–350.

Martin, R. A. (1981). On extinct hominid population densities. *J. hum. Evol.* **10**, 427–428.

Martin, R. D. (1981). Relative brain size and basal metabolic rate in terrestrial vertebrates. *Nature* **293**, 57–60.

Martin, R. D. (1983). *Human Brain Evolution in an Ecological Context*. New York: American Museum of Natural History.

Martin, R. D. (1984). Body size, brain size and feeding strategies. In (D. J. Chivers, B. A. Wood & A. Bilsborough, Ed.) *Food Acquisition and Processing in Primates*, 73–103. New York: Plenum Press.

McHenry, H. M. (1991*a*). Sexual dimorphism in *Australopithecus afarensis*. *J. hum. Evol.* **20**, 21–32.

McHenry, H. M. (1991*b*). Femoral lengths and stature in Plio-Pleistocene hominids. *Am. J. phys. Anthrop.* **85**, 149–158.

McHenry, H. M. (1991*c*). The petite bodies of the "robust" australopithecines. *Am. J. phys. Anthrop.* **86**, 445–454.

McHenry, H. M. (1992). Body size and proportions in early hominids. *Am. J. phys. Anthrop.* **87**, 407–431.

McNab, B. K. (1963). Bioenergetics and the determination of home range size. *Am. Nat.* **97**, 133–140.

Miller, J. A. (1991). Does brain size variability provide evidence of multiple species in *Homo habilis*? *Am. J. phys. Anthrop.* **84**, 385–398.

Milton, K. & May, M. L. (1975). Body weight, diet and home range area in primates. *Nature* **259**, 459–462.

Olson, T. R. (1981). Basicranial morphology of the extant hominids and Pliocene hominoids: The new material from the Hadar Formation, Ethiopia, and its significance in early human evolution and taxonomy. In (C. B. Stringer, Ed.) *Aspects of Human Evolution* **21**, 99–128. London: Taylor and Francis.

Olson, T. R. (1985). Cranial morphology and systematics of the Hadar Formation hominids and "*Australopithecus*" *africanus*. In (Eric Delson, Ed.) *Ancestors: The Hard Evidence*, 102–119. New York: Alan R. Liss, Inc.

Parker, S. T. (1990). Why big brains are so rare: Energy costs of intelligence and brain size in anthropoid primates. In (S. T. Parker & K. R. Gibson, Ed.) *"Language" and intelligence in monkeys and apes*, 129–154. Cambridge: Cambridge University Press.

Rightmire, G. P. (1993). Variation among early *Homo* crania from Olduvai Gorge and the Koobi Fora region. *Am. J. phys. Anthrop.* **90,** 1–33.

Ruff, C. B. & Walker, A. (1993). Body size and body shape. In (A. Walker & R. Leakey, Eds) *The Nariokotome Homo erectus Skeleton*, 234–265. Cambridge: Harvard University Press.

Senut, B. (1980). New data on the humerus and its joints in Plio-Pleistocene hominoids. *Coll. Anthrop.* **4,** 87–93.

Senut, B. & Tardieu, C. (1985). Functional aspects of Plio-Pleistocene hominoid limb bones: Implications for taxonomy and phylogeny. In (Eric Delson, Ed.) *Ancestors: The Hard Evidence*, 193–201. New York: Alan R. Liss, Inc.

Smith, E. A. (1992). Human behavioral ecology: I. *Evol. Anthrop.* **1,** 20–24.

Smith, H. (1993). The physiological age of KNM-WT 15 000. In (A. Walker & R. Leakey, Eds). *The Nariokotome Homo erectus skeleton*, 195–220. Cambridge: Harvard University Press.

Smith, R. J. (1992). Transformation bias of allometric equations results in systematic errors in fossil body weight estimates. *Am. J. phys. Anthrop. supplement* **14,** 153.

Smith, R. J. (1993). Logarithmic transformation bias in allometry. *Am. J. phys. Anthrop.* **90,** 215–228.

Stringer, C. B. (1986). The credibility of *Homo habilis*. In (B. Wood, L. Martin & P. Andrews, Ed.) *Major Topics in Primate and Human Evolution*, 266–294. Cambridge: Cambridge University Press.

Tobias, P. V. (1991). *Olduvai Gorge: The skulls endocasts and teeth of Homo habilis 4*, Cambridge: Cambridge University Press.

Tuttle, R. H. (1988). What's new in African paleoanthropology? *Ann. Rev. Anthrop.* **17,** 391–426.

Walker, A. (1993). Perspectives on the Nariokotome discovery. In (A. Walker & R. Leakey, Eds). *The Nariokotome Homo erectus skeleton*, 411–430. Cambridge: Harvard University Press.

White, T. D. (1985). The hominids of Hadar and Laetoli: An element-by-element comparison of the dental samples. In (Eric Delson, Ed.) *Ancestors: The Hard Evidence*, 138–152. New York: Alan R. Liss, Inc.

Wood, B. A. (1991). *Koobi Fora Research Project IV: Hominid Cranial Remains from Koobi Fora*. Oxford: Clarendon.

Wood, B. A. (1992). Origin and evolution of the genus *Homo*. *Nature* **355,** 783–790.

Wood, B. A., Li, Y. & Willoughby, C. (1991). Intraspecific variation and sexual dimorphism in cranial and dental variables among higher primates and their bearing on the hominid fossil record. *J. Anat.* **174,** 185–205.

Nicola Stern

Department of Archaeology, La Trobe University, Bundoora, Victoria 3083, Australia

Received 24 September 1993
Revision received 30 January 1994 and accepted 7 March 1994

Keywords: Lower Pleistocene, archaeology, Koobi Fora Formation, time resolution

The implications of time-averaging for reconstructing the land-use patterns of early tool-using hominids

Indications that shifts in diet, foraging strategies and land use patterns were driving forces in hominid evolution are largely responsible for archaeologists' longstanding interest in reconstructing the land use patterns of early tool-using hominids. Attempts to reconstruct past land use patterns involve a study of the distribution of material remains through sediments representing different portions of an ancient landscape. Depositional systems play a major role in structuring these types of regional archaeological records. There is, for example, an inverse relationship between the area of an ancient landscape being sampled, the quantity of archaeological debris available for study and the amount of time represented by that debris and its encasing sediments. The micro-stratigraphy and time resolution of a set of archaeological remains in the lower Okote Member of the Koobi Fora Formation are described. It is argued that this record comprises time-averaged palimpsests of debris spanning 65 ± 5 k.y.a. Only by ignoring the time dimension of these data is it possible to invoke interpretive theories that are based on ethnographic scale observations of the interactions between individuals and their environments. It is argued that serious consideration should be given to the suggestion that this is an ontologically singular record of hominid action with the potential to provide an entirely unique perspective on the history of the hominid lineage.

Journal of Human Evolution (1994) **27,** 89–105

Introduction

Archaeologists' interest in reconstructing the land use patterns of early hominids has been spurred by suggestions that changes in ranging patterns may have characterised the evolutionary history of the hominid lineage (e.g., Isaac, 1983; Binford, 1984; Foley, 1987; Sept, 1992). There are indications that shifts in diet, foraging strategies and ranging patterns both helped to initiate a minor radiation in the hominid lineage in the late Pliocene and to maintain species diversity between 2·5 and 1 m.y.a. (e.g., Foley, 1987). There have also been suggestions that the behaviours underlying the particular pattern of land-use that characterises the modern hunter gatherer adaptation may have been important in differentiating successive hominid grades and that at least some fundamental components of this adaptive strategy were in place 2 Ma (Isaac, 1976; 1978; 1983).

Thus, Palaeolithic archaeologists have been encouraged to seek a material record for the land use patterns of the earliest tool-using hominids. Isaac's (1981) suggestion that the distribution of archaeological debris through laterally extensive, but vertically discrete horizons will provide the information needed to assess alternative models of hominid movements across the landscape has been taken up by a number of researchers (e.g., Isaac & Harris, 1980; Potts, 1989; Stern, 1991; Blumenschine & Masao, 1991; Rogers & Harris, 1992). But while Isaac's vision of artefacts as "stone-age visiting cards" is an appealing metaphor, the distribution of stone tools through the sedimentary record does not translate readily into a map of the movements of early tool-using hominids across an ancient landscape (Stern, 1991).

Although Palaeolithic archaeologists have been willing to acknowledge that the archaeological record is not really a simple map of where hominids discarded things, let alone of where they went (Isaac, 1981), the notion that "it is a partial image . . . distorted and blurred" and that "with care and caution inferences can be drawn about the spatial configuration of daily

0047–2484/94/010089+17 $08.00/0

life and aspects of the use of a landscape" (Isaac, 1981) is very persistent. Ultimately all the analytical and interpretive strategies that archaeologists have employed to reconstruct past land use patterns are predicted on the assumption that it is possible to identify functionally integrated sets of material debris (e.g., Isaac, 1981), or at the very least, sets of material remains that preserve consistent sets of relationships with the ecological variables that influenced hominid foraging strategies (e.g., Sept, 1992; Blumenschine & Masao, 1991). This assumption needs investigation, particularly in the light of recent discussions about the implications of time resolution and time-averaging for the understanding of patterning in the fossil record (e.g., Behrensmeyer, 1987; Furisch & Aberhan, 1990; Kidwell, 1985, 1986; Peterson, 1977; Staff et al., 1986) and in view of the vastly improved understanding that archaeologists now have about the role of depositional processes in structuring the archaeological record.

There are no depositional contexts in which sedimentation is continuous both through time and across space (Sadler, 1981). This is a consequence of the fact that the geomorphic processes resulting in net sediment accumulation, and in the burial of fossil and archaeological debris, vary in magnitude and in frequency. Thus, the accumulation of sediment across a landscape tends to be intermittent and localised, resulting in the formation of time-averaged assemblages of material remains (Behrensmeyer, 1987, 1991).* The mode and rate of sediment accumulation that prevailed in any given depositional environment will have had a significant effect on the amount of time-averaging involved in the formation of a regional archaeological record. It is possible to estimate the amount of time-averaging involved in the formation of such a record and to use this information to pursue questions about the categories of palaeoecological information preserved at different time-scales. But no matter how much time-averaging was involved in the formation of a regional archaeological record the fact remains that ethnographic and ecological theory can only be applied to the interpretation of these data by ignoring their time dimensions. Despite occasional suggestions to the contrary the methodologies being used to characterise the palaeoecology of extinct tool-using hominids all draw on interpretive theories developed from observations of the interactions between individuals and between individuals and their environments. But the archaeological record is made up of material aggregates built up as a consequence of the actions and interactions of many different individuals. Before engaging in further discussion about the categories of palaeoecological information encapsulated in the Pleistocene archaeological record a debate is needed about whether or not the time dimension of these aggregates mark them out as an "ontologically singular" (Murray, 1987, 1993) record of human action.

In this paper I describe the geological context of archaeological debris preserved in the lower Okote Member of the Koobi Fora Formation with the aim of generating discussion about its ontological significance. The discussion focuses on aspects of site formation and time resolution, the two features of the archaeological record that distinguish it from the ethnographic and ecological records currently providing the interpretive frameworks applied to these data (e.g., Tooby & DeVore, 1987). The descriptions of the micro-stratigraphy of these archaeological remains and the estimates of their time resolution offered in this paper simply involve the application of standard geological methods to archaeology-bearing sediments. The rationale for their presentation in this volume is the desire to generate discussion about the behavioural and palaeoecological information that might be gleaned from this record if its time dimension were not ignored. Is the Pleistocene archaeological record a

---

*Other types of fossil assemblages, exhibiting very different attributes, may occur in specific locations on the landscape, for example, mass death deposits or lair deposits (Behrensmeyer, 1987, 1991). But most landscape assemblages are dispersed, fragmentary and time-averaged (ibid).

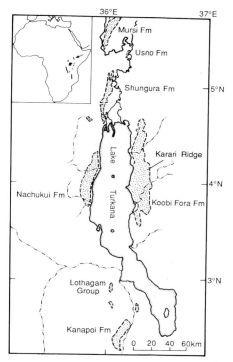

Figure 1. Map of the Lake Turkana Depression and surrounding in north-west Kenya, showing the distribution of the Koobi Fora Formation and the location of the Karari Ridge where the study area is situated.

uniquely structured record of hominid activities and does it have the potential to contribute to a unique archaeological perspective on the evolution of hominid behaviour? Recognising archaeological data as time-averaged accumulations of debris is only the first step in a long research endeavour ultimately aimed at developing interpretive strategies specifically designed to decode material aggregates spanning long segments of time. Although this discussion is about ontology, it is one stimulated by observations of the empirical structure of the record itself and it is the data themselves that are the focus of the remainder of this paper.

The geological setting

The Koobi Fora Formation, which outcrops over an area of 80 km by 40 km along the eastern shore of the modern Lake Turkana in northwest Kenya (Figure 1), preserves abundant fossil remains, including those of a variety of early human ancestors, and their archaeological traces (Coppens et al., 1976; Leakey & Leakey, 1978; Harris, 1982; Feibel, 1988; Isaac, n.d.). It comprises more than 560 metres of lake margin, lake, river and floodplain sediments deposited from approximately 4·2 Ma until 0·6 Ma in an asymmetric graben between the Kenyan and Ethiopian domes (Brown & Feibel, 1986; Feibel, 1988; McDougall, 1985; see Figure 2). A succession of rhylolitic tuffs punctuate the entire sequence providing the basis for correlating geographically separated localities, for dating and for palaeogeographic reconstructions of the region at specific time intervals (Brown & Feibel, 1988; Feibel, 1988).

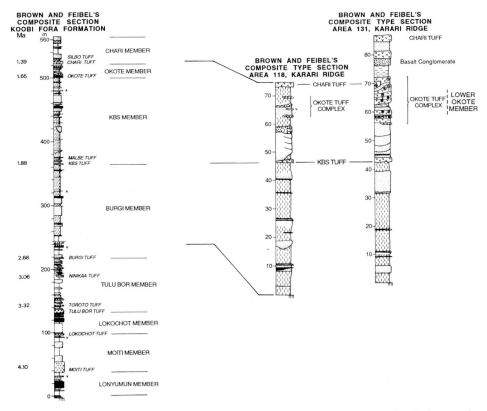

Figure 2. Composite stratigraphic section for the Koobi Fora Formation showing the bracketing ages for each member and the composite type sections for the Karari Ridge (after Brown & Feibel, 1986). The stratigraphic position of the Lower Okote Member is marked.

The archaeological record under discussion comes from a narrow stratigraphic interval within the Okote Member of the Koobi Fora Formation (Figure 2). The Okote Member outcrops most extensively along the Karari Ridge approximately 40 km to the east of Lake Turkana (Brown & Feibel, 1986). It comprises flat-lying silts, sands and tuffs representing a variety of channel, bank and floodplain deposits. These contain abundant archaeological remains dating to the time range 1·6–1·4 m.y.a. (Isaac & Harris, 1978; Harris, 1978; Isaac, n.d., see Figure 3). The particular stratigraphic interval under study is referred to informally as the lower Okote Member (Stern, 1991). It consists of up to 8 metres of inter-locking channel, bank and floodplain deposits (Kaufulu, 1983; Stern, 1991; see Figures 4 and 5). Archaeological remains occur as a variable density scatter throughout these sediments, although there tends to be more debris in some depositional contexts than in others and some beds contain localised, high density patches that punctuate the lower density scatter found in most beds (Stern, 1991).

The artefacts and fossils lying on the surface of lower Okote Member outcrops, and their geological context, were sampled in 1986 and 1987 during a field survey undertaken to describe the low density scatter in between two, well-known, high density patches on the western face of the Karari Ridge, FxJj20 and FxJj37 (Figure 4), building on methods pioneered

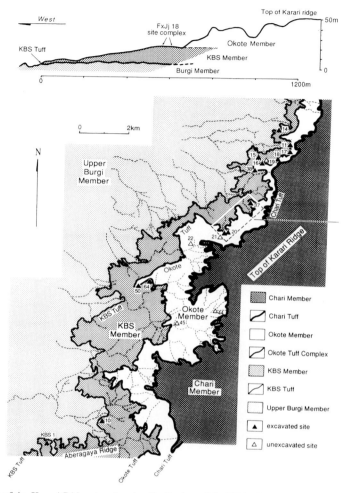

Figure 3. Map of the Karari Ridge showing the distribution of the KBS, Okote and Chari Members of the Koobi Fora Formation and the locations of the two archaeological sites defining the edge of the present study area, FxJj20 and FxJj37 (after Isaac, n.d.).

by Isaac (1981). Information about the sampling strategies, analytical and interpretive procedures employed in the study of these remains can be found in Stern (1991).

The minimum-archaeological stratigraphic unit

Any attempt to describe the land-use patterns of the hominids who left behind the debris that is now found scattered throughout the lower Okote Member involves documenting the distribution of archaeological remains in relation to palaeolandscape features (Isaac, 1981; Stern, 1991). To do this it is necessary to identify the "minimum archaeological-stratigraphic unit", the smallest sedimentary package that can be used to study the distribution of archaeological debris across an ancient landscape (Stern, 1991). The concept is useful because

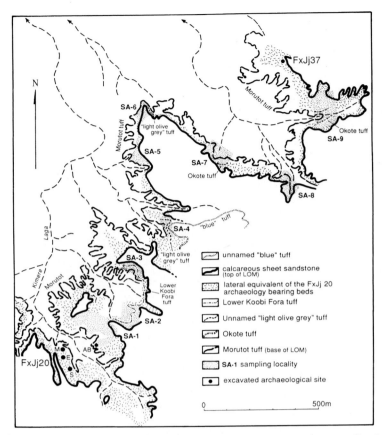

Figure 4. Geological map of the study area showing the distribution of the Morutot tuff and calcareous sandstones defining the boundaries of the Lower Okote Member, the lateral equivalents of the archaeological horizon at FxJj20 and the locations of sampling areas.

the idea that the archaeological debris recovered from one distinctive bed of sediment may not be derived from the same time plane is counter-intuitive for many archaeologists. Furthermore, it allows discussion of the trade-off that exists between the coverage of an ancient landscape, time resolution and the quantities of archaeological debris available for analysis.

Landscapes are made up of different depositional units, the boundaries of which shift over time. Thus, no single sedimentary horizon can provide information about the distribution of archaeological debris across an ancient landscape, and contiguous beds cannot be treated as though they were contemporaneous. In order to establish the configuration of a landscape at a specified time it is necessary to identify a time-line that cuts across facies boundaries. Time lines are most readily established by widespread horizons representing geologically instantaneous depositional events, like an air-fall ash spread over a large area by a single volcanic eruption. But because archaeological debris does not accumulate over the same time intervals, or at the same rate as the sediments encasing them there is often an inverse relationship

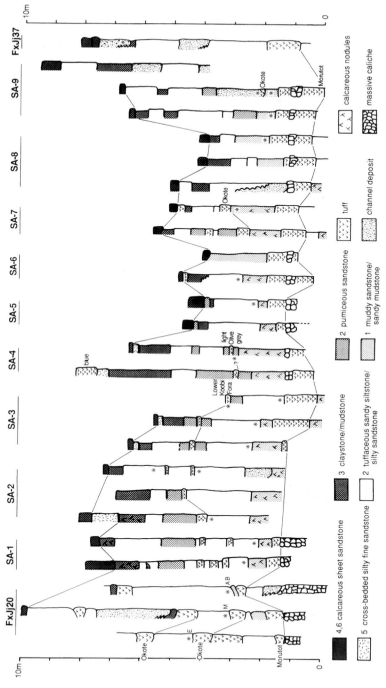

Figure 5. Set of correlated sections for the study area. The stars mark the stratigraphic origin of archaeological debris. Unit 2 is the lateral equivalent of the archaeological horizon at FxJj20. (Reproduced by permission of The University of Chicago Press, from *Current Anthropology* 34(2): 201–225)

between the rate of sediment accumulation and the amount of archaeological debris preserved. And although ash-beds represent geologically instantaneous events, they rarely contain much archaeological debris.

Depending on the stratigraphic sequence under investigation, one of two strategies may be used to investigate the distribution of material remains across an ancient landscape. In some sequences a depositional hiatus may be traced out around the modern erosion front and records made of the distribution of the archaeological debris encountered along the eroding outcrops. This horizon may be a distinctive soil horizon, like the one being studied by Potts (1989) at Olorgesailie, or it may be an isochronous bed on which a soil horizon has formed, like Tuff 1B at Olduvai Gorge, being investigated by Blumenschine & Masao (1991). In stratigraphic sequences without marked depositional hiatuses, but containing closely spaced marker horizons, it may be more appropriate to trace out a set of "pene-contemporaneous" sediments representing different parts of the same depositional system (e.g. Stern, 1991). The marker horizons may be isochronous units but the sediments between them will have accumulated intermittently in different parts of the landscape at different times.

The Koobi Fora Formation contains numerous widespread, fluvially re-worked ashes that serve as marker horizons (Findlater, 1976, 1978; Brown & Feibel, 1988). Three of these occur in the stratigraphic sequence exposed along the western face of the Karari Ridge, the KBS, the Morutot and the Chari tuffs, but none of them contains sufficient quantities of archaeological material to provide the basis for an investigation of its differential distribution across the palaeolandscape. There are at least another fifty chemically distinct tuffs in the southern Karari area, but most of these occur at a single section or outcrop over limited areas, do not provide anything approximating a transect across the ancient landscape, and not many of them contain any quantity of archaeological debris (Stern, 1991). The tuffs that lie within the lower Okote Member can be used to define facies relationships, but not to investigate what lay on the surface of the palaeolandscape at a geological instant in time. Reconstruction of the lower Okote landscape is an approximation of the main topographic features that accumulated sediment between two marker horizons over a quantifiable interval of time. Differences in the time span represented by a narrow band of sediments and by a soil horizon may be as great as one order of magnitude (e.g., Stern, 1991; Blumenschine & Masao, 1989). Both stratigraphic situations present archaeologists with a series of time-averaged assemblages, however, and raise questions about the behavioural and ecological "processes" or "relationships" that are preserved over periods of 1000–10 000 years.

In the study area between FxJj20 and FxJj37 (Figures 4 and 5) the minimum-archaeological-stratigraphic unit is defined by two marker horizons bracketing the archaeology-bearing beds at FxJj20. The lower boundary of this unit is defined by the Morutot tuff, a widespread tuff with a K/Ar age estimate of 1.64 ± 0.04 Ma. (McDougall, 1985; Stern, 1991). The upper boundary is defined by the next marker horizon in the stratigraphic sequence, a thin, laterally extensive and distinctive horizon of calcareous sandstone representing a change in depositional regime (Stern 1991). The sediments sandwiched between these two marker horizons represent the smallest thickness of sediment, and hence the smallest unit of time, that can be used to describe the distribution of material remains across the lower Okote Member landscape. They comprise massive and undifferentiated sands, silts and tuffs representing intercalated channel and floodplain deposits. Different sampling localities may have represented different, but contiguous, sedimentary environments (channel-proximal floodplain; proximal floodplain-distal floodplain) at different stages during this stratigraphic interval. However, unless a single horizon can be traced out between two geographically separate sections there is no way of correlating the sub-facies encountered at different sampling localities. The large number of chemically distinct tuffs with limited outcrop area illustrates the intermittent and localised character of sediment accumulation in this stratigraphic interval. It is highly unlikely that any two geographically separate localities accumulated sediment over precisely the same time interval or at the same rate. Contemporaneity and facies relationships can only be established with respect to the Morutot tuff and the calcareous sandstone capping the sequence.

The minimum-archaeological stratigraphic unit thus identifies a block of "pene-contemporaneous" sediments containing abundant material remains that can be traced around the modern erosion front for several kilometres providing a transect sample across a portion of the lower Okote landscape (Stern, 1991). It also provides a means of sampling the amount and type of archaeological debris that accumulated in different parts of that landscape over whatever time span is represented by the minimum archaeological stratigraphic unit.

These sediments were apparently laid down by two alternating depositional regimes, an axial and a marginal drainage (Feibel, 1988). The proto-Omo flowed from north to south along the axis of the sedimentary basin, carrying ashes and pumices that were originally erupted in the Ethiopian highlands. At times these ashes choked the river sytem, resulting in the formation of numerous braided channels (*ibid*). The proto-Bakate flowed from the volcanic highlands bordering the eastern margin of the basin, probably disgorging into the proto-Omo somewhere near the present study area (*ibid*). A succession of fluvial sub-facies make up the lower Okote Member. The lowermost depositional unit (1) comprises widespread alluvial floodplain deposits, including the Morutot tuff. Extensive caliche development points to prolonged periods of subaerial exposure and the development of a soil horizon on these deposits. These are overlain by a complex set of interlocking channel and floodplain deposits (unit 2), although overbank deposits make up most of the sequence. Sometimes channel segments can be traced short distances, but individual segments cannot be linked, though they may represent parts of the same drainage network, albeit at different stages in its history. These are the lateral equivalents of the archaeology-bearing horizon at FxJj20. Overlying these channel and overbank deposits are widespread distal floodplain deposits (unit 3). Abundant calcareous nodules indicate that these were part of a well-vegetated floodplain subject to intermittent deposition. A change in depositional regime is marked by the widespread sandsheets capping the stratigraphic sequence (units 4, 5, 6). These are interpreted as the deposits of frequently shifting, shallow, braided channels.

Archaeological debris is scattered throughout this stratigraphic sequence, though it is more abundant in the tuffaceous sandy mudstones and muddy sandstones of unit 2 than in the other subfacies. Much of the archaeological debris thus lies towards the base of the lower Okote Member, though this does not mean that it was all deposited at about the same time, since the time span of net sediment accumulation clearly varied from one locality to another (this argument is expanded below). But it does indicate that most of the archaeological debris is found in proximal floodplain settings (Stern, 1993).

Establishing the time resolution of the minimum archaeological stratigraphic unit
During the last decade palaeontologists have taken a particular interest in the problem of how to estimate the chronological resolution of fossiliferous portions of the rock record (e.g., Behrensmeyer & Schindel, 1983). Geologists have long recognised that the greater portion of geological time is represented by small-scale disconformities in the rock record (e.g., Barrell, 1917) and that the stratigraphic sequence at any one locality is inherently incomplete (Reineck, 1960; Newell, 1972). But in the early 1980s, attempts to quantify the completeness of the

record were hastened by the desire to resolve a decade-long debate about the mode and tempo of change in the earth's lifeforms (Eldredge & Gould, 1972; Gould & Eldredge, 1977; Gould, 1980; Eldredge, 1982) and a number of measures of incompleteness were proposed (Sadler, 1981; Schindel, 1980, 1982; Tipper, 1983; Badgely et al., 1986). Given ample illustration of the way in which sedimentary processes influence the distribution of fossil remains through the sedimentary record (e.g., Sadler & Dingus, 1982; Behrensmeyer, 1982, 1987) it is perhaps surprising that this literature has not been discussed more widely by archaeologists. Before trying to establish the time resolution of the lower Okote Member it is worth taking a brief digression to review the strategies that palaeontologists have used to estimate the time resolution of fossil assemblages.

Most of the measures of stratigraphic completeness proposed in the early 1980s (e.g., Schindel, 1980, 1982; Sadler, 1981; Tipper, 1983) were based on compilations of median sedimentation rates recorded over a range of time spans for a variety of depositional environments. These compilations were presented as plots of sedimentation rate against time span of observation to show that a persistent inverse relationship exists between rate and time-span of observation for all sedimentary environments, although the slope of the regression line varies from one depositional context to another (e.g., Sadler, 1981). Stratigraphic completeness was then defined as the ratio of long term sediment accumulation rate to the average rate of sediment accumulation for a specified time interval (ibid: 581) and calculated using the documented relationship beween short- and long-term sedimentation rates. Thus it represents a measure of the proportion of time intervals of a given duration that are represented by sediment at the sections under study. Initial applications of this measure to specific fossiliferous sequences resulted in the rather dismal conclusion that most vertebrate fossil sequences were too incomplete to resolve the evolutionary questions to which palaeontologists were seeking answers (e.g., van Andel, 1981; Sadler & Dingus, 1982; Gingerich, 1982).

But it was soon realised that median sedimentation rates do not provide meaningful estimates of the completeness of any individual section of the rock record, primarily because median sedimentation rates rarely reflect the actual rates of sediment accumulation at a specific locality and partly because they are artefacts of measurement techniques (Anders et al., 1987). Furthermore, estimates of completeness based on the relationship between long and short term sedimentation rates are complicated by the fact that plots of time versus rate produce spurious correlations (Badgely et al., 1986; Anders et al., 1987). Thus attempts to establish the stratigraphic completeness of any particular portion of the rock record (or chronological resolution) are most appropriately made on information derived from the stratigraphic sequence under study.

The particular measures of time resolution employed, and the way in which they are made, will depend on the problem under investigation and the time intervals being sought, as well as the types of sediments making up the sections under study and the types of fossil or archaeological assemblages and datable materials that they encapsulate (e.g., Behrensmeyer & Schindel, 1982). The mode and energy of sediment deposition, regional patterns of sedimentation and the taphonomic history of the material remains under investigation are crucial sources of information about the time resolution of any set of material remains encountered in the sedimentary record (see for example, Behrensmeyer, 1987; Kidwell, 1986; Kidwell & Bosence, 1991).

Time resolution may be defined in more than one way. It is often useful to distinguish the total time span taken to accumulate a stratigraphic section, or its scope, from the time span represented by each individual fossil sample in that section, its micro-stratigraphic acuity (Schindel, 1982). But time resolution has also been discussed in terms of stratigraphic completeness, a concept that has been variously defined as the proportion of time intervals of a given duration that are represented by sediment in a particular geological section (Sadler, 1981), the net duration of deposition represented by sediment in a given section (Tipper, 1983), and the temporal distribution of gaps in a stratigraphic sequence relative to the actual rock record (Badgely et al., 1986). Some of these measures are more useful than others for establishing the time resolution of the lower Okote Member (LOM) archaeological record.

The distinction between scope and micro-stratigraphic acuity is a useful one for understanding the empirical structure of the LOM archaeological record. Given the intermittent and localised nature of sediment build up in the Karari area during Okote Member times, it is highly unlikely that either the stone artefacts or the fossil remains accumulated at the same rate, or over the same time spans, as their encasing sediments. In fact, there is ample evidence that they did not. The Morutot tuff, for example, represents a single depositional event, but within the present study area no archaeological debris was incorporated into this flood deposit. However, during the depositional hiatus that followed the deposition of the Morutot tuff archaeological materials accumulated on its surface at a number of different localities (SA-8, SA-3; Figure 3). There are many lower Okote Member beds in the study area that do not contain archaeological debris and others that contain abundant debris at specific levels in specific locations (Stern, 1991).

It is difficult to obtain direct measures of the amount of time represented by a discrete cluster of archaeological debris from any single bed. Bone weathering stages have been used by a number of researchers to estimate the time span over which the faunal assemblages from individual high density patches of debris accumulated (e.g., Bunn, 1982; Potts, 1986). These estimates are based on actualistic studies documenting the patterns of surface modification that characterise bones exposed on the surface of a modern landscape for known periods of time (e.g., Behrensmeyer, 1978; Tappen, 1992). These studies indicate that most bones lying on the surfaces of savanna woodland environments disintegrate within 15 years unless they are buried. Bone is not well preserved in lower Okoke Member sediments in the study area, reflecting low rates of net sediment accumulation. The surface features of those bones that have survived suggest that they had lain on the surface of the landscape for 6 years or less before being buried (Bunn, 1982; Stern, 1991). A few, discrete clusters of bone thus have high micro-stratigraphic acuity, but most of the lower Okote Member faunal assemblage is made up of highly fragmented, dispersed remains (Stern, 1991).

Occasional sets of refitting artefacts and bones, like those from FxJj50 described by Bunn et al. (1980), provide another example of a set of material remains with high micro-stratigraphic integrity. In this case it is argued that these are functionally integrated sets of debris that result from a single, identifiable behavioural event. But for archaeologists interested in hominid land use patterns the existence of a few sets of remains representing short time spans is not particularly useful, since they do not establish the time resolution of most of the archaeological remains found within the band of sediments approximating the ancient landscape. This is because it is impossible to determine whether any two sets of refitting objects recovered from different sampling locations were deposited in the same time interval, or over the same time span. Only if the refitting sets were recovered from the same isochronous depositional unit would it be possible to argue that they had any functional inter-relationship. But as pointed out earlier, tuffs in the lower Okote Member rarely contain archaeological debris. Furthermore, these tuffs are all fluvially re-deposited and were originally deposited as air-fall ashes in the

Table 1   Estimates for the scope of the Lower Okote Member, using sedimentation rates from Feibel 1988

Stratigraphic interval	Mean sedimentation rate cm per 1000 years	Estimated scope of the LOM (units 1–6) in years	Estimated scope of the equivalents of FxJj20 (unit 2) in years
Okote Member	8·3	86 000	70 000
Morutot-Chari tuffs	9·23	78 000	65 000
White-Black Pumice tuffs	9·5	75 000	60 000

Ethiopian highlands; sometimes, as at FxJj20, the same source bed was re-worked downstream on more than one occasion (Brown & Feibel, 1985). Contemporaneity between archaeological assemblages recovered from different sampling localities, or from different beds within the lower Okote Member, can only be established in relation to the two marker horizons that define its stratigraphic boundaries.

For the purposes of palaeolandscape reconstruction the time resolution of the lower Okote Member is most appropriately defined in terms of the total time span taken to accumulate the deposits in this stratigraphic interval, i.e., its scope. Because the lower Okote Member is the smallest package of sediment that can be used to investigate the distribution of archaeological debris across the ancient landscape, its scope defines the time span over which the archaeological assemblages were accumulated. This establishes the time frame over which the patterned distributions of material remains being studied by archaeologists were formed. The interpretation and explanation of those patterns must be commensurable with the variables that contributed to patterns created at that time scale (e.g., Fletcher, 1992).

The Okote Member is bracketed by the Morutot and Chari tuffs, with K/Ar age estimates of 1·65 ± 0·03 Ma and 1·39 ± 0·01 Ma respectively (McDougall, 1985), setting an upper limit on estimates of the scope of the lower Okote Member. However, the actual scope of the lower Okote Member can be estimated using measured, long term average sedimentation rates for the Okote Member itself, or for the volcaniclastic interval straddling the upper portion of the KBS Member and the lower portion of the Okote Member, of which the lower Okote Member forms a part (see Stern, 1991). Depending on the stratigraphic interval used to calculate long term, average sedimentation rates, estimates of the scope of the LOM range from 75 000 to 86 000 years and from 60 000 to 70 000 years for the lateral equivalents of the archaeology-bearing beds at FxJj20 (Table 1). Feibel *et al.* (1989) have used stratigraphic scaling techniques to estimate an age of 1·55 ± 0·04 Ma for the Black Pumice tuff, which lies at least 3 metres above the top of the lower Okote Member. Thus, the estimated scope of the lower Okote Member is consistent with the estimated age of the Black Pumice tuff.†

These estimates underscore the implications of using a minimum archaeological strati-graphic unit, whose accumulation was intermittent and localised and spans 65 000 ± 5000 years, to reconstruct early hominid land use patterns. It is highly unlikely that the distribution of artefacts and bones through the lower Okote Member reflects a set of ecological relationships that prevailed throughout the time span of this stratigraphic horizon.

---

†A forthcoming K/Ar age estimate for a tuff immediately overlying the lower Okote Member will provide a useful check on these estimates of the total time span taken to accumulate the stratigraphic interval under investigation.

Is this a time-averaged assemblage?

Time-averaging is a term that was defined by invertebrate palaeontologists to refer to fossil assemblages "that accumulate from the live community during the time required to deposit the containing sediment" (Walker & Bambach, 1971). They pointed out that in situations where rates of sediment accumulation are slow by comparison with the life-span of an organism, the fossil assemblages created do not reflect the community structure that prevailed at any given time. Rather, they are composites, spanning long periods of time. Walker & Bambach (*ibid.*) argued that short term fluctuations in community structure cannot be discerned in these fossil assemblages, only persistent, long-term trends. Since then invertebrate palaeontologists have come to regard the identification and quantification of time-averaging as one of the major aims of taphonomic analysis (Kidwell & Bosence, 1991), given its impact on measures of diversity, relative abundance and other palaeoecological parameters (Staff *et al.*, 1986; Kidwell & Bosence, 1991). The focus of palaeontological research is on quantifying time-averaging, on generating criteria for identifying assemblages with comparable taphonomic histories, and on establishing the categories of palaeoecological information preserved in particular geological contexts (e.g., Kidwell, 1986; Kidwell & Bosence, 1991).

Clearly, the archaeological debris preserved in the lower Okote Member is time-averaged: it is the aggregated material residues of many hominid social groups, few of whom shared the same landscape. It is not possible to select from this record a set of material aggregates with high micro-stratigraphic acuity that were all deposited in the same 1000 year time-slice. Contemporaneity can be assessed only in terms of the marker horizons that define the upper and lower boundaries of the minimum archaeological-stratigraphic unit. Although there are some individual archaeological occurrences that do exhibit high micro-stratigraphic acuity they do not provide a basis for investigating patterns of palaeolandscape distribution. This problem cannot be overcome by focusing the analysis on two or more assemblages with fine temporal resolution for there is no way of assessing their contemporaneity, except within a 60–70 000 year time interval. Furthermore, by restricting the analysis to rare, individual occurrences that may represent functionally integrated sets of debris, the bulk of the record is being ignored. If attempts to establish the categories of behavioural and palaeoecological information encapsulated in this record are to be successful, the entirety of the time-averaged assemblage must be examined.

Having argued that the lower Okote Member preserves a time-averaged assemblage with a chronological resolution of 65 ± 5 k.y.a. it is worth stressing that "time-averaged" is not the same as "an average over time". In fact, the longer the time span of accumulation the greater the number of behavioural events and biological and geomorphic processes that will have contributed to the structure of a record. It is highly improbable that a single set of ecological relationships will have been maintained over the time span involved in the formation of the lower Okote Member (see for example Bradley, 1985). Micro-stratigraphic studies of the context in which archaeological debris is found (e.g. Kaufulu, 1983; Stern, 1991), and of the Okote Member as a whole (Brown & Feibel, 1985), certainly indicate that conditions conducive to the preservation of material remains were not always present and did not prevail in all parts of the landscape at the same time; this is reflected, for example, in the differential preservation of bone in the lower Okote Member (Stern, 1991). Given the myriad of variables affecting the behavioural events that result in the loss or discard of stone tools, and the formation of clusters of carcasses or carcass parts, the accumulation of debris on the LOM landscape was undoubtedly sporadic and variable in both quantity and character over time. Habitat boundaries, the structure of ecological communities, foraging strategies, population structure and dynamics and patterns of group interactions are all subject to changes over time and these will have influenced the quantities, the types and the locations of accumulating material residues. Neither the behavioural events resulting in the discard of material remains, nor the processes affecting their burial and preservation in the sedimentary record, were steady through time. The aggregations of debris encountered in lower Okote Member sediments do not reflect an average of all one thousand year time slices any more than they reflect the relationships that existed between hominid activities and specific landscape attributes in any given one thousand year time slice.

The patterned distributions of debris that can be identified in this record were generated over an 80 000 year time span. There is no reason to doubt that the factors influencing the ranging patterns of extant primates also influenced the movements of early hominids across the landscape. But these were not the only factors that contributed to the patterns and configurations that can be identified in the material aggregates preserved in the lower Okote Member. Thus, the issue that needs discussion is whether or not the processes and the length of time involved in the formation of these time-averaged assemblages can be understood in the same terms as the behaviour of extant primates.

Conclusions

Clearly, it is possible to identify stratigraphically discrete, but laterally extensive sedimentary horizons that contain sufficient archaeological debris that they can be used to study the differential distribution of material remains across an ancient landscape. However, the archaeological materials contained in these horizons are time-averaged palimpsests. The time resolution of the lower Okote Member indicates that the patterned distributions that characterise these data were the outcome of aggregating tens of thousands of years of debris. But while it is feasible to quantify the amount of time-averaging involved in the formation of this landscape assemblage it is not possible to draw a simple equation between the distribution of debris in relation to depositional environments and the movements of hominids across the palaeolandscape without ignoring the time dimension of these data.

This raises questions about whether or not aggregates of material remains can be understood using interpretive theories ultimately based on short term observations of the actions of individuals, their interactions with one another and with the ecological systems of which they are a part. The viability of attempting to reconstruct the way in which early hominids moved around the landscape from the configurations of debris that they left behind is thus open to serious investigation. A growing number of researchers have acknowledged the mis-match between theoretical constructs based on ethnographic scale observations and the material record of the past (e.g., Bailey, 1983; Cosgrove, 1991; Murray, 1992, 1993; Stern, 1991). But there is no consensus about the interpretive strategies that should be developed in response to this observation. A number of researchers have suggested that interpretive theories based on ethnographic scale data are just one rung of an explanatory hierarchy that incorporates material aggregates and behavioural processes and mechanisms at a variety of time scales (e.g., Bailey, 1983; Fletcher, 1992). There is, however, no consensus about the components of this behavioural hierarchy, its form, or the interdependence of its different levels.

It is difficult, however, to envisage behavioural or ecological processes that operate over the time intervals that it took to accumulate many Lower Pleistocene archaeological records. Behaviours are made up of individual actions and reactions and behavioural ecology of patterns of interactions between specific behaviours and ecological parameters maintained over relatively short spans of time (from hours, to seasons, to a few decades). The archaeological record allows the identification of some specific categories of behaviour in which early hominids engaged, like tool-use and meat-eating (e.g., Isaac, 1984; Potts, 1988). Although the archaeological record was built up through the performance of activities like these, patterns in the distribution of artefacts and bones across an ancient landscape are not simply the outcome of hominid activities. Patterns and configurations amongst these data may well have the potential to provide a unique archaeological perspective on the evolutionary history of hominid behaviour.

For archaeologists who concur that the time resolution of material aggregates is a potent reason for describing the archaeological record as "an ontologically singular record of human action" (Murray, 1987) the road ahead is uncharted. For whilst there is growing interest in developing a theory of human action applicable to patterns and configurations of material data generated over long time spans (Bailey, 1983, 1987; Fletcher, 1992; Murray, 1987, 1992, 1993, n.d.; Stern, 1991) there are no existing theoretical precedents. The rationale for this discussion is to stimulate interest in the development of theoretical constructs specifically aimed at understanding the material record of the past on its own terms. Solutions cannot be devised for problems whose existence are not acknowledged. Admittedly, not all archaeologists perceive this mis-match between artefact distributions and reconstructions of early hominid land-use patterns. Only those who do will want to engage in further discussion about the ontological significance of the time-averaged assemblages of material remains that characterise the Lower Pleistocene archaeological record. It is a long term research endeavour, but one that may ultimately provide a unique archaeological perspective on the evolution of human behaviour.

Acknowledgements

The field research on which this paper is based was undertaken with permission from the Office of the President of the Kenyan Government and under the sponsorship of the National Museums of Kenya. I thank Dr Simuyu Wandibba, then the Head of the Archaeology Division at the National Museums of Kenya and Mr Richard Leakey, then its Director/Chief Executive, for the support they gave this work. The field work was funded by grants from the National Science Foundation (Dissertation Improvement Grant), the L.S.B. Leakey Foundation, and the Department of Anthropology at Harvard University. Harry Merrick, Craig Feibel and Marsha Smith provided invaluable logistical support. Frank Brown generously undertook the analysis of the tuff samples, the cost of which was covered by a small grant from the American School of Prehistoric Research. I thank the editors of this volume for inviting me to participate in the symposium at which an earlier version of this paper was presented and three anonymous reviewers whose comments helped to sharpen the arguments presented.

References

Anders, M. H., Kruger, S. W. & Sadler, P. M. (1987). A new look at sedimentation rates and the completeness of the stratigraphic record. *J. Geol.* **95,** 1–14.
Badgely, C. (1986). Taphonomy of mammalian fossil remains from Siwalik rocks of Pakistan. *Paleobiol.* **12,** 119–142.
Bailey, G. N. (1983). Concepts in time in quaternary prehistory. *Ann. Rev. Anthrop.* **12,** 165–192.
Bailey, G. N. (1987). Breaking the time barrier. *Archaeol. Rev. Cambridge* **6,** 5–21.

Barrell, J. (1917). Rhythms and the measurement of geologic time. *Geol. Soc. Am. Bull.* **28**, 745–904.

Behrensmeyer, A. K. (1987). Taphonomy and hunting. In (M. H. Nitecki & D. V. Nitecki, Eds) *The Evolution of Human Hunting.* New York: Plenum Press, pp. 423–450.

Behrensmeyer, A. K. (1991). Terrestrial vertebrate accumulations. In (P. A. Allison & D. E. G. Briggs, Eds) *Taphonomy: Releasing the Data Locked in the Fossil Record.* New York: Plenum Press, pp. 291–335.

Behrensmeyer, A. K. & Schindel, D. (1983). Resolving time in paleobiology. *Paleobiol.* **9**, 1–8.

Binford, L. R. (1984). *Faunal Remains from Klasies River Mouth,* New York: Academic Press.

Blumenschine, R. J. & Masao, F. T. (1991). Living sites at Olduvai Gorge, Tanzania? Preliminary landscape archaeology results in the basal Bed II lake margin zone. *J. hum. Evol.* **21**, 451–462.

Bradley, R. S. (1985). *Quaternary Paleoclimatology: Methods of Paleoclimatic Reconstruction.* Boston: Allen and Unwin.

Brown, F. H. & Feibel, C. S. (1985). Stratigraphical notes on the Okote tuff complex of Koobi Fora, Kenya. *Nature* **316**, 794–797.

Brown, F. H. & Feibel, C. S. (1986). Revision of lithostratigraphic nomenclature in the Koobi Fora region, Kenya. *J. Geol. Soc.* **143**, 297–310.

Brown, F. H. & Feibel, C. S. (1988). "Robust" hominids and Plio-Pleistocene paleogeography of the Turkana basin, Kenya and Ethiopia. In (F. E. Grine, Ed.) *Evolutionary History of the "Robust" Australopithecines.* New York: Aldine de Gruyter, pp. 325–341.

Bunn, H. T. (1982). Meat-eating and human evolution: studies on the diet and subsistence patterns of Plio-Pleistocene hominid in East Africa. Ph.D. dissertation, Department of Anthropology, University of California, Berkeley.

Bunn, H. T., Harris, J. W. K., Isaac, G. Ll., Kaufulu, Z., Kroll, E., Schick, K., Toth, N. & Behrensmeyer, A. K. (1980). FxJj50: an early Pleistocene site in northern Kenya. *World Archaeol.* **12**, 109–136.

Coppens, Y., Howell, F. C., Isaac, G. Ll. & Leakey, R. E. F. (1976). *Earliest Man and Environment in the Lake Rudolf Basin.* Chicago: University of Chicago Press.

Eldredge, N. (1982). Evolutionary rates. *Syst. Zool.* **31**, 338–347.

Eldredge, N. & Gould, S. J. (1972). Punctuated equilibria: an alternative to phyletic gradualism. In (T. J. M. Schopf, Ed.) *Models in Paleobiology.* San Francisco: Freeman and Cooper, pp. 156–167.

Feibel, C. S. (1988). Paleoenvironments of the Koobi Fora Formation, Turkana Basin, Northern Kenya. Ph.D. Dissertation, Department of Geology and Geophysics, University of Utah.

Feibel, C. S., Brown, F. H. & McDougall, J. (1989). Stratigraphic context of fossil hominids from the Omo Group deposits: northern Turkana Basin, Kenya and Ethiopia. *Am. J. phys. Anthrop.* **78**, 595–622.

Findlater, I. C. (1976). Tuffs and the recognition of isochronous mapping units in the East Rudolf succession. In (Y. Coppens, F. C. Howell, G. Ll. Isaac & R. E. F. Leakey, Eds) *Earliest Man and Environments in the Lake Rudolf Basin.* Chicago: University of Chicago Press, pp. 94–104.

Findlater, I. C. (1978). Isochronous surfaces within the Plio-Pleistocene sediments east of Lake Turkana. In (W. W. Bishop, Ed.) *Geological Background to Fossil Man.* Edinburgh: Scottish Academic Press, pp. 415–420.

Fletcher, R. (1992). Time perspectivism, Annales and the potential of archaeology. In (B. Knapp, Ed.) *Ethnohistory. New Directions in Archaeology.* New Directions in Archaeology Series. Cambridge: Cambridge University Press.

Foley, R. (1987). *Another Unique Species.* Harlow: Longman Scientific and Technical.

Fürsich, F. T. and Aberhan, M. (1990). Significance of time-averaging for palaeocommunity analysis. *Lethaia* **23**, 143–152.

Gingerich, P. (1982). Time resolution in mammalian evolution: sampling, lineages and faunal turnover. *Proceedings of the 3rd North American Paleontological Convention Volume 1.* pp. 205–221.

Gould, S. J. (1980). Is a new and general theory of evolution emerging? *Paleobiology* **6**, 119–130.

Gould, S. J. & Eldredge, N. (1977). Punctuated equilibria: the tempo and mode of evolution reconsidered. *Paleobiology* **3**, 115–151.

Harris, J. M. Ed. (1983). *Koobi Fora Research Project. Volume 3. The Fossil Ungulates: Proboscidea, Perissodactyla and Suidae.* Oxford: Clarendon Press.

Harris, J. W. K. (1978). The Karari Industry: Its Place in East African Prehistory. Ph.D. Dissertation, Department of Anthropology, University of California, Berkeley.

Isaac, G. Ll. (1976). The activities of early African hominids: a review of archaeological evidence from the time span two and a half to one million years ago. In (G. Ll. Isaac & E. McCown, Eds) *Human Origins: Louis Leakey and the East African Evidence.* Menlo Park: Benjamin, pp. 483–514.

Isaac, G. Ll. (1978). The food-sharing behaviour of proto-human hominids. *Sci. Am.* **238**, 110–113.

Isaac, G. Ll. (1981). Stone Age visiting cards: approaches to the study of early land-use patterns. In (I. Hodder, G. Ll. Isaac & N. Hammond, Eds) *Pattern of the Past.* Cambridge: Cambridge University Press, pp. 131–155.

Isaac, G. Ll. (1983). Aspects of evolution. In (D. S. Bendall, Ed.) *Evolution from Molecules to Men.* Cambridge: Cambridge University Press, pp. 509–543.

Isaac, G. Ll. Ed. (n.d.). *Koobi Fora Research Project. Volume 3. The Archaeology.* Oxford: Clarendon Press.

Isaac, G. Ll. & Harris, J. W. K. (1978). Archaeology. In (M.G. & R. E. F. Leakey, Eds) *Koobi Fora Research Project. Volume 1. The Fossil Hominids and an Introduction to their Context 1968–1974.* Oxford: Clarendon Press, pp. 64–85.

Isaac, G. Ll. & Harris, J. W. K. (1980). A method for determining the characteristics of artefacts between sites in the Upper Member of the Koobi Fora Formation, East Lake Turkana. In (R. E. Leakey & B. Ogot, Eds) *Proceedings of the 8th Panafrican Congress of Prehistory and Quaternary Studies, Nairobi, 5 to 10 September 1977.* Nairobi: T.I.L.L.M.I.A.P., pp. 19–22.

Kaufulu, Z. M. (1983). The geological context of some early archaeological sites in Kenya, Malawi, and Tanzania: microstratigraphy, site formation and interpretation. Ph.D. Dissertation, Department of Anthropology, University of California, Berkeley.

Kidwell, S. M. (1985). Palaeobiological and sedimentological implications of fossil concentrations. *Nature* **318,** 457–460.

Kidwell, S. M. (1986). Modes for fossil concentrations: paleobiologic implications. *Paleobiology* **12,** 6–24.

Kidwell, S. M. & Bosence, W. J. (1991). Taphonomy and time-averaging of marine shelly faunas. In (P. A. Allison & D. E. G. Briggs, Eds) *Taphonomy: Releasing the Data Locked in the Fossil Record.* New York: Plenum Press, pp. 115–209.

Leakey, R. E. F. & Leakey, M. G. Ed. (1978). *Koobi Fora Research Project. Volume 1. The Fossil Hominids and an Introduction to their Context. 1968–1974.* Oxford: Clarendon Press.

McDougall, I. (1985). K-Ar and 40Ar/^{39}Ar dating of the hominid-bearing Pliocene-Pleistocene sequence at Koobi Fora, Lake Turkana, Kenya. *Geol. Soc. Am. Bull.* **96,** 159–175.

Murray, T. (1987). Remembrance of things present: appeals to authority in the history and philosophy of archaeology. Ph.D. Dissertation, Department of Anthropology, University of Sydney.

Murray, T. A. (1992). The Tasmanians and the constitution of the "Dawn of Humanity". *Antiquity* **66,** 700–743.

Murray, T. (1993). Dynamic modelling and a new social theory of the mid to longterm. In (C. Renfrew & S. Van der Leeuw, Eds) *Dynamic Modelling and the Study of Change in Archaeology.* Cambridge: Cambridge University Press.

Murray, T. (n.d.). Humanising the Palaeolithic. In (R. Jones & L. Hiatt, Eds) *Ethnography and European Thought.* Sydney: Oceania Monographs.

Newell, N. D. (1972). Stratigraphic gaps and chrono-stratigraphy. *24th Int. Geol. Congr.* **7,** 198–204.

Peterson, C. H. (1977). The paleontological significance of undetected short-term temporal variability. *J. Paleont.* **51,** 976–981.

Potts, R. B. (1986). Temporal span of bone accumulations at Olduvai Gorge and implications for early hominid foraging behavior. *Paleobiology* **12,** 25–31.

Reineck, H. E. (1960). Über zeitlucken in rezenten flachee-sedimenten. *Geologische Runschau* **49,** 149–161.

Rogers, M. & Harris, J. W. K. (1992). Recent investigations in landscape archaeology at East Turkana. *Nyame Akuma* **38,** 41–47.

Sadler, P. M. (1981). Sediment accumulation rates and the completeness of stratigraphic sections. *J. Geol.* **89,** 569–584.

Sadler, P. M. and Dingus, L. W. (1982). Expected completeness of sedimentary sections: estimating a time-scale dependent, limiting factor in the resolution of the fossil record. *Proc. Third N. Am. Paleont. Conv.* **2,** 461–464.

Schindel, D. E. (1980). Microstratigraphic sampling and the limits of paleontological resolution. *Paleobiology* **6,** 408–426.

Schindel, D. E. (1982). Resolution analysis: a new approach to the gaps in the fossil record. *Paleobiology* **8,** 340–353.

Sept, J. (1992). Archaeological evidence and ecological perspectives for reconstructing early hominid subsistence behavior. *Adv. Archaeol. Method Theory* **4,** 1–56.

Staff, G. M., Stanton, R. J., Powell, E. N. & Cummins, H. (1986). Time-averaging, taphonomy, and their impact on paleocommunity reconstruction: Death assemblages in Texas bays. *Geol. Soc. Am. Bull.* **97,** 428–443.

Staff, G. M. & Powell, E. N. (1988). The paleoecological significance of diversity: the effect of time averaging and differential preservation on macroinvertebrate species richness in death assemblages. *Palaeogeog., Palaeoclimat., Palaeoecol.* **63,** 73–89.

Stern, N. (1991). The scatters-between-the-patches: A study of early hominid land use patterns in the Turkana Basin, Kenya. Ph.D. Dissertation, Department of Anthropology, Harvard University.

Stern, N. (1993). The structure of the Lower Pleistocene archaeological record: a case study from the Koobi Fora Formation. *Curr. Anthrop.* **34,** 201–225.

Tipper, J. C. (1983). Rates of sedimentation and stratigraphical completeness. *Nature* **302,** 696–698.

Tooby, J. & DeVore, I. (1987). The reconstruction of hominid behavioural evolution through strategic modeling. In (W. G. Kinzey, Ed.) *The Evolution of Human Behavior: Primate Models.* Albany: State University of New York Press, pp. 183–237.

Van Andel, T. H. (1981). Consider the incompleteness of the geological record. *Nature* **294,** 397–398.

Walker, K. R. & Bambach, R. K. (1971). The significance of fossil assemblages from fine grained sediments: time-averaged communities. *Geol. Soc. Am. Abstr. Programs* **3,** 783–784.

Ellen M. Kroll

Department of Anthropology, University of Wisconsin, Madison, WI 53706, U.S.A.

Behavioral implications of Plio-Pleistocene archaeological site structure

Received 24 December 1993
Revision received 29 April 1994
and accepted 29 April 1994

Keywords: Plio-Pleistocene distributional data, archaeological excavations, horizontal density, vertical resolution, refitting, activity areas, mini-site, maxi-site, shade trees.

Current classifications of excavated Plio-Pleistocene archaeological distributions (mini-site, maxi-site, on-site, off-site) oversimplify the apparent incompleteness of the excavations and the variability in the vertical and horizontal distributions demonstrated by systematic spatial analysis. An alleged mini-site (FxJj 64) and maxi-site (FxJj 50) at Koobi Fora are illustrative of lithic and faunal distributions contained in consolidated, massive, fine-grained beds having geological evidence of vegetation (including trees), of biogenic and mechanical processes that can destroy internal bedding and vertically displace archaeological pieces, and of irregular paleotopography that can distort archaeological distributions. Horizontal distributions are subdivided into clusters and subclusters that have variable vertical resolution. Vertical distributions of refitted and unrefitted pieces are suggestive that the greater thickness of some distributions relates primarily to the reuse of particular places by hominids after some previously discarded refuse was partially or completely buried and secondarily to bioturbation and other post-depositional processes. Although the lesser thickness of some distributions (or vertical zones thereof) probably relates to the accumulation of refuse on one old land surface (or possibly two), the contents provide converging evidence that refuse accumulated during repeated (discontinuous) episodes of hominid activity. Repetition of hominid activities at particular places and refuse accumulated in clusters and subclusters are attributed in part to the physical characteristics of particular places, including trees that might have attracted hominids especially for refuge (shade, retreat from on-the-ground dangers) and enabled them to engage in activities more safely and unhurriedly than at treeless places. It is urged that Plio-Pleistocene archaeological excavations be deeper and wider and alternative interpretations be considered of each aspect of hominid behavior in developing comprehensive models.

Journal of Human Evolution (1994) **27,** 107–138

Introduction

Archaeologists interested in the behavior of Plio-Pleistocene hominids focus on the context, composition, and distribution of lithic artifacts and fossil animal bones. Distributional data integrate the composition and context in time and space. Distributions of lithics and fauna are documented and studied on modern erosional land surfaces and in archaeological excavations at localities in East Africa, including Koobi Fora, Kenya, and Olduvai Gorge, Tanzania (Figure 1). Excavated distributions (intrasite spatial data) are traditionally categorized into different types of sites based on the excavators' impressions and visual assessments of the vertical and horizontal density of the contents (e.g., living floor, site with diffused material). The contents themselves (assemblage data) are separately categorized into different types of sites based on qualitative and quantitative analyses of assemblage composition (e.g., type B or butchery site, type C or home-base site; Isaac, 1989; Isaac & Harris, 1978; Leakey, 1971). Surface and excavated distributions away from the traditional excavations are newly studied to understand the variability of larger-scale distributions in Plio-Pleistocene landscapes (Blumenschine & Masao, 1991; Bunn, 1994; Isaac & Harris, 1980; Rogers, 1993; Stern, 1993).

Now that surface and subsurface distributional data of differing spatial scales and densities are available, it is necessary to analyse them systematically and quantitatively for regulated comparisons and to consider the current constraints on interpretations of hominid behavior

0047-2484/94/010107+32 $08.00/0

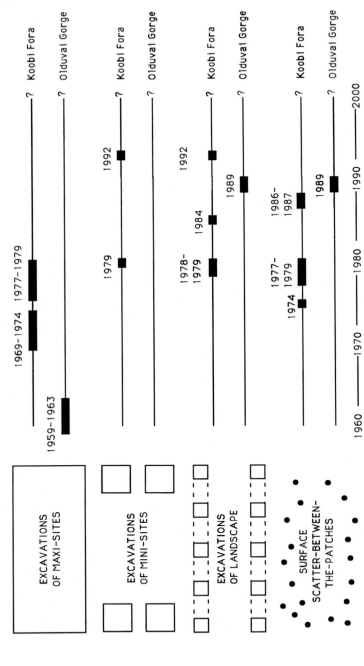

Figure 1. Field methods and periods for recovering Plio-Pleistocene distributional data in the Koobi Fora Formation at Koobi Fora, Kenya, and in Bed I at Olduvai Gorge, Tanzania, during the 35 years since the discovery of the FLK *Zinjanthropus* skull at Olduvai. Left side of chart identifies four field methods for mapping distributions of lithics and fauna, starting with the traditional excavation of maxi-sites. Right side shows relatively short periods devoted to each method from 1959–1992. See also Bunn (1994) about study of fauna-only distributions not included in this figure.

imposed by premature classifications (e.g., maxi-site, mini-site, on-site, off-site). This paper offers some observations about the recovery, analysis, classification, and interpretation of Plio-Pleistocene distributional data based on systematic spatial analysis of many of the traditionally excavated subsurface distributions at Koobi Fora and Olduvai Gorge (Kroll, n.d., in press) excavated and reported as archaeological "sites" by Isaac (in press) and by Leakey (1971). An alleged mini-site (FxJj 64) and limited comparison to a maxi-site (FxJj 50) illustrate some of the variability in the vertical and horizontal distributions at Koobi Fora and the importance of basing comparisons and classifications on empirical data regarding the density, composition, spacing, and demonstrated interconnectedness of continuous and discontinuous vertical and horizontal distributions of lithics and fauna. There is a convergence of different kinds of archaeological and geological evidence that some hominids repeatedly located certain activities at desirable places in the Plio-Pleistocene landscape.

Interpretations of hominid behavior from distributional data

The anthropological significance of Plio-Pleistocene distributions of lithics and fauna needs to be evaluated regarding the physical and behavioral characteristics of the different hominid taxa, the opportunities and constraints of the Plio-Pleistocene environments in which they lived, and the processes and periods of formation. Archaeologists need to consider alternative interpretations of each particular aspect of those (e.g., diet, social organization, location of activities, timing of activities) and incorporate a selection of the successful interpretations into alternative integrated models of hominid behavior to be revised as new data and interpretations become available. The different behavioral models instead emphasize different aspects of those. The home-base model focuses on the social function of a "site" as a domestic locus for the delayed consumption and sharing of transported food enabled by a gender-based division of labor (Isaac, 1978, 1981). Revised as the central-place model (Isaac, 1983, 1984), the emphasis is on the place as a safe locus for the delayed consumption of transported food. The stone-cache model focuses on the energetics of hominid behavior and emphasizes the efficient transport and use of lithics for the quick processing of faunal resources to minimize conflicts caused by competitors (Potts, 1984, 1988, 1991). The routed-feeding model considers the distribution of exploited resources in the landscape and identifies the excavated "sites" as magnet midday-rest locations where scavenged faunal resources could be safely consumed diurnally (Binford, 1984).

Representative interpretations of hominid behavior need to consider distributions from traditional excavations (e.g., Isaac, in press; Leakey, 1971) and from newly studied landscape distributions (e.g., Blumenschine & Masao, 1991; Bunn, 1994; Isaac & Harris, 1980; Isaac et al., 1981; Potts, 1991:168; Rogers, 1993; Stern, 1991). Demonstration of the contributing agents and processes ("integrity") and the episodes of activity ("resolution") is a prerequisite to interpreting archaeological distributional data (terms from Binford, 1981:19–20). One agent or process produces a distribution of high integrity. Several agents or processes produce a distribution of low integrity. One episode of activity produces a distribution of high resolution. Independent (discontinuous) or protracted (continuous) episodes of activity produce a distribution of low resolution. Actualistic research demonstrates that archaeological distributional data can be studied for a variety of anthropological and non-anthropological evidence, but the questions that are answerable depend on the integrity and resolution of a distribution (e.g., Binford, 1983; Ebert, 1992; Kent, 1987; Kroll & Price, 1991; Schiffer, 1987; Whallon, 1984:243). Despite the interpretive ease ascribed to distributions of high integrity and high resolution (although precise interpretations are still often unattainable), it is probable that many archaeological distributions consist of the reverse, about which anthropologically interesting questions can still be answered (Binford, 1987:19–20; Ebert, 1992). The refuse produced from discrete activities or behavioral events may be discernible within a distribution of low integrity and/or low resolution, although the relationship among discard events, as well as other information valued by archaeologists, may be elusive. Distributions, whether accumulated on one old land surface (during one or multiple episodes of activity), or on superimposed old land surfaces, can be informative, if they are recognized and interpreted accordingly (e.g., Bordes, 1975, in Straus, 1979; Brooks & Yellen, 1987).

Recognition of the varied integrity and resolution of the traditionally excavated Plio-Pleistocene archaeological distributions ("sites") at Koobi Fora and Olduvai Gorge is an outcome of nearly 20 years of intensive analysis and debate by numerous researchers (e.g., Binford, 1981; Bunn, 1982; Isaac, 1983, 1984; Kaufulu, 1983; Kroll, n.d.; Potts, 1982; Schick, 1984; Stern, 1993; Toth, 1982). There is now consensus among researchers that (1) hominids transported lithics (raw materials and tools) and carcass parts (meat and marrow bones), used lithic tools to process various natural products, and abandoned lithic refuse and faunal refuse together and separately, (2) various artificial and natural agents and processes contributed in varying degrees to the accumulation and burial of particular lithic and faunal distributions, (3) hominids were the predominant accumulators and arrangers of at least some of the traditionally excavated distributions (see reviews by Bunn, 1991, and Potts, 1991). Disagreement continues, however, about various aspects of hominid behavior and site formation that are already incorporated into alternative interpretations of specific excavated distributions and broader models of hominid behavior (Binford, 1984, 1988; Blumenschine & Masao, 1991; Bunn, 1991, 1994; Bunn & Kroll, 1986, 1987, 1988; Isaac, 1978, 1983, 1984; Isaac et al., 1981; Kroll & Isaac, 1984; Potts, 1984, 1987, 1988, 1991; Schick, 1987; Stern, 1991, 1993; Toth, 1987). Relevant herein are differing interpretations (or lack thereof) of the resolution of the distributions. It is generally acknowledged that

> Getting more definite information on the number and duration of discard episodes represented at sites is an urgent need in the development of research. (Isaac, 1984:154).

The alternative site interpretations and behavioral models vary in the attention given to the duration and frequency of activity episodes and the elapsed period of site formation as illustrated by the following examples (see also review of seasonal site formation by O'Brien, n.d.)

Interpretations differ concerning the singularity or repetition of episodes of hominid activity. For the excavated distributions in Beds I and II at Olduvai Gorge, high resolution, resulting from a continuous period of diurnal and nocturnal hominid activity, is implicit in interpreting a vertically concentrated distribution (10 cm thick) on an old land surface (paleosol) as a "living floor" or "occupation site" (e.g., FLK *Zinjanthropus*) analogous to a campsite of modern hunter-gatherers, although it is acknowledged that "there is, unfortunately, no means [either] of assessing the length of time during which any campsite was occupied . . ." (Leakey, 1971:258–262). The implication is that "sites with diffused material" (e.g., FLK North Levels 1–5) are of low resolution*

---

*The traditionally excavated distributions ("sites") at Koobi Fora are classified as vertically thin in channel contexts (e.g., FxJj 16) and vertically thick in floodplain contexts (e.g., FxJj 20) (Harris, 1978), although there is vertical variability within all distributions (Kroll, n.d., in press). The floodplain vertical distributions until now are attributed to diagenetically-caused post-depositional movement of pieces that accumulated on one old land surface (Bunn et al., 1980; Harris, 1978:422, 478; Kaufulu, 1983; Kroll & Isaac, 1984), rather than to the mixing of pieces that accumulated on different old land surfaces, except for two horizons at FxJj 18G and a minor overlying horizon at FxJj 20E (Harris, 1978; see reanalysis by Kroll, n.d., in press).

... where repeated accumulations of debris occur throughout a considerable thickness of deposit, the position is not clear, although the most likely explanation appears to be that these sites were reoccupied on successive occasions. (Leakey, 1971:260)

Binford (1987) suggests alternatively that the living floors and the sites with diffused material at Olduvai Gorge are all of low resolution resulting from independent (discontinuous) episodes of activity. He bases this on the similarity of lithic tool types (choppers, polyhedrons, discoids, light-duty scrapers, spheroids) in assemblages from the two kinds of sites, although the example of a "living floor" assemblage (HWKE level 1) used by Binford is one that Leakey (1971:89) herself does not consider as a typical lithic assemblage from an occupation site.

At Olduvai Gorge, as at Isimila, the only condition consistently associated with "living floors" is a geologically stable land surface ... This characteristic, coupled with the similarity between the temporally collapsed assemblages from the diffuse deposits and the assemblages on the "living floors", leads one to conclude that a stable land surface was subject to the same episodal utilization as the many surfaces in the diffuse deposits. Both are the consequence of many discrete, nonintegrated events of tool manufacture, use, and discard, but on the stable land surfaces this palimpsest is vertically undifferentiated in the archaeological record. (Binford, 1987:26)

The traditionally excavated distributions at Olduvai Gorge and Koobi Fora have been compared to buried surface distributions forming from many different and unrelated episodes of hominid activity (Binford, 1987:20; Stern, 1993:214; see also Ebert, 1992). The possibility that they might have formed from repeated, unrelated episodes of hominid activity motivated Isaac to distinguish between dense "maxi-sites" (e.g., FxJj 50) and less dense "mini-sites" (e.g., FxJj 64) (Isaac, 1984; Isaac et al., 1981).†

This term [mini-site] refers to situations in which a cluster of up to a few dozen artefacts occurs in minimally disturbed context in an area with a diameter of only five to ten metres ... it is our hope that these may often represent the residue from a very restricted part of early protohuman life: perhaps at the scale of a single toolmaking bout, a single feeding event or some such. (Isaac et al., 1981:264)

Isaac died in 1985 before resolving whether maxi-sites formed from the "additive combination" (palimpsests of refuse from successive episodes of activity) of the materials found at individual mini-sites (Isaac et al., 1981:265), or whether there are distinctive compositional and behavioral differences between the two densities of sites (Isaac et al., 1981:265; see also Isaac, 1984:153–154).

... one way in which sites showing the peak level of artefact density could form is through the repeated occurrence at one spot of the same kinds of behavioural events that otherwise formed the scattered, separate mini-sites. (Isaac et al., 1981:263) If the large, dense sites are composite, long-term palimpsests, then a series of mini-sites may eventually provide the best way of distinguishing the components that have become tangled up at maxi-sites (Isaac, 1984:154).

Interpretations also differ regarding the time of day and the duration of each episode of hominid activity. Interpreting some of the traditionally excavated distributions at Olduvai Gorge and Koobi Fora as "home-base" sites implicitly requires one period or several periods of daily continuity in hominid activities enabling foraging away from a place (group fission) and returning to it (group fusion) for gender-based sharing of transported resources (Isaac,

---

†Another type of site is noted by Leakey, analogous to what Isaac et al. (1981) refer to as mini-sites "artefacts sparsely scattered on a former surface or palaeosol ..., but none has been excavated" (1971:258).

1978). The interpretation of "central-place" sites does not explicitly involve overnight activities (sleeping) by hominids, but does suggest that resources were transported to safe places (Isaac, 1983) where use of them would not necessarily be hurried. A main point in reinterpreting some of the living floors at Olduvai Gorge as "stone-cache" sites is that repeated brief episodes of hominid activity were restricted to the making and using of lithic tools for processing meat and marrow quickly to minimize conflicts with carnivores (Potts, 1984:343–346, 1988). The same "sites" at Olduvai Gorge interpreted by one researcher as stone caches are interpreted (along with the excavated sites at Koobi Fora) by several other researchers as favorite or desirable places in the landscape beside trees (potential sources of shade and refuge) to which hominids transported lithics and fauna at least diurnally enabling safer and unhurried activities (e.g., Bunn et al., 1980:250, 256; Cavallo, 1993; Isaac, 1978, 1983; Kroll, 1986, 1987; Kroll & Isaac, 1984:28). This is similar to interpreting the sites at Olduvai Gorge as "midday-rest locations" near a water source (Binford, 1984:206–208, 249, 253), although there are still differences to be resolved among researchers concerning the importance of meat and not only marrow in the diet of hominids.

Finally, there are different interpretations of the total time during which an archaeological distribution accumulated. Bone weathering data are reported and used differently by different researchers for particular Plio-Pleistocene assemblages to estimate the period of aerial exposure of bones prior to their burial (see review by Lyman & Fox, 1989). It is debated whether bones were exposed prior to burial for (a) up to 5–10 years at each of the Olduvai "sites", indicating that length period of refuse accumulation (Potts, 1982, 1984:342; 1986, 1987, 1988:256–257; note: exposure of stone caches prior to burial needed to be long enough for deliberate caching to be worthwhile), or (b) for less than a year up to several years at each of the Olduvai and Koobi Fora "sites", indicating that length of aerial exposure and not necessarily of refuse accumulation (Bunn, 1982; Bunn & Kroll, 1986:434, 1987:97–98; Bunn et al., 1980:241, 244, 256). A geologically-short period of less than ten years, however, contrasts markedly with a time estimate based on average sedimentation rates at Koobi Fora: the traditionally excavated distributions in the Okote Member of the Koobi Fora Formation are each interpreted as time-averaged palimpsests that accumulated over tens of thousands of years and that cannot be resolved within a period of 65 000–75 000 years (Stern, 1991, 1993:214). It is concluded that it is not possible to "identify a subset of artefacts . . . all of which derive from the same 100- or 1000 year slice" (Stern, 1993:214). Another landscape study attributes archaeological materials (the excavated "living floor" at HWKE level 1 and the landscape distribution excavated elsewhere) contained in a tuff horizon in Bed II at Olduvai Gorge to " 'landscape time', rather than the finer 'ethnographic time' (<100 years) or coarser 'geological time' (>10 000 years)" (Blumenschine & Masao, 1991:454).

Plio-Pleistocene distributional data

Plio-Pleistocene distributional data are collected in four ways at Olduvai Gorge and Koobi Fora (Figure 1): (1) traditional archaeological excavations of relatively large dense distributions, initially interpreted as representative examples of hominid sites (Bunn et al., 1980; Harris, 1978; Isaac, in press; Isaac & Harris, 1978; Isaac et al., 1971; Isaac, Harris & Crader, 1976; Leakey, 1971), and subsequently interpreted (e.g., Isaac et al., 1981) as unrepresentative maxi-sites after the discovery of some of the following different kinds of distributions; (2) surface surveys of the scatter between the patches which is alleged to be ubiquitous (Isaac & Harris, 1980; Rogers, 1993; Stern, 1991, 1993), (3) archaeological

Table 1   Area and boundaries of some traditional Plio-Pleistocene archaeological excavations at Koobi Fora and Olduvai Gorge

Excavation	Total area (sq m)	Contiguous trench areas (sq m)	Boundaries of largest trench			
			N	E	S	W
FxJj 1	65·0	65·0	E	E, O	O	E
FxJj 3	44·5	34·5, 10·0	E	O	O	E
FwJj 1	28·5	28·5	E	E	E	O
FxJj 17	25·0	25·0	E	E	O	O
FxJj 20 Main East, South	290·0	(See below)				
FxJj 20 Main	135·0	126·0, 9·0	E	E	O	O
FxJj 20 East	140·0	132·0, 4·0, 4·0	E, O	E	O	E
FxJj 20 South	15·0	15·0	E, O	E	O	O
FxJj 50	193·0	177·0, 4·0, 4·0, 4·0, 4·0	P	E	D	O
FxJj 64	39·5	39·5	O	O	E	E
DK level 3	231·0	231·0	O	E	E	W
FLKN levels 1–2	100·0	100·0	E	O	O	O
FLKNN level 3	230·0	230·0	E	E	O	O
FLK "Zinjanthropus"	300·0	300·0	E	O	O	D, O

Boundaries of largest trench: N=north, E=east, S=south, W=west; D=decreased horizontal density of pieces, E=erosion, O=overburden, P=paleochannel.

excavations of relatively small, less dense mini-sites which are hoped to be from a single episode of hominid activity (Isaac *et al.*, 1981; Rogers, 1993), and (4) widely dispersed, small-scale excavations (and surface surveys) of allegedly continuous off-site landscape distributions (Blumenschine & Masao, 1991; Rogers, 1993). Because the traditional excavations of hominid behavioral sites occurred first, distributions discovered subsequently are interpreted relative to the traditionally excavated "sites", resulting in some premature and limited classifications and interpretations (see Dunnell & Dancey, 1983; Ebert, 1992; Foley, 1981*a*, 1981*b*; and Thomas, 1975, about the site concept and archaeological distributional data).

It is recognized by many researchers that the area (horizontal distribution) and depth (vertical distribution) at four traditional excavations at Olduvai Gorge (Bed I) and at more than a dozen excavations at Koobi Fora (KBS and Okote Members) are constrained physically by erosion and overburden and logistically by funding and time (Table 1). Excavation boundaries at the partially exposed "sites" are not systematically defined, and none of the excavations completely exposes the horizontal and vertical limits (decreased density or disappearance of pieces) of a Plio-Pleistocene archaeological distribution (Kroll, in press; n.d.; Kroll & Isaac, 1984). At Koobi Fora and Olduvai Gorge, larger excavations (100–300 m^2) usually incompletely expose several clusters and subclusters of lithics and/or fauna (some of which have demonstrated interconnectedness by the refitting of pieces and less definitively by the taxonomic representation; Kroll, in press; n.d.). Smaller excavations (16–100 m^2) usually incompletely expose one cluster of lithics and sometimes an additional cluster (or subclusters) of fauna. Each of the clusters and subclusters has distinctive vertical distributions as well. Non-contiguous excavations (20 m apart) of 135 m^2 (FxJj 20M) and 140 m^2 (FxJj 20E) at Koobi Fora demonstrate the probable vastness of Plio-Pleistocene distributions (Kroll & Isaac, 1984). Certain untested usages of "off-site" consequently are premature and/or potentially misleading (e.g., Blumenschine & Masao, 1990:456; Toth,

1987:764, 783), considering that the number of excavated large "sites" is small, the field time devoted to traditional excavations at Olduvai and Koobi Fora is relatively short-term (Figure 1), the width and depth of the excavated distributions have not been systematically or completely exposed, the interconnectedness of adjacent distributions (clusters and scatters) at particular excavations has not been fully studied, and the resolution of site formation episodes is still debated.

One purpose in studying mini-site, scatter-between-the-patches, and landscape distributions is to discover rather than presume the spatial scale of Plio-Pleistocene distributions. There appears to be premature downplaying of the spatial focusing of activities by hominids at behavioral "sites" and premature downsizing of the excavations that are intended to expose alternative distributional data. First, it is concluded, for example, that more of the Plio-Pleistocene lithics and fauna occur in scatters (off-site) rather than in patches (on-site) within the Okote Member of the Koobi Fora Formation (Isaac & Harris, 1980:21; Isaac *et al.*, 1981:263; Stern, 1991:341, 1993:204) and within the preliminarily-studied "target clay" horizon in Bed II at Olduvai (Blumenschine & Masao, 1991:456).

> The [scatter-between-the-patches] study confirmed that vast numbers of artefacts are scattered between sites as conventionally defined. Probably much less than 10% of all artefacts occur within excavatable concentrations which could be called sites. (Isaac & Harris, 1980:21)
> In general, the data in Figure 2 are consistent with our expectation that the vast majority of archaeological remains, and hence behavioral information derivable from them, are located "off-site". The HWKE level 1 excavation reveals a small segment of a continuous archaeological record distributed over large areas in which hominids were discarding artifacts in often higher densities. (Blumenschine & Masao, 1991:456)

Is the surface distribution at Koobi Fora and the surface-surveyed and excavated landscape distribution at Olduvai Gorge a sufficient basis for those conclusions? Surface pieces have indefinite stratigraphic origins and might themselves have eroded from intermediate-density or high-density patches (Bunn, 1994; Bunn & Kroll, 1993). The excavated landscape distribution at Olduvai consists of lithics and fauna excavated within 17 "widely dispersed" $1·5 \times 1 m^2$ trenches within approximately $1 km^2$ (Blumenschine & Masao, 1991:455). It demonstrates a widespread (rather than continuous) horizontal distribution of varying density, and presumably a varying vertical distribution, although the latter is not reported. That sample is then compared to the number of pieces incompletely exposed at a "living floor" (HWKE level 1). HWKE level 1 is mainly classified as a "living floor" because of the vertical concentration of pieces on a paleosol, despite the scarcity of fauna and uncertainty about the context and about the composition of lithic artifacts (Leakey, 1971:89). The latter

> . . . do not appear to be a typical assemblage of material from an occupation site . . . proportions [of different artefact types] are similar to those on the outskirts of the FLK occupation floor, although the material is not so densely concentrated. (Leakey, 1971:89)

Yet it is further concluded about the landscape distribution that

> The available results are adequate to invite consideration of an alternative working hypothesis to the living site interpretation of HWKE level 1, and, by extension, to other Type C Oldowan sites from similar depositional environments in Bed I and Lower Bed II . . . The apparent continuous distribution of artifacts and associated bones in the portion of the mudflat examined, and their often

higher densities in areas lateral to the HWKE level 1 excavation, suggest that home bases, or repeatedly visited focal locations for *multiple* hominid activities, have not been shown to exist during basal Bed II times. (Blumenschine & Masao, 1991:458)

A widespread landscape distribution away from a somewhat arbitrary traditionally excavated distribution need not mean that focal locations for hominids did not exist, any more than it automatically means that hominid behavioral "sites" were ubiquitous. Instead of forcing new distributional data into either the vaguely defined concept of "site" (e.g., "a cluster of clusters" in Isaac, 1978:213, 225) or the existing categories of site types (e.g., maxi-site, mini-site, on-site, off-site), it is prudent first to develop consistent descriptive categories that can portray the variety of distributional data. To do this, the excavations need to be large enough to study the variability in vertical and horizontal distributions.

Second, the methods of archaeological surface survey and excavation involve a gradual reduction in the size of individual excavations (Figure 2 top to bottom): (1) maxi-site excavations of approximately 300 m^2 (Leakey, 1971), (2) mini-site excavations up to approximately 40 m^2 (Isaac, in press; Isaac et al., 1981), (3) landscape excavations of $1 \cdot 5 \times 1$ m^2 (Blumenschine & Masao, 1991), 2×2 m^2 at FxJj 50 (Bunn et al., 1980; Kroll & Isaac, 1984), and up to 15 m^2 near FxJj 20E (Harris & Isaac, in press; Kroll & Isaac, 1984), and (4) scatter-between-the-patches surface survey without excavation (Isaac & Harris, 1980; Stern, 1991, 1993). Excavations need to be wider and deeper to investigate the variability in the probable widespread distribution of Plio-Pleistocene archaeological materials and to enable systematic comparisons of the different kinds of distributional data. The need for large-scale exposures by archaeologists is indicated by the now apparent grand scale of the archaeological distributions themselves and also by actualistic observations in ethnographic, ethological, and other contexts (e.g., Binford et al., 1988; Gould & Yellen, 1987; O'Connell, 1987; Schick, 1984; Sept, 1992).

Goals and methods of spatial analysis

My analysis of the Plio-Pleistocene distributions exposed by traditional excavations involves visual inspection of point pattern, frequency per unit of excavation, and density contour data (Kroll, n.d., in press; Kroll & Isaac, 1984). The excavated distributions have vertical and horizontal dimensions that are viewed in sections and plans, respectively. The vertical dimensions are of variable thickness in massive deposits, and the apparent distributions of lithics and fauna may not accurately represent the number of separate old land surfaces on which pieces originated or the number of separate episodes of refuse-producing activities. The horizontal dimensions consist of one or several clusters, subclusters, and scatters of pieces that vary in composition, density, location, shape, and size. The episode or episodes of accumulation even on one old land surface need to be addressed in order to evaluate the relationship of the pieces within and among the individual clusters, subclusters, and scatters.

Separate vertical distributions of lithics and fauna are analysed visually in 1-meter-wide longitudinal and transverse sections, in 1·5-meter-wide diagonal sections, and in tables of the frequency of pieces in the individual excavation squares and spits (1 m$^2 \times 5$ cm, respectively; Kroll, n.d., in press). The difference in depth (range) of the lithics and fauna is calculated in the individual 1-meter squares to minimize the slope effect, and it is presented separately for continuous and discontinuous vertical distributions as 0–10 cm (very low), 10–20 cm (low), 20–30 cm (moderate), 30–40 cm (high), and ≥ 40 cm (very high) (see Kroll, n.d., in press,

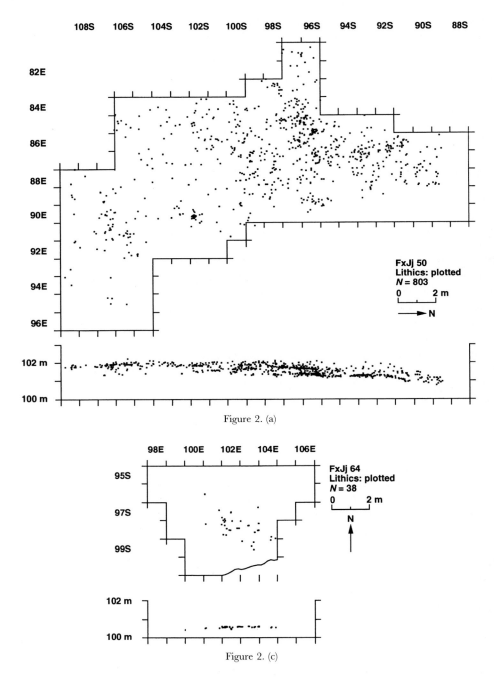

Figure 2. (a)

Figure 2. (c)

about maximum and minimum differences in depth using spit data). The vertical distribution is evaluated relative to the position and slope of the contacts between lithological units, the depositional and post-depositional conditions of individual lithological units, the extent of erosion, the top and bottom depths of excavation in individual 1-meter squares, and the

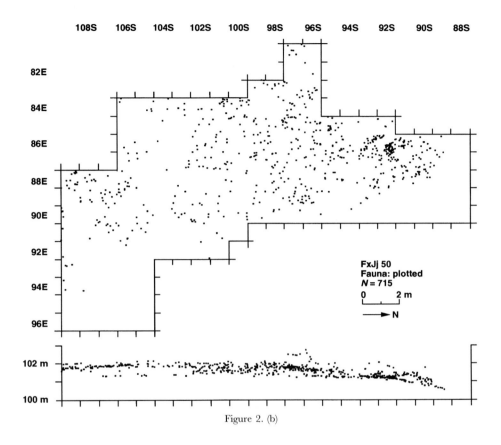

Figure 2. (b)

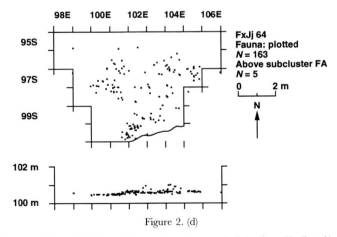

Figure 2. (d)

Figure 2. Distribution of plotted lithics and fauna at FxJj 50 and FxJj 64 (from Kroll, n.d.). Not shown are the southern trenches (four 4-sq-m trenches) at FxJj 50 or distributions of screened pieces, which are available in Kroll (n.d., in press).

vertical distribution of refitted pieces. Vertical distributions are analysed twice: (1) prior to horizontal spatial analysis to separate pieces into vertical zones that obviously accumulated on separate old land surfaces (e.g., FxJj 64 fauna); (2) vertical zones that cannot be consistently separated across each excavation are analysed as a horizontal unit after which the vertical distribution is evaluated within each cluster, subcluster, and scatter (e.g., FxJj 50 lithics and fauna, FxJj 64 fauna).

Horizontal distributions of lithics and fauna are each divided into clusters, subclusters, and scatters by density contour analysis. There is averaging of the number of pieces in every four interior excavation squares and in every two perimeter or three-inward-corner perimeter squares (note: outward-corner perimeter squares excluded; Kroll, n.d., in press). This method is advocated by Whallon (1984; see criticism of density contour analysis by Ebert, 1992:173–185). Density contour analysis is appropriate to the spatial data at Koobi Fora because of the high percentage of screened pieces having square (1 m^2) and spit (5 cm) coordinates and the long-term interest in different densities of lithics and fauna in the Plio-Pleistocene landscape. Contour intervals are 4 (very-low density), 8 (low density), 16 (medium density), 32 (medium-high density), and 64 (high density). Distributions less than contour interval "4" are scatters. Distributional data between excavations can be consistently evaluated and compared. Comparing the distributions in different excavations alternatively by average density per m^2 (total pieces divided by m^2 area of excavation) or average density per m^3 (total pieces divided by m^3 area of excavation) masks the variability in the density and thickness of distributions within each excavation. It also ignores that the vertical and horizontal dimensions of each excavation are not determined systematically and that the starting and stopping depths are uneven within many individual excavations (data available in Kroll, n.d.). Contents are compared within and among clusters, subclusters, and scatters as frequency representations of different categories of pieces (refitting, piece size, surface modifications, lithic artifact type and raw materials, faunal taxon and skeletal part).

Distributional data at FxJj 64

The excavated 39·5 m^2 was a stream-channel distal floodplain (an estimated 15–20 m from an undetermined watercourse) when the lithics and fauna accumulated (Figure 2; Kaufulu, 1983; Marshall et al., in press). The stratigraphic context is massive fine sandy siltstone or mudstone (lithological unit 1) which underlies massive silty claystone (lithological unit 3). There is no evidence of fluvial rearrangement (Kroll, n.d., in press). The lithics consist of 37 surface-collected pieces (44·6%) and 46 excavated pieces (55·4%), all of which are detached pieces (flakes and flake fragments). The fauna consists of surface pieces and 576 excavated pieces of which 230 pieces are excluded from this analysis (see note to Table 2). The percentage of excavated plotted pieces, relative to screened pieces having square and spit coordinates, is high compared to other excavations (lithics, 82·6%; fauna, 73·1%). There are fauna in the lower exposure of lithological unit 3, from 101·20–100·85 m, only where the excavated deposits were sieved (not shown in Figure 2 having screened coordinates only; available in Kroll, n.d.). In the lower exposure of lithological unit 1, lithics occur from depths 100·65 to 100·46 m and fauna from 100·85 to 100·44 m.

In Figure 3, there are one lithic cluster ($_LA$=97·7% of excavated lithics) and three faunal subclusters ($_FAa$=49·1% of excavated fauna, $_FAb$=22·1%, $_FAc$=27·0%) in lithological unit 1. Varying from 5 to 40 cm above subcluster $_FAa$ in lithological unit 3 is 35·7% of the

Table 2   Taxonomic representation in the faunal subclusters at FxJj 64

	Size group	MNI	N	Cluster and subclusters				<2 m contour	Excluded pieces*
				FAa above	FAa	FAb	FAc		
Identifiable (counted in MNI):									
Bovidae gen. et. sp. indet.	3A	1	1	—	1	—	—	—	—
Proximal sesamoid									
Bovidae gen. et. sp. indet.	3B	1	3	1	2	—	—	—	—
Mandible frag., metapodial proximal epiphysis and shaft frag, distal sesamoid									
Hexaprotodon karumensis	5	1	2	2	—	—	—	—	—
Teeth									
Proboscidea gen. et sp. indet.	6	1	75	—	12	22	41	—	—
Rib shafts									
Mammalia gen. et sp. indet.	1–2	1	4	2	2	—	—	—	—
Limb shafts									
Claris sp.	—	1	4	—	4	—	—	—	—
Cranial plate frags (3), pectoral spine									
Less identifiable:									
Mammalia gen. et sp. indet.	3	—	9	—	5	3	1	—	—
Rib shafts (3), humerus shafts (2), phalanx distal epiphysis frag, limb shafts (4)									
Mammalia gen. et sp. indet.	3–4	—	5	2	2	—	—	1	—
Mandible frags (2), teeth frags (3), cranium frag									
Mammalia gen. et sp. indet.	5	—	9	—	1	7	1	—	—
Petrosal frags (4), cranium frags (2), non-identifiable frags (3)									
Mammalia gen. et sp. indet.	—	—	460	116	79	14	17	4	*230
Non-identifiable frags (460)									
Pisces	—	—	4	—	1	3	—	—	—
Non-identifiable frags (4)									
Total		6	576	123	109	49	60	5	*230
Surface modifications:									
Cut-marked rib shaft frag of Proboscidean			1	—	—	—	1	—	—
Refitted fragments:									
Limb			2	—	2	—	—	—	—
Rib			4	—	—	4	—	—	—

Sources: Identifications from Bunn (1982, in press).

*Excluded from spatial analysis are 230 recently broken screened pieces smaller than 20 mm presumably broken from larger pieces in the assemblage during or after the excavation.

N=number of pieces.

fauna excavated at FxJj 64. Each cluster and subcluster is probably incompletely exposed vertically and laterally. Cluster $_L$A, for example, is incompletely exposed at top (erosion increasing southeastward), at bottom (insufficient sterile spits), and laterally (erosion eastward and southward), as indicated by three surface pieces refitted with excavated pieces in two lithic groups (Figures 4, 5). Pieces in cluster $_L$A and in subclusters $_F$Aa–$_F$Ac are at similar depths in the lower exposure of lithological unit 1. The range in depths of the lithics is very low to low (0–11 cm and mostly 0–4 cm in the individual excavation squares) and of the

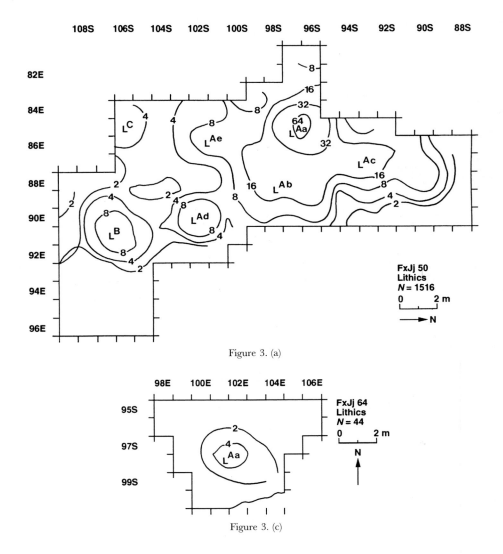

Figure 3. (a)

Figure 3. (c)

fauna very low to moderate (0–30 cm in the individual excavation squares). There are some sterile gaps of 5 to 15 cm in the vertical distribution of the fauna, below which there are lithics and fauna including all of the Proboscidean and most of the other identifiable and less identifiable taxa, and above which there are fauna in the upper exposure of lithological unit 1. There is very low vertical separation of refitted pieces of 2 cm or less in two refitted lithic groups having depth coordinates and in two refitted faunal elements (shafts of limb and rib).

Three faunal subclusters of low density surround one lithic cluster of very low density (Figure 3). The distribution of cluster $_L$A overlaps subclusters $_F$Aa and $_F$Ac. A high proportion of lithics (27·9%) are refitted in five groups, and the average lateral separation of pieces in the individual refitted groups is less than 1 meter (Figure 4). Very small (<10 mm) to large (40–80 mm) detached pieces are represented in cluster $_L$A, and most are of medium

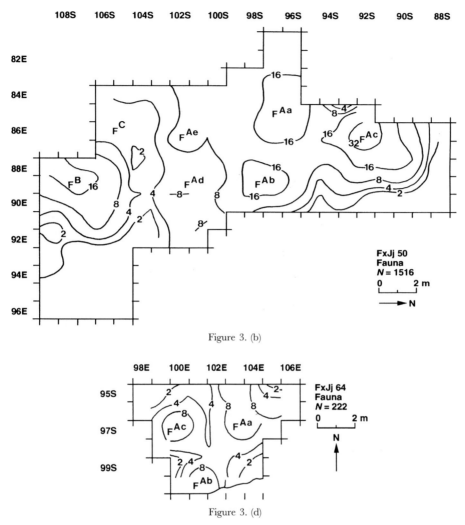

Figure 3. (b)

Figure 3. (d)

Figure 3. Density contour distributions of lithics and fauna excavated at FxJj 50 and FxJj 64 (from Kroll, n.d., in press).

(20–40 mm) and large size. The fauna represent an estimated minimum number of six individuals of different taxa and restricted skeletal parts (Table 2; Bunn, 1982, in press). Unidentifiable mammalian fragments predominate overall but vary in their representation within the faunal cluster and subclusters (94·3% of pieces above subcluster $_F$Aa; $_F$Aa, 72·5%; $_F$Ab, 28·6%; $_F$Ac, 28·3%). The identifiable taxonomic representation in the three faunal subclusters is dominated by Proboscidean rib shaft fragments ($_F$Aa, 11·0% of total pieces, 40·0% of identifiable and less identifiable mammalian pieces; $_F$Ab, 44·9% of total pieces, 68·8% of identifiable and less identifiable mammalian pieces; $_F$Ac, 68·3% of total pieces, 95·3% of identifiable and less identifiable pieces), the breakage of which is more likely by hominids than by other processes (Bunn, 1982). Subcluster $_F$Aa has the greatest taxonomic diversity, including pieces of three other identifiable mammalian taxa and three less identifiable mammalian taxa. Subclusters $_F$Ab and $_F$Ac also have pieces of two of the less identifiable mammalian taxa that are in subcluster $_F$Aa. Fauna above subcluster $_F$Aa represent one taxon not represented below (*Hexaprotodon karumensis*) and three taxa that are also represented below (bovid size 3B, mammal size 1–2, mammal size 3–4). The fragmentation of the identifiable and less identifiable pieces in the three subclusters and in the overlying cluster is more likely by hominids than by other processes (Bunn, 1982).

FxJj 64 is the only distribution interpreted as a mini-site at Koobi Fora (Isaac, in press). The fauna overlying subcluster $_F$Aa, if interpreted as a hominid accumulation, is evidence of repeated activities at one place after the refuse from previous activities was already buried below. The absence of lithics could be from incomplete excavation, erosion, or removal of lithics by hominids similar to the fauna-only distributions reported by Bunn (1994). The lithic distribution vertically and horizontally, especially the clustering of refitted pieces which is analogous to flake scatters replicated by modern lithnic knappers, is initially suggestive of one episode of hominid activity on one old land surface. It is possible, alternatively, that the incompleteness of the refitted reduction sequences results from higher turnover of cores during repeated episodes of hominid activity (see Discussion). The vertical distribution in faunal subclusters $_F$Aa–$_F$Ac is also initially suggestive of the accumulation of most faunal pieces on the same old land surface as the lithics. The minimal vertical separation of refitted bones and the sterile gaps in some excavation squares, however, are suggestive that at least a small number of pieces accumulated on an overlying old land surface also in lithological unit 1. An interconnectedness among the subclusters is suggested by the taxonomic and skeletal part composition in each of them, but the total taxonomic representation is not necessarily consistent with the interpretation of FxJj 64 as a mini-site formed from a "single feeding event" (Isaac *et al.*, 1981:264), unless it is argued that hominids had (1) one bountiful episode of butchery involving skeletal parts from different animals, (2) one prolonged (continuous) use of the place for repeated butcheries, or (3) involvement in processing the skeletal parts of only one taxon despite the diagnostic evidence otherwise. If there were several episodes of butchery during repeated visits, then FxJj 64 is not a "mini-site" as defined by Isaac *et al.* (1981), although it still is an important example of a lower density distribution relative to other excavated Plio-Pleistocene distributions (see for comparison Aridos I in Santonja & Villa, 1990).

It was obviously not intended that FxJj 64 be used to test the interpretive potential of mini-sites until other comparable sites were also excavated.

> A significant part of the rationale in pursuing the study of a series of mini-sites is that, while the behavioural interpretation of each individual instance may be ambiguous, the recurrence of some features may make certain lines of interpretation more secure. FxJj 64 has been reported here simply as an example of the kind of data that may derive from following this line of approach. (Isaac *et al.*, 1981:265)

My comments about FxJj 64 are meant to improve not impede the mini-site method. Can the mini-site method succeed if (1) excavations are too small to show that the alleged mini-sites are not part of incompletely exposed, larger-scale distributions?, (2) interpretations precede analyses of the context, composition, and vertical and horizontal distributions of lithics and fauna?, (3) behavioral interpretation of each individual instance is ambiguous? Would the information gained from FxJj 64 actually be more "informative" had the lithics been associated with the skeletal parts of only one carcass?

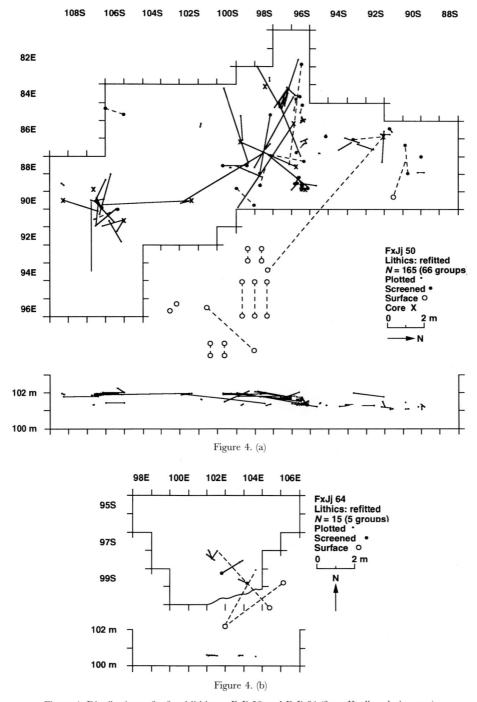

Figure 4. (a)

Figure 4. (b)

Figure 4. Distributions of refitted lithics at FxJj 50 and FxJj 64 (from Kroll, n.d., in press).

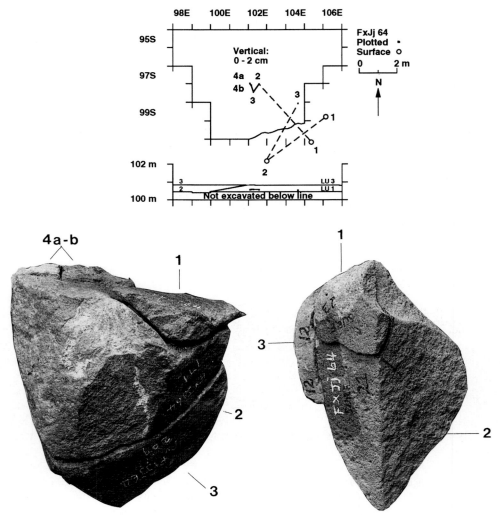

Figure 5. Distribution of two refitted lithic groups at FxJj 64. Artifacts actual size, 1, 2, 3, 4=flakes removed in refitted sequence. a, b=split flakes of an individual reconstituted flake. Screened and surface pieces positioned randomly in square and spit. LU=lithological unit.

Discussion

By contrast with [FxJj] 64, major sites such as the FLK Zinj. site at Olduvai (M. D. Leakey, 1971) or Koobi Fora sites like FxJj 20 (Harris, 1978) or FxJj 50 (Bunn *et al.*, 1980) involve thousands of stone and bone specimens packed into a limited area. (Isaac *et al.*, 1981:265)

The lithics and fauna at FLK *Zinjanthropus*, FxJj 20, and FxJj 50 are not, of course, "packed into a limited" space vertically or horizontally (see Isaac, 1981, in press; Leakey, 1971; Kroll, n.d., in press; Kroll & Isaac, 1984). There is variability in the density and spacing of the vertical and horizontal distributions. At FxJj 50, for example, there are several clusters and subclusters of lithics and fauna having different vertical distributions (Figures 2 and 3). The southern clusters of lithics ($_L$B) and fauna ($_F$B) are similar to the distribution at FxJj 64 in density, composition, and context (data available in Kroll, n.d.; the distribution at FxJj 50 is described preliminarily in Bunn *et al.*, 1980, and Kroll & Isaac, 1984, and more fully in Kroll, n.d.; see also Bunn *et al.*, in press; Kroll, in press).

Interpreting FxJj 50 as a maxi-site and FxJj 64 as a mini-site is largely related to the total occurrence at FxJj 50 of more lithics and fauna than at FxJj 64 (average density as defined above), appearance of a larger area of hominid activity at FxJj 50 than at FxJj 64 (spatial scale), and presumption that FxJj 64 formed during a shorter period of hominid activities (moment in time) compared to a longer period of hominid activities at FxJj 50 (moments in time). It is presumed that lower densities of pieces are from fewer hominid activities or from fewer episodes of activity. If that were so, then FxJj 64 would have a higher incidence of associated pieces having formed from fewer activities than FxJj 50. But is the site formation scenario actually easier to understand at the alleged mini-site? Or is the southern distribution at FxJj 50 also a mini-site because of the lower density of pieces there than northward? There is a convergence of spatial, lithic, faunal, and contextual data indicating that hominids repeatedly visited FxJj 50 and FxJj 64 and that refuse from separate episodes of activity accumulated sometimes on one old ground surface and other times on superimposed old ground surfaces.

1. Vertical distribution

The difference in depth of the lithics in the individual one-meter squares is very low to low at FxJj 64 and very low to very high at FxJj 50, and the difference in depth of the fauna is very low to moderate at FxJj 64 and very low to very high at FxJj 50. Because some of the individual groups of refitted lithics and fauna at FxJj 50 have absolute differences in depth up to 60 cm and the refitted groups together form an overlapping network (Kroll & Isaac, 1984), the thicker vertical distribution was preliminarily and provisionally interpreted as evidence of vertical dispersal of pieces that had accumulated on one old land surface (Bunn *et al.*, 1980; Kroll & Isaac, 1984), noting the possibility that subsequent analysis might alter that interpretation.

> We are thus inclined to treat the entire [vertical] occurrence as an indivisible archaeological entity . . . nearly all of the material was laid down on a single irregular and gently sloping surface . . . Subsequently bioturbation processes, especially termite action, seem to have dispersed the material to varying degrees in different parts of the site . . . On the other hand, it is possible that in some parts of the site bioturbation has caused an upper (major) archaeological horizon to merge with a lower (minor) horizon . . . We intend to test these hypotheses . . . (Bunn *et al.*, 1980:235, 238)

Most of the individual groups of refitted lithics and fauna actually have minimal differences in depth at FxJj 64 (less than 5 cm) and at FxJj 50 (less than 10–15 cm) (Figure 4; data tables available in Kroll, n.d., in press). The difference in depth is less than 10 cm in 81% of the individual refitted groups and 10–60 cm in 19% of them. The refitted groups having greater differences in depth at FxJj 50 usually consist of pieces that are either (1) laterally separated with one piece being upslope and the other piece being downslope, or (2) laterally close together in areas of increased topographic irregularities or in areas of increased bioturbation evidenced by root casts, termite nest concretions, and absence of bedding structures. In the northernmost trench, for example, exactly where there are two groups of refitted fauna, each occurring in a less than 1 meter diameter area and having differences in depth of 7 cm and 14 cm, a topographic irregularity (hollow) is recorded in the field records.

In Figure 5, the distributions of several refitted groups illustrate the pattern. Within less than 1 meter diameter laterally and 2 centimeters vertically are five whole and broken flakes of lava representing a four-piece reduction sequence without the core. In Figure 6, 0·7 to 1·2 m apart laterally and at the same depth are a polyhedron and two whole flakes of lava representing a two-piece reduction sequence with the core present, and 2.7 m apart laterally and differing in depth by 14–19 cm (range due to screened piece) are a chopper and whole flake of ignimbrite. In Figure 7, 0·5 to 1·7 m apart laterally and differing in depth by 14 cm (difference in depth varies from 1 to 10 cm by comparing the depth of each piece in the sequence of core reduction) are ten whole flakes, broken flakes, and flake fragments of lava representing a seven-piece reduction sequence without the core present. In Figure 8, less than 1·0 m apart (same excavation square and spit) to 3·3 meters laterally and differing in depth by less than 5 cm (same spit) to 60 cm (difference in depth varies from <5 cm to 36 cm comparing the depth of each piece only to depth of polyhedron-core) are a polyhedron, core-cobble fragment, and two whole flakes of lava representing a three-piece reduction sequence.

The use of refitting in interpreting archaeological vertical distributions is prevalent (e.g., Cziesla & Eickhoff, 1990; Hofman, 1986; Cahen & Moeyersons, 1977; Van Noten *et al.*, 1980; Villa, 1982). If the excavated pieces accumulated during repeated episodes of hominid activity on superimposed old land surfaces, but the latter are indistinguishable in archaeological context because of both small-scale mixing of the separate depositional units and minimal vertical displacement of the archaeological pieces, then there might be minimal vertical separation among the individual refitted groups, although all of the groups together might form an overlapping network across the full thickness of an archaeological distribution. If the pieces instead accumulated during one or more episodes of hominid activity on a single old land surface after which they were extensively dispersed vertically by post-depositional processes, then there might be some refitting among pieces across the full thickness of an archaeological distribution (Kroll, n.d., in press; see also Kroll & Isaac, 1984:12, 14; Villa, 1983:65–75; see Gifford-Gonzalez *et al.*, 1985; McBrearty, 1990; Stockton, 1973; Villa & Courtin, 1983, about the vertical movement of archaeological pieces by depositional and post-depositional processes). These predictions are the basis for interpreting the vertical distribution of refitted pieces at FxJj 50, which has a pattern similar to the first prediction.

The difference in depth of the total lithics at FxJj 50, in contrast to the difference in depth among refitted pieces, is less than 10 cm in a minority of excavation squares, 10–60 cm in a majority of squares, and 60–120 cm in an equivalent minority of squares. Because the difference in depth of the total lithics exceeds the difference in depth in the majority of refitted groups and many of the groups having greater differences in depth can be linked to topographic irregularities, it is probable that some of the pieces at FxJj 50 accumulated on more than one old land surface of uneven paleotopography, after which diagenetic processes obliterated the bedding structures and vertically mixed some of the pieces. In some excavation squares, it is possible to separate the vertical distribution into at least two vertical zones, as is described in the alternative interpretation of FxJj 50 already quoted (Bunn *et al.*, 1980:238), but this cannot be done consistently in each excavation square, partly because of the unevenness of the bottom depth of excavation (data available in Kroll, n.d.). The clusters and subclusters described in the next section, however, do not consist predominantly of vertically-scattered pieces collapsed into a horizontal distribution. The individual clusters and subclusters consist of definite vertical zones defined by refitted pieces. If the zones could be consistently resolved throughout the excavation, there would be superimpositioning of at least two vertical zones containing horizontal clusters and subclusters.

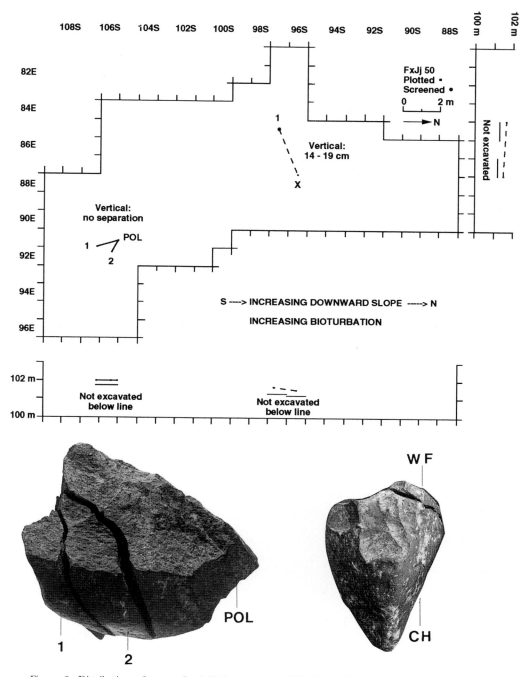

Figure 6. Distribution of two refitted lithic groups at FxJj 50. Artifacts actual size. CH=chopper. POL=polyhedron. WF=whole flake. 1, 2=flakes removed in refitted reduction sequence. Screened piece positioned randomly in square and spit.

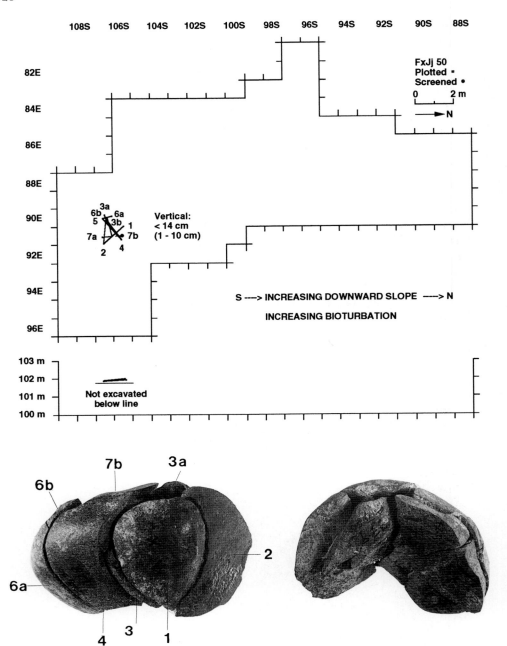

Figure 7. Distribution of one refitted group at FxJj 50. Artifacts actual size. Screened piece positioned randomly in square and spit. < =less than.

2. *Lateral distribution*

Density contours are used to identify the lateral clustering (variable density) of lithics and fauna at FxJj 50 and FxJj 64 (Figure 3). Northward at FxJj 50 are subclusters of lithics (${}_L$Aa–${}_L$Ae) and

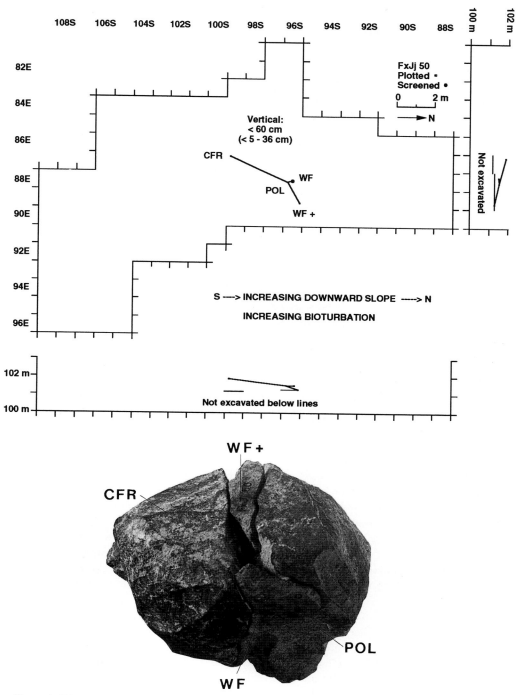

Figure 8. Distribution of one refitted lithic group at FxJj 50. Artifacts actual size. CFR=core-cobble fragment. POL=polyhedron. WF, WF+=flakes of uninterpreted reduction sequence. Screened piece positioned randomly in square and spit. <=less than.

fauna ($_F$Aa–$_F$Ae) of varying density, and southward are clusters of lithics ($_L$B) and fauna ($_F$B) of densities similar to FxJj 64. The southern clusters represent 6·9% of the lithics and 15·3% of the fauna excavated at FxJj 50. That the low density clusters southward at FxJj 50 are similar to the lithic cluster and faunal subclusters at FxJj 64 is a further reminder of the prematurity in interpreting the distribution exposed by the small excavation at FxJj 64 as a discrete or isolated mini-site when it might be the exposed area of a larger-scale hominid site. Compared to the nearly 200 m^2 excavation at FxJj 50, it is equally possible that the 39·5 m^2 excavation at FxJj 64 incompletely exposes (1) an isolated relatively low-density distribution, or (2) a very-low-density lithic cluster and low-density faunal subclusters adjacent to other currently unexposed clusters. FxJj 64 is a small excavation and not necessarily a discrete small site (see similar interpretation of FxJj 17 by Harris, 1978:423, 454). That can be tested by partial expansion of the excavation (see boundaries in Table 1).

The spatial scale and spacing of the clusters and subclusters at FxJj 50 and FxJj 64 are also interesting. What were the natural and behavioral constraints on the distribution of refuse-producing activities and refuse, despite the open-air, seemingly unconstrained space? What effect did vegetation (bushes, distribution of shade) have on the spatial scale and discreteness of activities and of the resultant lithic and faunal distributions (Figure 9; e.g., see Bartram *et al.*, 1991; Gould & Yellen, 1987; O'Connell, 1987; Yellen, 1977, for ethnographic observations)? In studying some of the archaeological distributions at Olduvai Gorge, including FLK *Zinjanthropus*, Leakey (1971:262) concluded ". . . the existence of some factor affecting the horizontal diffusion of debris on the living floors is indicated". This is also so for the exposed distributions at FxJj 50 and FxJj 20 and needs to be tested further in future larger-scale distributional research in different vegetational contexts (e.g., plains, open woodlands, riparian woodlands).

. . . patches of refuse as have been recovered at FxJj 50 at Koobi Fora and FLK *Zinjanthropus* at Olduvai Gorge would all fit perfectly well into the shade areas afforded by trees growing in comparable situations in modern East Africa. (Kroll & Isaac, 1984:28)

3. Lithics

The differences between the lithic assemblages at FxJj 50 and FxJj 64 are the greater number of pieces at FxJj 50, the presence of cores and unmodified cobbles and pebbles at FxJj 50 (four cores and 21 unmodified cobbles and pebbles in cluster $_L$B) and their absence at FxJj 64, the diversity of raw materials at FxJj 50 (e.g., detached pieces in cluster $_L$B are 93·5% lava, 1·3% chert, 2·6% ignimbrite, 2·6% indeterminate between lava and ignimbrite) compared to only lava at FxJj 64, and the higher percentage of refitted pieces in the FxJj 64 assemblage (26% of the detached pieces) than in the FxJj 50 assemblage (9·9% of the detached pieces; 13·2% of the flaked pieces, 5·6% of the pounded pieces, 1·5% of the allegedly unmodified cobbles and pebbles). The refitting composition in cluster $_L$B at FxJj 50 and in cluster $_L$A at FxJj 64 is similar (five refitted groups and half of two other groups representing 30·0% of detached pieces, 25·0% of flaked pieces, 5·3% of unmodified cobbles and pebbles in cluster $_L$B at FxJj 50; compared to five refitted groups representing 27·9% of the detached pieces in cluster $_L$A and three additional surface-collected pieces at FxJj 64). Cluster $_L$B at FxJj 50 and cluster $_L$A at FxJj 64 each represents less residue from lithic knapping than subclusters $_L$Aa–$_L$Ac at FxJj 50. The distributions of refitted lithics at FxJj 50 and FxJj 64 in Figure 5 are similar to those identified as lithic knapping areas at younger archaeological sites (e.g., Villa, 1983; Cziesla, 1991) and to those replicated by modern lithic knappers.

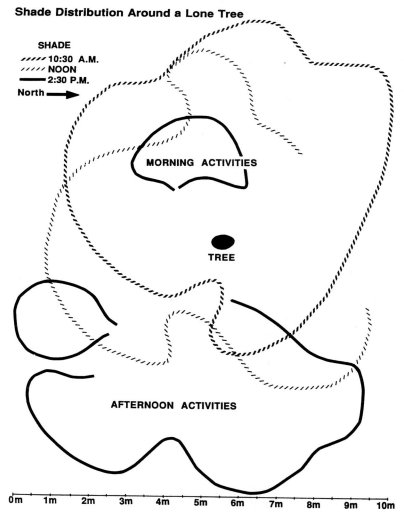

Figure 9. Location of Kua San activities in tree shade in Botswana made in 1985 (from Kroll, 1986, 1987, n.d.).

The lithics in the very-low-density cluster $_L$A at FxJj 64 are attributed to the flaking of a few cobbles (Isaac, 1984; Isaac *et al.*, 1981), although none of the cores was excavated and individual reduction sequences are incompletely refitted. The lithics in cluster $_L$B at FxJj 50 and in the northern subclusters include cores, but again the reduction sequences are incompletely refitted (Bunn *et al.*, 1980). It is probable that further refitting analysis will yield additional refitted pieces from FxJj 50 and FxJj 64. Refitting the majority of pieces is unlikely, however, considering the numbers of distinctive flake scars to which no pieces have been refitted, distinctive flake shapes which have not yet been refitted to other pieces, and cores and flakes of distinctive raw materials for which there are no mates or no refittable mates (Kroll, personal observation 1977–79). The incompleteness of the reduction sequences at FxJj 50 as evidence that hominids carried cores and flakes to and from the sites (Bunn *et al.*, 1980; Schick, 1987; Toth, 1987).

The incompleteness of the refitted lithics is a pattern in twelve excavated Plio-Pleistocene assemblages (Kroll, n.d., in press). If individual hominids carried at least a hammerstone and a core from which were knapped a limited number of flakes for each particular activity, then highly mobile hominids returning repeatedly to preferred localities might create an admixture of lithic pieces that cannot be refitted into long reduction sequences because of the small numbers of flakes that were usually knapped from each core. If the low percentages of refittable pieces in the excavated lithic clusters are largely because the pieces come from many different cores and are consequently unrefittable, then the spatial distribution of refitted pieces—especially the low number of vertically-separated pieces—must be used cautiously because pieces that were knapped during one episode of activity and that might have been vertically displaced by post-depositional processes would not be refittable.

Are there enough refittable pieces to choose definitively between (1) the greater vertical movement of pieces that originated mainly on one ground surface and (2) the lesser vertical movement of pieces that originated on different ground surfaces that are now unidentifiable in massive bioturbated deposits? The available evidence in the different assemblages having refitted pieces indicates that the prevalent spatial pattern consists of minimal vertical separation among the pieces in the individual refitted groups in consolidated massive fine-grained contexts. The accumulation of refuse during hominid revisits on multiple old ground surfaces is likely considering the abundance of minimally dispersed refitted pieces. The weakness in this interpretation is that refitting itself may not be the ultimate measure of vertical dispersion in Plio-Pleistocene archaeological distributions: if the pieces come from many different cores, then it is not possible to use the small number of vertically-separated refitted pieces as evidence against vertical movements.

4. Fauna

The differences between the faunal assemblages at FxJj 50 and FxJj 64 are the greater number of pieces at FxJj 50, the higher minimum number of individuals at FxJj 50, and the greater number of cut marks at FxJj 50 (see Bunn, 1982, in press; Bunn *et al.*, 1980, in press). Similar to the lithics, the fauna can be evaluated among the clusters and subclusters rather than between the whole FxJj 50 and FxJj 64 assemblages (data available in Kroll, n.d.). The southern low-density faunal cluster at FxJj 50 and three low-density faunal subclusters at FxJj 64 are similar in taxonomic representation, skeletal parts, and refitting, although there are more taxa representing a higher number of carcasses in cluster $_F$B at FxJj 50. Isaac (1984; Isaac *et al.*, 1981) concludes that it is difficult to establish the functional association between the lithics and fauna at individual mini-sites, but the repetition of associations at multiple mini-sites is the strength of the mini-site method. The functional association between the lithics and fauna in the individual clusters at FxJj 50 and FxJj 64 is demonstrated by cut marks and hammerstone fractures on the fauna (Bunn, 1982, 1991, in press; Bunn *et al.*, 1980) and by lithic microwear (Keeley & Toth, 1981). The excavated faunal clusters and subclusters at FxJj 50 and FxJj 64 are places where hominids defleshed and broke bones for meat and marrow using lithic tools. What is difficult to establish not only at FxJj 50 (maxi-site), but also at FxJj 64 (alleged mini-site), is the functional and temporal relationships among lithics, among fauna, and among lithics and fauna. At FxJj 50, Bunn's MNI estimate is 29 (22 mammals) compared to six at FxJj 64 (Figure 8). The same interpretive choices as were presented for FxJj 64 apply to FxJj 50. Ecologically diverse taxa in the clusters and subclusters indicate that if hominids were the principal accumulators of most of the fauna, then the clusters probably formed during repeated episodes of activity at FxJj 50 and FxJj 64. Discontinuity of the episodes of hominid activity is consistent with the paucity of epiphyses in both assemblages. The latter is interpreted as evidence that between episodes of hominid activity carnivores scavenged still-desirable faunal refuse abandoned by hominids (Bunn, 1987, 1991; Bunn et al., 1980; Bunn & Kroll, 1986, 1988).

5. Context

It is necessary to consider why and for how long hominids sometimes returned repeatedly to the same places and engaged in lithic and faunal refuse-producing activities? Were the activity episodes quick and restricted only to butchering and consumption activities as proposed in the stone-cache model (Potts, 1984, 1988)? There are many reasons and many aspects of hominid behavior to consider. One factor that should be incorporated into interpretations of particular excavated distributions and into broader models of hominid behavior is the presence of trees. Obviously many other behavioral (e.g., social organization) and environmental (e.g., water, food resources) factors need to be incorporated as well. Ethnoarchaeological and ethological studies have demonstrated the magnetic attraction of trees for shade and refuge (Bartram et al., 1991; Cavallo, 1993, personal communication 1992; Gould, 1969; Hayden, 1979; Kroll, 1986, 1987, n.d.; Kroll & Isaac, 1984; Yellen, 1977; see also Binford, 1984). Root casts and pollen of trees at FxJj 50, and proximity to a stream channel at FxJj 50 and FxJj 64, are ample evidence that trees existed at both localities (Bunn, personal communication; Bunn et al., 1980, in press; Isaac, 1983; Kaufulu, 1983; Vincens, 1979; see also evidence of trees at FLK Zinjanthropus in Leakey, 1971:49). Trees might have attracted hominids for food, lookouts, and especially refuge (shade, shelter, protection from on-the-ground danger). Trees might have provided hominids a place to conduct activities unhurriedly because of potential refuge from carnivores. The importance of various resources, including trees and water, is emphasized in some interpretations of particular archaeological distributions and in some behavioral models (e.g., Binford, 1984; Bunn et al., 1980; Cavallo, 1993; Isaac, 1978, 1983; Kroll, 1986, 1987; Kroll & Isaac, 1984). Trees are de-emphasized specifically in the stone-cache model (Potts, 1988:262–64). Yet, the presence of trees suggests that the quickness of hominid activities, which is inferred in the stone-cache model and attributed to on-the-ground dangers, may be unwarranted. Perhaps social activities occurred at the desirable places too?

> The fact that a 3-kg carcass can be distributed among as many as 15 chimpanzees who spend up to 9 hours sharing and consuming the meat contributed greatly to the premise that predatory behavior among these apes has a social as well as a nutritional basis (Teleki, 1973a:171–173). (Teleki, 1981:333).

Regarding the total time during which episodes of hominid activity occurred at a particular place, there is not sufficient geological evidence to conclude a period of tens of thousands of years (compare Bunn & Kroll, 1993; and Stern, 1993). The average rate of regional sedimentation as used by Stern need not represent the rate of sedimentation at individual localities as Stern acknowledges (1993:214). It is probable at individual localities that the massive beds containing the archaeological materials accumulated from repeated deposition of thin layers of sediments after which a variety of natural processes obliterated the bedding structures (Kaufulu, 1983). Based on bone weathering data, the bone refuse from individual episodes of activity was not exposed for long periods prior to burial (Bunn & Kroll, 1987; Potts, 1987). The period of use of a particular location by hominids might have been governed by dynamic microtopographic and vegetational features that were attractive for perhaps several decades or the lifespan of a good shade tree. The time-span represented at each place need not lessen the intensity or behavioral significance of hominids repeatedly returning to particular places. The excavated sample of those kinds of places is unfortunately relatively minimal in areal exposure and relatively small in number.

The studied distributions have implications for the recovery, analysis, and interpretation of Plio-Pleistocene distributional data. Lithics and fauna occur regionally and locally in different densities. Archaeological excavations so far incompletely expose distributions that vary in their vertical and horizontal dimensions. Integrity (agents and processes of formation) and resolution (periods of formation) vary among the distributions. Horizontal clusters and subclusters of lithics and fauna define places where hominids at least made and used lithic tools and butchered and consumed carcasses. The excavated distributions might not reflect or represent the full suite of hominid behavior. There is evidence that some lithics and fauna in one place probably accumulated on different old land surfaces. Other pieces at the same place probably accumulated on one old land surface during repeated (discontinuous) episodes of activity. The accumulation of refuse in clusters that were created by the repeated use of particular places during geologically short periods is consistent with the evidence that some hominids deliberately positioned at least their tool-making, butchering, and consumption activities beside trees that might have provided shade and a reliable refuge. Trees might have enabled hominids to engage in activities more safely and unhurriedly than at treeless places. Other factors in addition to trees contributed to the location and timing of activities and to the reuse of particular places. That a desirable place might have been preferentially reused during a geologically short period needs to be considered as an alternative interpretation of one aspect of hominid behavior in broader behavioral models. Attention is focused currently on why hominids transported resources and created dense (and less dense) concentrations of lithics and fauna (e.g., Potts, 1988; Schick, 1987). Future research needs to consider whether and why (1) hominids *repeatedly* transported resources to particular places and (2) created a *variety* of horizontal and vertical distributions of lithics and fauna at different places in different micro-habitats.

It is further suggested that the variety of archaeological distributional data now available will be beneficial to interpretations of hominid behavior in the Plio-Pleistocene landscape only if (1) all excavations are deeper and wider, (2) premature classifications (especially oversimplified dichotomies) of site types are restrained, (3) distributions are classified empirically based on systematic descriptions of vertical and horizontal distributional data, and (4) analyses both evaluate the density, composition, and vertical resolution of each individual cluster, subcluster, and scatter of pieces and document the spacing and relationships (interconnectedness) among discontinuous vertical distributions and adjacent horizontal clusters, subclusters, and scatters.

Summary and conclusions

Four points are made in this paper:
  (1) Think big. Single-trench small excavations are not informative of probable large-scale distributions.
  (2) Interpretive dichotomies based on incompletely excavated distributions are premature and misleading.
  (3) There is archaeological evidence that hominids revisited large areas of the landscape and repeatedly engaged in a variety of activities before and after refuse from previous activities was buried.

(4) There was preferential location by hominids at trees providing shade and refuge and probably enabling a variety of unhurried activities.

Points (1) and (2) emphasize that the size of all archaeological excavations needs to be greatly increased as part of current field strategies aimed at investigating the scale of lithic and faunal distributions in the Plio-Pleistocene landscape. Plio-Pleistocene distributions were probably much vaster laterally and much deeper vertically than has been exposed by archaeological excavations to date. Large-scale excavations of contiguous trenches and of discontinuous trenches at and away from presumed hominid behavioral sites are imperative to document and understand landscape distributions. Landscape archaeology is sampling locations at the spatial scale necessary to resolve major questions about hominid land use, but the discontinuous excavations need to be large enough (minimally 2×2 m^2) to evaluate the variable density and spacing of archaeological materials. It is possible that the recent landscape methods of collecting distributional data will bring us full circle back to large-scale excavations of behavioral sites.

Points (3) and (4) need to be further evaluated and perhaps incorporated into interpretations of particular distributions and into broader models of hominid behavior. They are not presented as mutually exclusive alternatives to previously offered explanations and are compatible with some of the interpretations and models discussed in the second section of the paper. I have not addressed how often or how many hominids were revisiting particular places. The time represented by an individual cluster or palimpsest-cluster is probably neither a moment in time as hoped for in mini-site research nor thousands-of-time-averaged years as concluded in scatter-between-the-patches research. A main point of this paper is the convergence of archaeological evidence that hominids revisited particular places for as long as there were physically attractive features. I have tried to relate the spatial analysis of the traditionally excavated distributions to the recent trend toward landscape distributions. Although the former were available first, we may need to view them differently because of new knowledge of the latter. Archaeologists studying the Plio-Pleistocene need to prepare the terminology and interpretive framework for the new kinds of distributional data that will be available in the rapidly approaching next century (Figure 1).

Acknowledgements

I thank: G. Isaac for supporting my research; National Museums of Kenya, Nairobi, for supporting my refitting and other analyses of Plio-Pleistocene lithic assemblages; H. Bunn for sharing faunal identifications and refitting data and for continued support during all phases of the research; J. W. K. Harris (FxJj 50) and F. Marshall (FxJj 64) for some lithic attribute data; N. Toth and K. Schick (FxJj 50, FxJj 64) and G. Isaac and F. Marshall (FxJj 64) for some lithic refitting data; L. Bartram for assisting with Figure 9 in Botswana; H.-P. Blankholm and R. Whallon for advice about density contour analysis; M. Rogers for sharing his typescript for the Paleoanthropology Society meetings in 1993; H. Bunn, C. O'Brien, N. Pierzina, and L. Torgerson for assisting in the completion of this paper; H. Bunn, L. Bartram, A. K. Behrensmeyer, J. Cavallo, D. Dechant Boaz, I. Findlater, F. C. Howell, J. W. K. Harris, A. B. Isaac, G. Isaac, Z. Kaufulu, F. Marshall, C. O'Brien, R. Potts, J. Speth, P. Villa, and A. Vincent for discussions; J. Oliver, N. Sikes, and K. Stewart for inviting me to participate in the AAPA symposium; K. Stewart for exemplary management of my contribution to this volume; H. Bunn and three anonymous reviewers for constructive comments on an earlier version of this paper.

References

Bartram, L. E., Kroll, E. M. & Bunn, H. T. (1991). Variability in camp structure and bone food refuse patterning at Kua San hunter-gatherer camps. In (E. M. Kroll & T. D. Price, Eds) *The Interpretation of Archaeological Spatial Patterning*, pp. 77–148. New York: Plenum Press.

Binford, L. R. (1981). *Bones: Ancient Men and Modern Myths*. New York: Academic Press.

Binford, L. R. (1983). *In Pursuit of the Past: Decoding the Archaeological Record*. New York: Thames and Hudson.

Binford, L. R. (1984). *Faunal Remains from Klasies River Mouth*. London: Academic Press.

Binford, L. R. (1987). Searching for camps and missing the evidence?: another look at the Lower Paleolithic. In (O. Soffer, Ed.) *The Pleistocene Old World: Regional Perspectives*, pp. 17–31. New York: Plenum Press.

Binford, L. R. (1988). Fact and fiction about the Zinjanthropus floor: data, arguments, and interpretations. *Curr. Anthrop.* **29**, 123–135.

Binford, L. R., Mills, M. G. L. & Stone, N. M. (1988). Hyena scavenging behavior and its implications for the interpretation of faunal assemblages from FLK 22 (the Zinj floor) at Olduvai Gorge. *J. Anthrop. Archaeol.* **7**, 99–135.

Blumenschine, R. J. & Masao, F. T. (1991). Living sites at Olduvai Gorge, Tanzania? Preliminary landscape archaeology results in the basal Bed II lake margin zone. *J. hum. Evol.* **21**, 451–462.

Bordes, F. (1975). Sur la notion de sol d'habitat en prehistoire paleolithique. *Bull. Soc. Prehist. Franc.* **72**, 139–144.

Brooks, A. S. & Yellen, J. E. (1987). The preservation of activity areas in the archaeological record: ethnoarchaeological and archaeological work in northwest Ngamiland, Botswana. In (S. Kent, Ed.) *Method and Theory for Activity Area Research: an Ethnoarchaeological Approach*, pp. 63–106. New York: Columbia University Press.

Bunn, H. T. (1982). Meat-eating and Human Evolution: Studies on the Diet and Subsistence Patterns of Plio-Pleistocene hominids in East Africa. Ph.D. Dissertation, Department of Anthropology, University of California, Berkeley.

Bunn, H. T. (1987). Paper presented at UISPP meetings, Mainz, Germany.

Bunn, H. T. (1991). A taphonomic perspective on the archaeology of human origins. *Ann. Rev. Anthrop.* **20**, 433–467.

Bunn, H. T. (1994). Early Pleistocene hominid foraging strategies along the ancestral Omo River at Koobi Fora, Kenya. *J. hum. Evol.* **27**, 247–266.

Bunn, H. T. (in press). Bone assemblages from the excavated sites. In (G. Ll. Isaac, Ed.) *Koobi Fora Research Project, Archaeology Volume, Chapter 8*. Oxford: Clarendon Press.

Bunn, H. T., Harris, J. W. K., Isaac, G., Kaufulu, Z., Kroll, E., Schick, K., Toth, N. & Behrensmeyer, A. K. (1980). FxJj 50: An early Pleistocene site in northern Kenya. *Wld Archeol.* **12**, 109–136. Reprinted in (B. Isaac, Ed.) *The Archaeology of Human Origins: Papers by Glynn Isaac*, pp. 228–257. Cambridge: Cambridge University Press (1989).

Bunn, H. T. & Kroll, E. M. (1986). Systematic butchery by Plio-Pleistocene hominids at Olduvai Gorge, Tanzania. *Curr. Anthrop.* **27**, 431–452.

Bunn, H. T. & Kroll, E. M. (1987). Reply to R. Potts. *Curr. Anthrop.* **29**, 96–98.

Bunn, H. T. & Kroll, E. M. (1988). Fact and fiction about the Zinjanthropus floor: data, arguments, and interpretations. Reply to L. R. Binford. *Curr. Anthrop.* **29**, 135–155.

Bunn, H. T. & Kroll, E. M. (1993). Comment on "The Structure of the Lower Pleistocene archaeological record: a case study from the Koobi Fora Formation" by N. Stern. *Curr. Anthrop.* **34**, 216–217.

Bunn, H. T., Kroll, E. M., Kaufulu, Z. M. & Isaac, G. Ll. (in press). FxJj 50. In (G. Ll. Isaac, Ed.) *Koobi Fora Research Project, Archaeology Volume, Chapter 4*. Oxford: Clarendon Press.

Cahen, D. & Moeyersons, J. (1977). Subsurface movements of stone artefacts and their implications for the prehistory of central Africa. *Nature* **266**, 812–815.

Cavallo, J. (1993). Paper presented at the Paleoanthropology Society meetings, Toronto.

Cziesla, E. & Eickhoff, E. (Eds) (1990). *The Big Puzzle: International Symposium on Refitting Stone Artefacts*. Holos: Verlag Bonn.

Dunnell, R. C. & Dancey, W. S. (1983). The siteless survey: a regional scale data collection strategy. In (M. B. Schiffer, Ed.) *Advances in Archaeological Method and Theory, volume 6*, pp. 267–287. New York: Academic Press.

Ebert, J. I. (1992). *Distributional Archaeology*. Albuquerque: University of New Mexico Press.

Foley, R. (1981a). *Off-site Archaeology and Human Adaptation in Eastern Africa*. Cambridge Monographs in African Archaeology 3. Oxford: British Archaeological Reports (International Series) 97.

Foley, R. (1981b). Off-site archaeology: an alternative approach for the short-sited. In (I. Hodder, G. Ll. Isaac & N. Hammond, Eds), *Patterns of the Past: Studies in Honour of David Clarke*, pp. 131–155. Cambridge: Cambridge University Press. In (B. Isaac, Ed.) *The Archaeology of Human Origins: Papers by Glynn Isaac*, pp. 206–227. Cambridge: Cambridge University Press (1989).

Gifford-Gonzalez, D. P., Damrosch, D. B., Damrosch, J. P. & Thunen, R. L. (1985). The third dimension in site structure: an experiment in trampling and vertical dispersal. *Am. Antiq.* **50**, 803–818.

Gould, R. A. (1969). *Yiwara: Foragers of the Australian Desert*. New York: Charles Scribner's Sons.

Gould, R. A. & Yellen, J. E. (1987). Man the hunted: determinants of household spacing in desert and tropical foraging societies. *J. Anthrop. Archaeol.* **6**, 77–103.

Harris, J. W. K. (1978). The Karari Industry: its place in East African prehistory. Ph.D. Dissertation, Department of Anthropology, University of California, Berkeley.

Harris, J. W. K. & Isaac, G. Ll. (in press). FxJj 20 Main, FxJj 20E, FxJj 20S. In (G. Ll. Isaac, Ed.) *Koobi Fora Research Project, Archaeology Volume, Chapter 4*. Oxford: Clarendon Press.

Hayden, B. (1979). *Palaeolithic Reflections*. Canberra: Australian Institute of Aboriginal Studies.

Hofman, J. L. (1986). Vertical movement of artifacts in alluvial and stratified deposits. *Curr. Anthrop.* **27**, 163–171.

Isaac, B. (Ed.) (1989). *The Archaeology of Human Origins: Papers by Glynn Isaac*, pp. 228–257. Cambridge: Cambridge University Press.

Isaac, G. Ll. (1978). The food sharing behavior of protohuman hominids. *Sci. Am.* **238**, 90–108. Reprinted in (B. Isaac, Ed.) *The Archaeology of Human Origins: Papers by Glynn Isaac*, pp. 289–311. Cambridge: Cambridge University Press (1989).

Isaac, G. Ll. (1981). Stone age visiting cards: approaches to the study of early land-use patterns. In (I. Hodder, G. Ll. Isaac, & N. Hammond, Eds), *Patterns of the Past: Studies in Honour of David Clarke*, pp. 131–155. Cambridge: Cambridge University Press. Reprinted in (B. Isaac, Ed.) *The Archaeology of Human Origins: Papers by Glynn Isaac*, pp. 206–227. Cambridge: Cambridge University Press (1989).

Isaac, G. Ll. (1983). Bones in contention: competing explanations for the juxtaposition of Early Pleistocene artefacts and faunal remains. In (J. Clutton-Brock & G. Grigson, Eds), *Animals and Archaeology: Hunters and Their Prey*. Oxford: British Archaeological Reports (International Series) 163: 3–19. Reprinted in (B. Isaac, Ed.) *The Archaeology of Human Origins: Papers by Glynn Isaac*, pp. 325–336. Cambridge: Cambridge University Press (1989).

Isaac, G. Ll. (1984). The archaeology of human origins: studies of the Lower Pleistocene in East Africa. In (F. Wendorf & A. Close, Eds) *Advances in Old World Archaeology, Volume 3*, pp. 1–87. New York: Academic Press. Reprinted in (B. Isaac, Ed.) *The Archaeology of Human Origins: Papers by Glynn Isaac*, pp. 120–187. Cambridge: Cambridge University Press (1989).

Isaac, G. Ll. (Ed.) (in press). *Koobi Fora Research Project, Archaeology Volume*. Oxford: Clarendon Press.

Isaac, G. Ll. & Harris, J. W. K. (1978). Archaeology. In (M. G. Leakey & R. E. Leakey, Eds) *Koobi Fora Research Project, Volume 1: The Fossil Hominids and Introduction to Their Context, 1968–1974*, pp. 64–85. Oxford: Clarendon Press.

Isaac, G. Ll. & Harris, J. W. K. (1980). A method for determining the characteristics of artefacts between sites in the Upper Member of the Koobi Fora Formation, East Lake Turkana. In (R. E. Leakey & B. A. Ogot, Eds) *Proc. 8th Panafr. Congr. Prehist. Quat. Stud.*, pp. 19–22. Nairobi: The International Louis Leakey Memorial Institute for African Prehistory.

Isaac, G. Ll., Harris, J. W. K. & Crader, D. (1976). Archaeological evidence from the Koobi Fora Formation. In (Y. Coppens, F. C. Howell, G. L. Isaac & R. E. F. Leakey, Eds) *Earliest Man and Environments in the Lake Rudolf Basin: Stratigraphy, Paleoecology and Evolution*, pp. 533–551. Chicago: University of Chicago Press.

Isaac, G. Ll., Harris, J. W. K. & Marshall, F. (1981). Small is informative: the application of the study of mini-sites and least effort criteria in the interpretation of the early Pleistocene archaeological record at Koobi Fora, Kenya. In (J. D. Clark & G. Ll. Isaac, Eds) *Las Industrias mas Antiguas* X Congreso Union Internacional de Ciencias Prehistoricas y Protohistoricas, Mexico, D.F.: 101–119. Reprinted in (B. Isaac, Ed.) *The Archaeology of Human Origins: Papers by Glynn Isaac*, pp. 258–268. Cambridge: Cambridge University Press (1989).

Isaac, G. Ll., Leakey, R. E. & Behrensmeyer, A. K. (1971). Archaeological traces of early hominid activities east of Lake Rudolf, Kenya. *Science* **173**, 1129–1134.

Kaufulu, Z. M. (1983). The Geological Context of Some Early Archaeological Sites in Kenya, Malawi, and Tanzania: Microstratigraphy, Site Formation and Interpretation. Ph.D. Dissertation, Department of Anthropology, University of California, Berkeley, California.

Keeley, L. H. & Toth, N. (1981). Microwear polishes on early stone tools from Koobi Fora, Kenya. *Nature* **293**, 243–267.

Kent, S. (Ed.) (1987). *Methods and Theory for Activity Area Research: an Ethnoarchaeological Approach*, pp. 63–106. New York: Columbia University Press.

Kroll, E. M. (1986). The home base revisited. Paper presented at the American Anthropological Association meetings, Philadelphia.

Kroll, E. M. (1987). Help, no hearths. Paper presented at the Society for American Archaeology meetings, Toronto.

Kroll, E. M. (n.d.). The Anthropological Meaning of Spatial Configurations at Plio-Pleistocene Archaeological Sites. Ph.D. Dissertation, Department of Anthropology, University of California, Berkeley, California.

Kroll, E. M. (in press). Lithic and faunal distributions at eight archaeological excavations. In (G. Ll. Isaac, Ed.) *Koobi Fora Research Project, Archaeology Volume, Chapter 9*, Oxford: Clarendon Press.

Kroll, E. M. and Isaac, G. Ll. (1984). Configurations of artifacts and bones at early Pleistocene sites in East Africa. In (H. J. Hietela, Ed.) *Intrasite Spatial Analysis in Archaeology*, pp. 4–31. Cambridge: Cambridge University Press.

Kroll, E. M. & Price, T. D. (Eds) (1991). *The Interpretation of Archaeological Spatial Patterning*. New York: Plenum Press.

Leakey, M. D. (1971). *Olduvai Gorge: Excavations In Beds I and II, 1960–1963*. Cambridge: Cambridge University Press.

Lyman, R. L. & Fox, G. L. (1989). A critical evaluation of bone weathering as an indication of bone assemblage formation. *J. Archaeol. Sci.* **16**, 293–317.

Marshall, F., Isaac, G. L., Harris, J. W. K. (in press). FxJj 64. In (G. Ll. Isaac, Ed.) *Koobi Fora Research Project, Archaeology Volume, Chapter 4*. Oxford: Clarendon Press.

McBrearty, S. (1990). Consider the humble termite: termites as agents of post-depositional disturbance at African archaeological sites. *J. Archaeol. Sci.* **17**, 111–143.

O'Brien, C. J. (n.d.). Determining Seasonality and Age in East African Archaeological Fauna: an Ethnoarchaeological Application of Cementum Increment Analysis. Ph.D. Dissertation, University of Wisconsin, Madison.

O'Connell, J. F. (1987). Alyawara site structure and its archaeological implications. *Am. Antiq.* **52,** 74–108.

Potts, R. (1982). Lower Pleistocene Site Formation and Hominid Activities at Olduvai Gorge, Tanzania. Ph.D. Dissertation, Harvard University.

Potts, R. (1984). Home bases and early hominids. *Am. Sci.* **72,** 338–347.

Potts, R. (1987). On butchery by Olduvai hominids. *Curr. Anthrop.* **28,** 95–96.

Potts, R. (1988). *Early Hominid Activities at Olduvai.* New York: Aldine de Gruyter.

Potts, R. (1991). Why the Oldowan? Plio-Pleistocene toolmaking and the transport of resources. *J. Anthrop. Res.* **47,** 153–176.

Rogers, M. (1993). An examination of varying archaeological densities at East Turkana, Kenya, and its implications for interpreting "sites". Paper present at the Paleoanthropology Society meetings, Toronto.

Santonja, M. & Villa, P. (1990). The Lower Paleolithic of Spain and Portugal. *J. Wld Prehist.* **4,** 45–94.

Schick, K. D. (1984). Processes of Paleolithic Site Formation: An Experimental Study. Ph.D. Dissertation, Department of Anthropology, University of California, Berkeley, California.

Schick, K. D. (1987). Modeling the formation of Early Stone Age artifact concentrations. *J. hum. Evol.* **16,** 789–807.

Schiffer, M. B. (1987). *Formation Processes of the Archaeological Record.* Albuquerque: University of New Mexico Press.

Sept, J. M. (1992). Was there no place like home? *Curr. Anthrop.* **33,** 187–207.

Stern, N. (1991). The Scatter-Between-the-Patches: A Study of Early Hominid Land Use Patterns in the Turkana Basin, Kenya. Ph.D. Dissertation, University of California, Berkeley.

Stern, N. (1993). The structure of the Lower Pleistocene archaeological record: a case study from the Koobi Fora Formation. *Curr. Anthrop.* **34,** 201–225.

Stockton, E. D. (1973). Shaw's Creek shelter: human displacement of artifacts and its significance. *Mankind* **9,** 112–117.

Straus, L. G. (1979). Caves: a paleoanthropological resource. *Wld Archeol.* **10,** 331–339.

Teleki, G. (1981). The omnivorous diet and eclectic feeding habits of chimpanzees in Gombe National Park, Tanzania. In (R. S. O. Harding & G. Teleki, Eds) *Omnivorous Primates: Gathering and Hunting in Human Evolution,* pp. 303–343. New York: Columbia University Press.

Thomas, D. H. (1975). Nonsite sampling in archaeology: up the creek without a site? In (J. W. Mueller, Ed.) *Sampling in Archaeology,* pp. 61–81. Tucson: University of Arizona Press.

Toth, N. P. (1982). The Stone Technologies of Early Hominids at Koobi Fora, Kenya: An Experimental Approach. Ph.D. Dissertation, University of California, Berkeley.

Toth, N. P. (1987). Behavioral inferences from early stone artifact assemblages: an experimental model. *J. hum. Evol.* **16,** 763–787.

Van Noten, F. L., Cahen, D., Keeley, L. H. (1980). A Paleolithic campsite in Belgium. *Sci. Am.* **252,** 48–55.

Villa, P. (1982). Conjoinable pieces and site formation processes. *Am. Antiq.* **47,** 276–290.

Villa, P. (1983). *Terra Amata and the Middle Pleistocene Archaeological Record of Southern France.* University of California Publications, Anthropology, 13. Berkeley: University of California Press.

Villa, P. & Courtin, J. (1983). The interpretation of stratified sites: a view from underground. *J. Archaeol. Sci.* **10,** 267–281.

Vincens, A. (1979). Analyse palynologique du site archeologique FxJj 50, Formation du Koobi Fora, Est Turkana (Kenya). *Bull. Soc. Geol. France* **21,** 343–347.

Whallon, R. J. (1984). Unconstrained clustering for the analysis of spatial distributions in archaeology. In (H. Hietala, Ed.) *Intrasite Spatial Analysis in Archaeology,* pp. 242–277. Cambridge: Cambridge University Press.

Yellen, J. (1977). *Archaeological Approaches to the Present: Models for Reconstructing the Past.* New York: Academic Press.

Michael J. Rogers &
John W. K. Harris
Department of Anthropology, Rutgers University, Douglass Campus, New Brunswick, New Jersey, 08903-0270 U.S.A.

Craig S. Feibel
Department of Geology and Geophysics, University of Utah, 84112-1183 U.S.A.

Received 14 March 1994
Revision received 6 April 1994
and accepted 6 April 1994

Keywords: paleogeographic reconstruction, land use, Early Stone Age archaeology, Turkana Basin, *Homo erectus*, Plio-Pleistocene.

Changing patterns of land use by Plio-Pleistocene hominids in the Lake Turkana Basin

The first step in a behavioural ecological study of stone-tool-using hominids involves the description of the character of lithic discard and the context within which the discard occurred. We examine and put into paleogeographic context the known archaeological traces in the Turkana Basin at three successive time intervals: 2·3 million years ago (Ma), 1·9–1·8 Ma, and 1·7–1·5 Ma. At 2·3 Ma, hominid use of stone appears restricted to small areas on the landscape where many resources such as water, shade, and stone are juxtaposed. In contrast, archaeological traces at 1·6 Ma are found in a variety of settings, which may in part be explained by the paleogeographic changes taking place at that time. This change coincides with the emergence of *Homo erectus*. The hominid fossil and archaeological records are shown to complement each other in the generation of ecological hypotheses of *H. erectus* behaviour.

Journal of Human Evolution (1994) **27**, 139–158

Introduction

To study a species' behavioural ecology is to study a species as a part of its ecosystem, which includes the physical environment around it as well as its relationships with other plants and animals (Foley, 1987). In short, the goal of behavioural ecology is to understand a species' niche—variables relevant to the adaptive strategies used by a species to survive over the long term. Variables such as habitat, home range size, diet breadth, group size, locomotion, food dispersion and body size, are all important to examine in this science, but the relationship between these variables are really what behavioural ecologists are trying to understand. Indeed, established ecological relationships can help to elucidate the niches of early hominids and how they may have changed through time (e.g., see Winterhalder, 1980, 1981; Blumenschine *et al.*, 1994).

In order for the study of early hominid behavioural ecology to be a viable endeavour, all possible avenues for obtaining information about the past must be considered. In this paper, we consider aspects of the geological, paleontological, and archaeological records in the Turkana Basin, northern Kenya, from 2·3 Ma (million years ago) to 1·5 Ma. All three records change dramatically during this time interval—a critical time period for understanding the emergence of both stone tool manufacture and the genus *Homo*. We emphasize how the archaeological record fits into recent (revised) paleogeographic reconstructions of the Turkana Basin, so that early hominid adaptive strategies can be examined in a broader, paleoenvironmental context, the first step towards a behavioural ecological understanding of human evolution. Finally, we make an attempt to integrate the hominid paleontological and archaeological records with ecological principles in a way that generates testable hypotheses of hominid behaviour.

Archaeological record in stratigraphic context

The Plio-Pleistocene strata of the Lake Turkana Basin are known collectively as the Omo Group, which includes the Koobi Fora Formation east of modern Lake Turkana, the

0047–2484/94/010139+20 $08.00/0

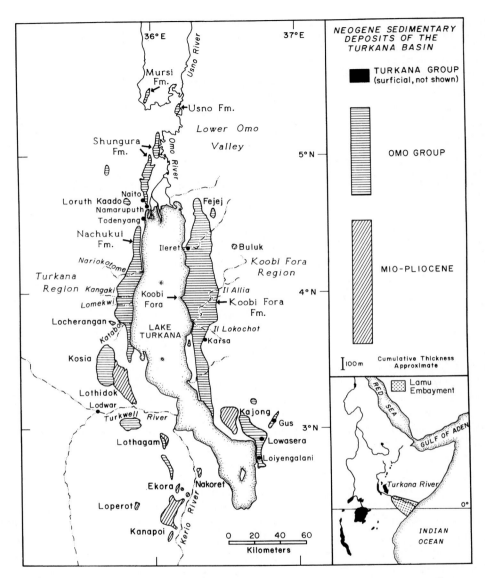

Figure 1. Map of the Turkana Basin, showing areas of outcrop for the Plio-Pleistocene sedimentary sequences (modified from Feibel *et al.*, 1989).

Shungura Formation in the lower Omo Valley to the north, and the Nachukui Formation on the west side of the lake (Figure 1; Brown & Feibel, 1986; de Heinzelin, 1983; Harris *et al.*, 1988). Each of these sedimentary sequences preserves abundant evidence of the activities of early hominids. Stratigraphic terminology and the levels of archaeological sites discussed here are shown in Figure 2.

The oldest archaeological traces recorded thus far in the basin are known from 2·3 Ma in Member F of the Shungura Formation (Chavaillon, 1976; Merrick, 1976; Howell *et al.*, 1987)

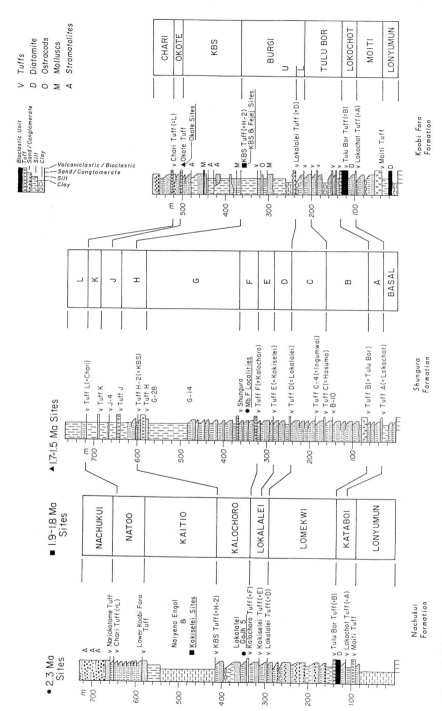

Figure 2. Schematic sections of artefact-bearing formations in the Turkana Basin showing stratigraphic placement of archaeological localities (modified from Feibel et al., 1989).

and the Kalochoro Member of the Nachukui Formation (Roche, 1989; Kibunjia, 1990; Kibunjia *et al.*, 1992; Roche & Kibunjia, 1994). Sediments of this age are not known from Koobi Fora, where an unconformity marks a period of uplift and erosion. The archaeological localities from the Shungura Formation, FtJj 1, 2, and 5, and Omo 57 and 123, consist of small numbers and densities of predominantly quartz artefacts, a locally available raw material. Lokalalei, from the Nachukui Formation on the west side of the modern lake, has thus far produced a similar low density of artefacts, but the raw material there is fine-grained lava. These localities are only sparsely associated with faunal remains. Figure 3 shows the artefact densities from these sites, as well as those from all the localities in the two succeeding time intervals discussed below.

The second period of archaeological evidence examined here occurs between 1·9 and 1·8 Ma. The oldest localities in this interval, FxJj 1 and 3, are documented from Koobi Fora at 1·88 Ma, the age of the KBS Tuff (McDougall, 1985). Other localities slightly younger in age are known from this interval in the Koobi Fora Formation (FxJj 10), the Kaitio Member of the Nachukui Formation (Kokiselei and Naiyena Engol), and at Fejej. The Koobi Fora localities are low density occurrences of lava artefacts and associated fauna, with density values similar to those of the 2·3 Ma localities (Isaac & Harris, 1978). The two localities reported from the Nachukui Formation have higher densities of fine-grained lava artefacts, but the preliminary nature of the investigations there, with consequent small excavated areas (two and five square meters), may be exaggerating these values (Kibunjia *et al.*, 1992). The preliminary report from Fejej also does not allow one to calculate a density figure (Asfaw *et al.*, 1991). Further field studies at Fejej led by Dr Y. Beyene yielded an *in situ* occurrence of quartz artefacts and fragmentary fauna (Beyene, pers. comm. to JWKH). This new round of investigations has therefore confirmed the utilization of quartz as the raw material for stone tool manufacture. The highly localized nature of the choice of raw material has important implications. Only 25 km to the south in the corresponding time interval are the KBS Member sites where artefacts were made almost exclusively on basalt. This feature suggests highly localized ranging patterns with respect to selection of raw materials and use of stone tools at this time.

The third, and most intensively-studied, stratigraphic interval with evidence of hominid activities ranges from about 1·7 to 1·5 Ma. All of the reported archaeological traces from this interval derive from the Koobi Fora region, and they include localities such as FwJj 1, FxJj 64, 50, 18, 17, and 20 (Harris, 1978; Isaac & Harris, 1978; Bunn *et al.*, 1980; Isaac *et al.*, 1981). Several of these localities contain artefacts characteristic of the Karari Industry, a local variant of the Developed Oldowan, namely, single platform cores formerly labeled "Karari scrapers" (Harris, 1978). Most of the artefacts at these localities are made of fine-grained lava, with small numbers of quartz, chert, and chalcedony pieces. With this time interval we see a much greater range of artefact densities, as well as a corresponding increase in raw density values at some localities, e.g., FxJj 18IH and 20AB. This pattern is not simply due to different taphonomic processes, since the high density localities are found in low-energy, distal floodplain contexts (Kaufulu, 1987; Rogers, in press).

From 2·3 Ma to 1·5 Ma, then, we see the archaeological record become much more variable in terms of artefact density and artefact types. In addition, the localities within the 1·7–1·5 Ma interval do not seem to be as tied to sources of raw material as the earlier localities were, suggesting that a greater diversity of habitats were being used by stone-tool-using hominids by 1·6 Ma. This statement is supported and contextualized by the recent paleogeographic reconstructions of the Turkana Basin to which we now turn.

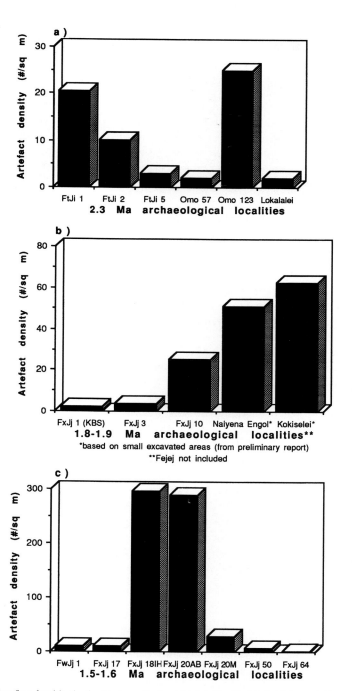

Figure 3. Artefact densities in the Turkana Basin at (a) 2·3 Ma, (b) 1·8–1·9 Ma, and (c) 1·5–1·6 Ma.

Introduction to the paleogeography

Throughout the Plio-Pleistocene, the Turkana Basin acted as a single, large depositional system. The main contributor to the basin, in terms of both water and sediment supply, was the ancestral Omo River. This was a perennial river arising in the Ethiopian Highlands to the north. Several tributary rivers, notably the Kerio and Turkwell Rivers in the south, joined the Omo before it exited the basin. These major rivers form the axial drainage system, flowing along the length of the basin, throughout the Plio-Pleistocene. Each had a large headwater catchment which lay a considerable distance from the depositional basin, often in a very different climatic regime. Many smaller drainages headed along the margins of the depositional basin. Their contribution was relatively minor in terms of quantity, but often significant. These were commonly high-gradient streams which transported very coarse debris into the basin, the raw material for stone-tool manufacture.

The interval from roughly 2·5–1·5 Ma was a period of intense activity in the Turkana Basin. The tectonic setting of the basin, which had been relatively stable for nearly 2 million years, experienced a major structural event about 2·5 Ma. The northeastern corner of the basin was uplifted, relative to the remainder of the region. This produced several immediate effects. Sedimentation in the Koobi Fora region ceased, and some incision began to take place. Farther to the northeast, the volcanic and metamorphic terrain in the Lake Stephanie region began to undergo active erosion. The effects of this erosion would not be recorded for another half million years, but at that time would show an influx of volcanic material to the Koobi Fora region, which would be the foundation of the archaeological record there. In the southeastern corner of the Turkana Basin, the main constructional phase in the development of the shield volcano Mt Kulal also began about 2·5 Ma. This activity likely produced a series of volcanic dams and eventually the edifice of Mt Kulal, which deflected the basin's outlet to the north. In the process, this volcanic accumulation dammed the basin's outlet and caused the formation of a major lake about 2 Ma. The infilling of this lake by sediments from the Omo River would be a major control on paleogeographic evolution in the interval 2·0–1·5 Ma. A final area of activity in this time span centered on the silicic volcanic centers of the Ethiopian Highlands. These volcanoes formed a part of the headwater region of the Omo River, and explosive volcanism associated with their development mantled this region with volcanic ash. Ensuing rainy seasons, produced by the Indo-African monsoon, flushed this ash from the headwaters into the depositional basin. At 2·5 Ma, the Ethiopian Highlands were experiencing a cyclical peak in explosive volcanism, as recorded in the tephra of the Turkana Basin. This was followed by a dramatic drop in activity, with a gradual buildup to an even greater peak of eruptive activity around 1·6 Ma. In this latter peak of activity, the large volumes of tephra washed into the depositional basin caused significant clogging of the fluvial systems, and changes in channel pattern. This would have important consequences for vegetation distribution and hominid habitats (Feibel, 1993).

Paleogeographic reconstructions

The record of changing hominid habitats and land use patterns is best illustrated on a basinal scale by looking at the changing paleogeographic context (Figures 4–7). These maps show the regional setting for each of the three intervals of the archaeological record outlined above. As with earlier reconstructions of basinal paleogeography (Feibel, 1988; Brown & Feibel, 1988, 1991; Feibel & Brown, 1993), modern Lake Turkana is included for

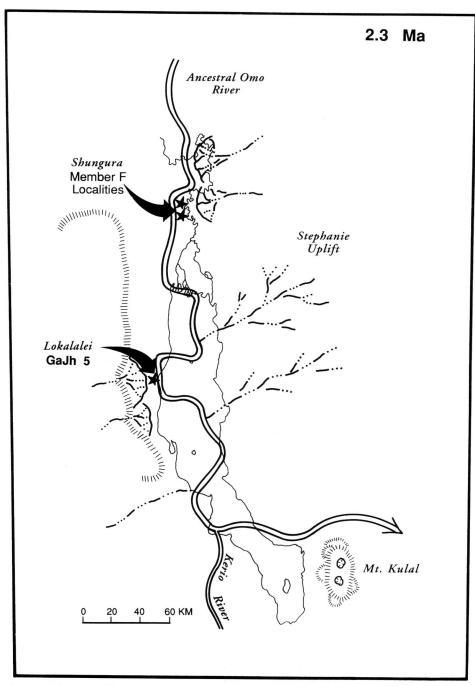

Figure 4. Paleogeographic reconstruction of the Turkana Basin at 2·3 Ma showing locations of archaeological localities. Outline of modern Lake Turkana shown for reference.

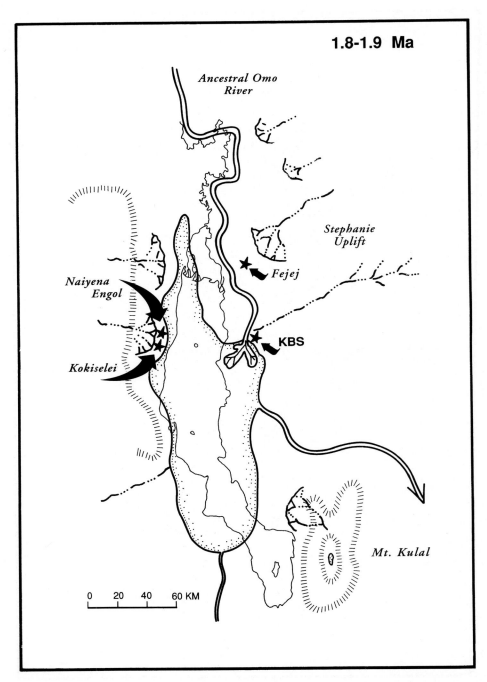

Figure 5. Paleogeographic reconstruction of the Turkana Basin at 1·8–1·9 Ma showing location of archaeological localities. Outline of modern lake shown for reference.

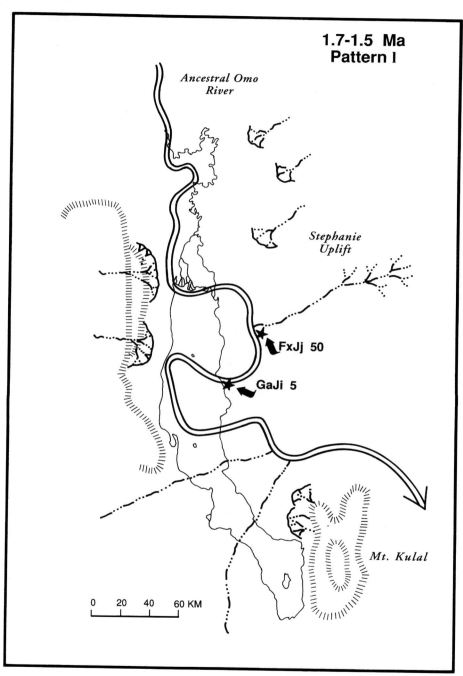

Figure 6. Pattern I paleogeographic reconstruction of the Turkana Basin at 1·7–1·5 Ma showing location of archaeological localities. Outline of modern lake shown for reference.

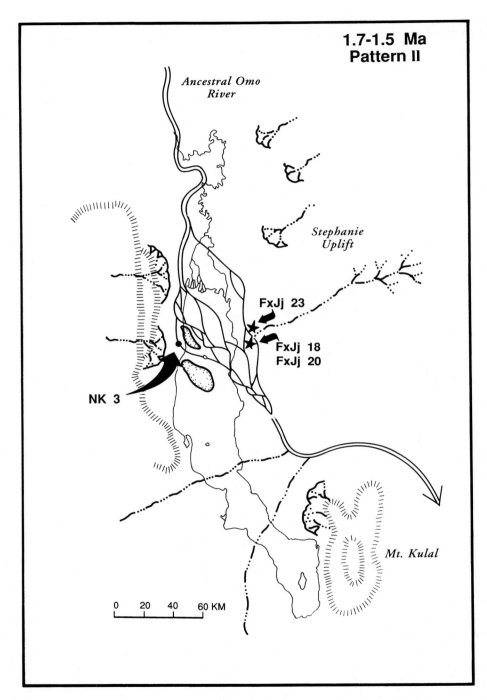

Figure 7. Pattern II paleogeographic reconstruction of the Turkana Basin at 1·7–1·5 Ma showing location of archaeological localities and the Nariokotome skeleton site (NK 3). Outline of modern lake shown for reference.

locational purposes only; the modern lake did not come into existence until the last several hundred thousand years. The maps show the ancestral Omo River flowing down from the Ethiopian Highlands into the depositional basin, and a second river, the Turkana River, exiting the basin. The exiting river accounts for the open character of the basin, and probably existed in some form for most of the past 4 million years. In the period of interest here, it reached to the Indian Ocean, as documented by the immigration of the stingray *Dasyatis africana* (Feibel, 1994). The character of the depositional basin varies considerably, depending on the presence or absence of a major lake, and the position and pattern of the fluvial system.

At 2·3 Ma there was no lake present in the basin (see Figure 4). The Omo River crisscrossed the basin in a pattern that appears to have been determined by structural control. Because of uplift in the northeast probably still underway at this time, the Koobi Fora region shows only erosional activity, and deposition along the axial system is restricted to a relatively narrow, north–south belt. Both sets of archaeological occurrences at this time lie between the toes of active alluvial fans from the marginal drainage system and the axial river system. Both systems are apparent in the sedimentary sequences associated with the sites. In the lower Omo Valley, de Heinzelin (1983) and Howell *et al.* (1987; also see Howell, 1976) both describe the situation in lower Member F as including strata indicative of a large meandering fluvial system, as well as deposits characteristic of a braided stream. They attributed both types of deposits to the same river, but a simpler explanation is that two systems left deposits of distinct character. Although the strata are not described in detail, Kibunjia *et al.* (1992) interpret the context of the Lokalalei site in the same terms. The main difference between the two localities is that the marginal drainage at Lokalalei was headed in the volcanic strata of the Murua Rith range to the west, while the marginal system of the lower Member F localities originated in the metamorphic terrain of the west side of the Stephanie uplift. It is likely that the lateral migration of the marginal system in the lower Omo Valley relates directly to the uplift going on in that region. The increased gradient caused by uplift would push the alluvial fans farther out into the basin, until they verged on the axial system. The ephemeral streams coming off the toes of those fans contained coarse metamorphic detritus which became the raw material for artefact manufacture. It is not currently known whether a similar pulse of uplift in the Murua Rith range may have been responsible for the progradation of the Lokalalei fans to verge on the axial systems there. Proximity to both the axial and marginal systems does seem to be critical at this early stage, for in other localities of the same temporal interval and with a proximity to only one or the other system, artefact occurrences have not been recorded.

By the next temporal interval of interest, 1·8–1·9 Ma, the paleogeographic picture has changed dramatically (Figure 5). The large lake which formed ca. 2 Ma is being filled with sediment by the delta of the Omo River along its eastern margin (Feibel, 1988). The early history of this lake records a stable system, but beginning about 1·89 Ma climatic fluctuations are first recorded, and the lake level becomes unstable. Archaeological traces of hominid activities are recorded in three different settings. The KBS site at Koobi Fora (FxJj 1) lies in a fluvial setting, close to the channel of the ancestral Omo River. The context of this site differs from earlier occurrences in that the raw material source appears to be the Omo itself. A tributary marginal drainage flowing off the Stephanie Uplift at this time joined the Omo a short distance upstream from the archaeological site. The volcanic gravels provided by the marginal source were presumably concentrated on gravel bars for a short distance downriver before they were dispersed to the point of being rare components.

Although its precise stratigraphic and paleogeographic setting is unclear (see Asfaw *et al.*, 1991; Haileab & Feibel, 1993; and Asfaw *et al.*, 1993), the Fejej locality seems to be located in an alluvial situation more similar to the 2·3 Ma localities. The axial system is in evidence from the occurrence of tephra, and a marginal source is indicated by the coarse metamorphic raw material (quartz). The third setting in this temporal interval is by far the most unique. The West Turkana localities of Kokiselei and Naiyena Engol are both situated between the alluvial fans draining the western hills and the lake margin. These are the only Plio-Pleistocene sites within the Turkana Basin known with a lake margin association.

The paleogeographic setting at 1·7–1·5 Ma records a complex situation, hence the two reconstructions (Figures 6 & 7). These figures portray two extremes of variability in a dynamic landscape, not necessarily a smooth chronological transition from one reconstruction to another. By this time the Omo River had completed filling the lake formed at 2 Ma, but it is likely that the compaction of the thick deltaic sediment package produced high subsidence rates. At its most stable, the fluvial regime produced a pattern similar in character to the earlier fluvial systems (Pattern I, Figure 6). At the other extreme, a combination of subsidence and clogging of channels by tephra resulted in a broken pattern of smaller channels and abundant floodplain marshes and ponds. The best example of a site associated with a Pattern I setting may be FxJj 50 (Bunn *et al.*, 1980). This site lies on the edge of a paleotopographic feature interpreted to represent the bank of a major river channel. Lateral to the site is a large gravel bar which might occur just downstream of a tributary gravel source. Although tephra are associated with this site (the Koobi Fora Tuff is present), this represents the waning stages of the eruptive peak at about 1·6 Ma. A second example of a Pattern I setting is GaJi 5, the cut-marked bone locality at Koobi Fora (Bunn, 1981). This locality lies in the stratigraphic interval below the eruptive peak, in sediments characterized by typical meandering channel patterns and overbank soil development.

Most of the 1·7–1·5 Ma interval sediments, however, were deposited in strata strongly influenced by the influx of large volumes of tephra (Pattern II, Figure 7). In this depositional setting, channels were much smaller, shallower, and straighter than the typical meandering form. The high influx of tephra caused these channels to fill rapidly and become unstable, at which time the channel would switch its position, abandoning the former channel to cut a new one at a different position on the floodplain. The small channels overtopped their banks easily in flood stage, and this resulted in a high proportion of poorly-sorted sediments, primarily tuffaceous silts, being deposited across the floodplain. The FxJj 18 site complex is one example of an archaeological locality in this setting. It has four distinct archaeological horizons representing channel, gravel bar, overbank, and floodplain depositional environments, probably reflecting the rapid filling by the Lower Koobi Fora Tuff and consequent migration of the associated channel. The Nariokotome hominid skeleton, KNM-WT 15 000, was also buried in a Pattern II setting (Feibel & Brown, 1993).

Changing patterns of land use

By considering the archaeological record within the paleogeographic reconstructions described above, we can begin to assess the possible factors that influenced hominid ranging and land use.

At 2·3 Ma the availability of resources such as stone, water, shade, shelter, and presumably food seems to play a large role in determining where stone tools were discarded. Only in certain places where all of these resources were available do we find accumulations of stone artefacts, and then only in small numbers. For example, all the archaeological localities within this time period were most likely located along the toes of alluvial fans where a marginal drainage containing lithic raw material fed into the axial Proto-Omo, at this time a large meandering river with a perennial water supply. Because of its restricted and sparse nature, stone tool use at this time seems sporadic, possibly seasonal, and most likely expedient (Harris & Capaldo, 1993). It has also been suggested that the artefacts of this time period show less technological skill than those from less than 2 Ma (Roche, 1989).

Even at 1·9–1·8 Ma, archaeological localities still seem to be tied to raw material sources to some extent. FxJj 1 and Fejej are only about 25 kilometers apart, yet the lithic remains at each locality are made only from the locally-available raw material—lava at FxJj 1 and quartz at Fejej. At this time, though, there is a hint that places other than the intersection of axial and marginal streams may have been used for stone-tool-related activites. The West Turkana localities within this time period are located on a lake margin, which suggests an expansion of the locational repertoire of stone tool use and discard. However, it is possible that these lake margin localities were essentially the same as the closed, mesic habitats utilized 0·5 million years earlier, with a lake substituting for the Proto-Omo as the perennial water supply. Future work by the West Turkana team (see Kibunjia et al., 1992) will of course provide more detailed information on these localities.

By 1·6 Ma, many different settings are being used for stone tool activities, and lithic traces are not necessarily tied to sources of raw material. Stone artefacts are found in practically all fluvial depositional environments (Kaufulu, 1983, 1987; Stern, 1991), including both proximal (see Stern, 1991) and distal floodplain (Rogers, in press) settings which would have been at least some distance away from the stream beds whose cobbles were used for lithic raw material. This increase in the variety of archaeological settings may in part be explained by the change in the depositional regime from Pattern I to Pattern II (see Figures 6 and 7). A system of rapidly-changing braided channels not only has the effect of widening the axial drainage, but it also might have reduced the area of stable canopy forest that would be associated with a perennial meandering river. Paleoenvironmental reconstructions of the Turkana Basin based on stable isotopes of paleosol carbonates corroborate this trend and point to a change towards drier, more open conditions at about 1·8–1·7 Ma (Cerling et al., 1988; Cerling, 1992). Bovid remains point to an increase in drier more open conditions sometime within the KBS Member (Feibel et al., 1991). Interestingly, studies that have investigated hominid habitat preferences using large mammal remains from the Turkana Basin have hinted that while Australopithecus may have been fairly restricted to closed, wet habitats, early Homo may have been more of a habitat generalist, able to survive in dry, open habitats as well as wet, closed areas (Behrensmeyer, 1985; Shipman & Harris, 1988; but see White, 1988).

The bovid and other large mammal remains left at Okote Member archaeological localities at East Turkana represent a mix of habitats, but there is also some indication that drier, more open habitats were now being utilized for lithic-related purposes (Rogers & Sorkowitz, 1992). Other aspects of the archaeological record also seem to have changed. For example, the character of archaeological traces increases in variety at this time, in terms of both lithic density (see Figure 3; Rogers & Harris, 1992; Rogers, in prep.) and types of lithic end products (Harris, 1978). Also by 1·6–1·5 Ma, there is secure evidence that both bone and stone were being transported across the landscape (Bunn, 1982; Toth, 1982, 1985) and that some locations were sometimes used repeatedly (Harris, 1978). There is even mounting evidence for hominid use of fire (Harris, 1978; Bellomo, 1990; see Bellomo, 1994).

Behavioural implications of the archaeological record and early *Homo erectus* anatomy

In order to approach a behavioural ecological understanding of early hominids, all possible lines of evidence must be examined. The archaeological and geological/paleogeographical records discussed above complement anatomical features noted in the hominid fossil record in the search for evidence of hominid behaviour. In the Turkana Basin, juxtaposing all three records is accomplished best with early *Homo erectus*, *Homo ergaster* to some (e.g., Wood, 1991, 1992). This is because of the spectacular fossil finds of this species, such as the KNM-ER 3733 and KNM-ER 3883 skulls (Leakey & Walker, 1985) and the KNM-ER 1808 partial skeleton (Leakey, 1974) found at East Turkana dated to 1·78 Ma, 1·57 Ma, and 1·70 Ma, respectively (Feibel *et al.*, 1989), and the nearly-complete Nariokotome skeleton found at West Turkana, KNM-WT 15 000, dated to 1·60 Ma (Brown *et al.*, 1985; Leakey & Walker, 1993). Also, with the emergence of *Homo erectus* at about 1·7–1·8 Ma, there is less ambiguity concerning the authorship of the archaeological record. By 1·6 Ma, the only other hominid species existing in the Turkana Basin was *Australopithecus boisei*. Although it is possible that this species used and perhaps had the capability to make stone tools, we feel that it is almost certain that the *changes* seen in the archaeological record at this time were due to an *emergent* hominid species, namely *Homo erectus*, not one that may have already existed for at least half a million years (i.e., *A. boisei*). A recent study of brain asymmetries on the Nariokotome endocast supports the view that early *H. erectus* was probably the sole creator of Early Pleistocene stone tools in the Turkana Basin (see Begun & Walker, 1993).

If one accepts, then, that *H. erectus* is responsible for the lithic record at 1·6 Ma, we can examine both the fossil remains of this species and the archaeological record at this time in the Turkana Basin for clues of its behaviour. The fossil specimens mentioned above, especially the Nariokotome skeleton, allow us to make fairly confident statements about *H. erectus* anatomy. Figure 8 lists some of the anatomical changes that occur with the emergence of this species. Our purpose here is not to provide an exhaustive compendium of *H. erectus* physical traits, as this has been done with great precision elsewhere (Leakey & Walker, 1993; Rightmire, 1990), but to illustrate how the fossil and archaeological records can complement each other in generating behavioural hypotheses.

Although brain size increases with the emergence of *H. erectus* (Holloway, 1982), overall encephalization increases just slightly over the earlier *Homo* condition, mainly because of the great increase observed in *H. erectus* body size (Begun & Walker, 1993; Ruff & Walker, 1993; McHenry, 1988, 1991). This jump in size and stature is accompanied by relatively narrow hips with subsequent small birth canal (Walker & Ruff, 1993) and extreme linearity in body shape (Ruff & Walker, 1993; Ruff, 1991, 1993). The final anatomical change considered here is the slight decrease in jaw and tooth size, although this characteristic may not distinguish *H. erectus* from *H. habilis sensu lato* (Brown & Walker, 1993; Wood, 1991).

From these anatomical changes observed in the fossil record we can confidently infer some accompanying physiological changes adopted by *H. erectus* that cannot be directly observed (see Figure 8). The combination of an enlarged brain and relatively narrow hips and small birth canal implies that early *H. erectus* infants were probably born before their brains reached their full gestational size, a condition shared with modern humans called secondary altriciality (Martin, 1983; Walker & Ruff, 1993). Of course, a longer period of maturation would also be required, especially given the large size of an adult. We can also infer from the increase in body size that early *H. erectus* would be in greater need of high quantity and/or quality foods (Cachel

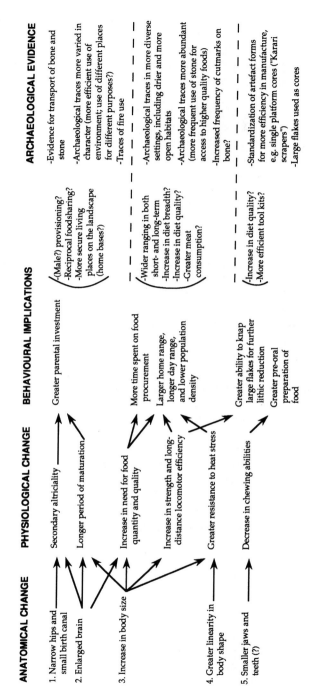

Figure 8. Summary of anatomical and physiological changes reflected in the earliest *Homo erectus* specimens 1·78–1·5 Ma, changes seen in archaeological traces at 1·6 Ma, and behavioural implications of both the archaeological and paleontological records in the Turkana Basin.

& Harris, 1993), stronger and more efficient in locomotion over long distances (Walker, 1993), and better able to withstand heat stress (Wheeler, 1991, 1992). The tall, thin body shape underscores the importance of thermoregulatory constraints that were placed on *H. erectus* (Ruff, 1991, 1993; Ruff & Walker, 1993; Wheeler, 1993). Finally, the slight reduction in the size of the dentition continues the trend that began with the earliest *Homo*, and implies even less need for heavy chewing.

The behavioural implications of these physiological changes listed in Figure 8 require a higher level of inference, but the first column of behavioural implications are derived from logical deductions and/or behavioural ecological principles and therefore seem fairly secure. In order to ensure the survival of helpless neonates, greater parental investment was probably needed; especially critical would have been the health of the mother and infant for the first few months of the infant's life (Martin, 1983). Larger brain and body sizes probably would have required more time spent on the search and procurement of food resources (Foley, 1984). The time required for manufacturing sharp-edged stone flakes for the ultimate purpose of acquiring food would be included here. If there was a decrease in masticatory abilities with the emergence of *H. erectus*, greater pre-oral preparation of food may have been necessary. The large, strong, but narrow body of *H. erectus* implies that this species had larger home and day ranges (Clutton-Brock & Harvey, 1977), lower population densities (Foley, 1984), and greater ability to create large stone flakes than its antecedents (also see McHenry, 1994).

The second column of behavioural implications in Figure 8 moves us from the realm of probability to that of plausibility. The behaviours listed with question marks are possible responses to the behavioural implications we extrapolated from the fossil record, but have not yet been empirically demonstrated. For example, one possible way for *H. erectus* to increase parental investment would be to adopt a home base type of foraging strategy, with provisioning and food-sharing as described by Isaac (1978). Although the archaeological evidence is consistent with this model, it cannot yet demonstrate without doubt that central-place foraging was indeed the dominant foraging strategy used by early *H. erectus*.

The archaeological record, however, does indicate that *H. erectus* was engaging in more complex foraging behaviours than earlier hominids. Archaeological traces in the Okote Member are found in most habitats, including drier, more open areas, and adjacent to marginal, seasonally-flowing streams. This contrasts sharply with the archaeological record in earlier times where stone tool use was restricted to closed habitats next to perennial water sources (Harris & Capaldo, 1993). The great variation in artefact densities at 1·6 Ma suggests that the process of stone tool manufacture and discard was more complex than it had been. Certain places on the landscape were used repeatedly, perhaps seasonally, illustrated by stratified multiple archaeological horizons at "site complexes" such as FxJj 18 and 20 (Harris, 1978). These favoured locations may have been in ecotones between riparian forest and open grassland, places from which hominids could have exploited a great variety of animal and plant food resources (Cachel & Harris, 1993).

The archaeological record strongly supports the second group of behavioural implications in Figure 8, particularly the suggestion that this hominid was ranging over much greater areas. At Ileret, about 15–20 kilometers to the northwest of the Karari Escarpment where most Okote Member localities are found, smaller, low-density artefact and fossil bone occurrences indicate some degree of hominid curation and the minimum distance over which hominids ranged (Harris, 1978; Cachel & Harris, 1993). The higher density and overall abundance of stone artefacts at 1·6 Ma suggests that by this time *H. erectus* was frequently using stone, probably to gain access to food resources it would not otherwise have been able to exploit.

Both diet quality and diet breadth would have increased as both meat (and marrow?) were more habitually part of a *H. erectus* meal. Further bone modification studies looking at changes in mark frequency and character through time may be able to test this idea further.

The beginnings of lithic standardization by the reduction of single platform cores (Karari "core scrapers") and the use of large flakes for cores (Karari "flake scrapers") both demonstrate that stone was extremely important to the lifeway of *H. erectus* (Ludwig, 1992); they also provide hard evidence for the motor sequencing skill perceived in the Nariokotome endocast (Begun & Walker, 1993). Moreover, the ability of early hominids to efficiently reduce cobblestones into the form of single platform cores and produce numerous flakes through this process provided an important advantage in terms of lithic conservation. When taken together with long-distance stone transport (curation), this indicates more energy-efficient use of the overall ancient landscape, particularly in places where stone raw material was scarce, e.g., in the Ileret region—see above (Ludwig & Harris, 1994).

On the whole, the evidence discussed above suggests that *H. erectus* was less constrained than earlier hominids by the natural distribution of resources across the Turkana Basin. The dangers of venturing out more often into open habitats would have been mitigated by *H. erectus*'s larger body size, which would have given this hominid a better chance in competitive interactions with other carnivores (Walker, 1993). Increased competition with various mammalian carnivores over ungulate carcasses may have selected for more complex sociality, larger group size, and possibly the development of cooperative foraging and food-sharing (Cachel & Harris, 1993).

Summary

Changing patterns of land use by stone-tool-using hominids in the Turkana Basin are discernible through an examination of the paleogeographic context of archaeological localities in the 2·3–1·5 Ma time interval. The earliest archaeological traces have small numbers of artefacts and are restricted to places where the axial river system is intersected by alluvial fans of marginal drainages, places where stone tool raw material, water, shade, shelter, and food are all likely to be present. By 1·9–1·8 Ma, there is a hint that hominids are beginning to use other places for lithic-related activities, but not until 1·6 Ma is there solid evidence for abundant hominid use of many different paleogeographic settings, depositional environments, and microhabitats. The emergence of wide-ranging *Homo erectus*, with its large and narrow body, is probably responsible for this pattern. Both anatomical characteristics of this hominid and aspects of the archaeological record suggest that *H. erectus* was able to secure places on the ground, expand its diet breadth, increase its diet quality, range over larger areas, forage in a more complex manner, and create standardized, efficient lithic forms.

Acknowledgements

We would like to thank Dr Richard Leakey and the Kenya Wildlife Service, Drs Mohammed Isahakia, Meave Leakey, and Hélène Roche and the National Museums of Kenya, and Dr Harry Merrick and the Koobi Fora Field School for their continued support. Funding for the authors' research in the Turkana Basin has been generously provided by the National Science Foundation (MJR, CSF, & JWKH), L.S.B. Leakey Foundation (MJR & JWKH), Holt Family Foundation (JWKH), Rutgers University Faculty of Arts & Science Dean's Fund (MJR & JWKH), the Boise Fund (MJR), and North Arm Resources (MJR). Thanks go to Drs Alison

Brooks and John Shea for their reviews on an earlier draft of the manuscript. Helpful comments were also provided by Dr Kathy Stewart and Jennifer Bishop. Finally, we would like to thank the organizers of the Symposium on Early Hominid Behavioural Ecology, Jim Oliver, Nancy Sikes, and Kathy Stewart, firstly for their foresight in providing a venue for the discussion of pioneering ideas in behavioural ecology and secondly for their kindness in inviting us to submit a paper even though we did not participate in the original symposium.

References

Asfaw, B., Beyene, Y., Semaw, S., Suwa, G., White, T. & WoldeGabriel, G. (1991). Fejej: a new paleoanthropological research area in Ethiopia. *J. hum. Evol.* **21,** 137–143.

Asfaw, B., Beyene, Y., Semaw, S., Suwa, G., White, T. & WoldeGabriel, G. (1993). Tephra from Fejej, Ethiopia—A reply. *J. hum. Evol.* **25,** 519–521.

Begun, D. & Walker, A. (1993). The Endocast. In (A. C. Walker & R. E. Leakey, Eds) *The Nariokotome* Homo erectus *Skeleton*, pp. 326–358. Cambridge, MA: Harvard University Press.

Behrensmeyer, A. K. (1985). Taphonomy and the paleoecologic reconstruction of hominid habitats in the Koobi Fora Formation. In (Y. Coppens, Ed.) *L'Environnement des Hominidés au Plio-Pléistocène*, pp. 309–324. Paris: Masson.

Bellomo, R. V. (1990). Methods for documenting unequivocal evidence of humanly controlled fire at an early Pleistocene archaeological site in East Africa: the role of actualistic studies. Ph.D. Dissertation, University of Wisconsin, Milwaukee.

Bellomo, R. V. (1994). Methods of determining early hominid behavioural activities associated with the controlled use of fire at FxJj 20 Main, Koobi Fora, Kenya. *J. hum. Evol.* **27,** 173–195.

Blumenschine, R., Cavallo, J. & Capaldo, S. (1994). Competition for carcasses and early hominid behavioral ecology: a case study and conceptual framework. *J. hum. Evol.* **27,** 197–213.

Brown, B. & Walker, A. (1993). The Dentition. In (A. C. Walker & R. E. Leakey, Eds) *The Nariokotome* Homo erectus *Skeleton*, pp. 161–192. Cambridge, MA: Harvard University Press.

Brown, F. H. & Feibel, C. S. (1986). Revision of lithostratigraphic nomenclature in the Koobi Fora region, Kenya. *J. Geol. Soc., Lond.* **143,** 297–310.

Brown, F. H. & Feibel, C. S. (1988). "Robust" hominids and Plio-Pleistocene paleogeography of the Turkana Basin, Kenya and Ethiopia. In (F. E. Grine, Ed.) *Evolutionary History of the "Robust" Australopithecines*, pp. 325–341. New York: Aldine de Gruyter.

Brown, F. H. & Feibel, C. S. (1991). Stratigraphy, depositional environments and paleogeography of the Koobi Fora Formation. In (J. M. Harris, Ed.) *Koobi Fora Research Project, volume 3*, pp. 1–30. Oxford: Clarendon Press.

Brown, F. H., Harris, J. M., Leakey, R. E. F. & Walker, A. (1985). Early *Homo erectus* skeleton from west Lake Turkana, Kenya. *Nature* **316,** 788–792.

Bunn, H. (1981). Archaeological evidence for meat-eating by Plio-Pleistocene hominids from Koobi Fora and Olduvai Gorge. *Nature* **291,** 574–577.

Bunn, H. (1982). Meat-Eating and Human Evolution: Studies on the Diet and Subsistence Patterns of Plio-Pleistocene Hominids of East Africa. Ph.D. Dissertation, University of California, Berkeley.

Bunn, H., Harris, J. W. K., Isaac, G., Kaufulu, Z., Kroll, E., Schick, K., Toth, N. & Behrensmeyer, A. K. (1980). FxJj50: an early Pleistocene site in northern Kenya. *World Archaeol.* **12,** 109–136.

Cachel, S. & Harris, J. W. K. (1993). Ranging Patterns, Land-Use, and Subsistence in *Homo erectus* from the Perspective of Evolutionary Ecology. Paper presented at the International Pithecanthropus Centennial Conference, Leiden, The Netherlands.

Cerling, T. (1992). Development of grasslands and savannas in East Africa during the Neogene. *Palaeogeogr., Palaeoclimat., Palaeoecol.* **97,** 241–247.

Cerling, T., Bowman, J. & O'Neil, J. (1988). An isotopic study of a fluvial-lacustrine sequence: The Plio-Pleistocene Koobi Fora sequence, East Africa. *Palaeogeogr., Palaeoclimat., Palaeoecol.* **63,** 335–356.

Chavaillon, J. (1976). Evidence for the technical practices of Early Pleistocene hominids, Shungura Formation, Lower Omo Valley, Ethiopia. In (Y. Coppens, F. C. Howell, G. L. Isaac & R. E. Leakey, Eds) *Earliest Man and Environments in the Lake Rudolph Basin*, pp. 565–573. Chicago: University of Chicago Press.

Clutton-Brock, T. H. & Harvey, P. H. (1977). Primate ecology and social organization. *J. Zool.* **183,** 1–39.

Feibel, C. S. (1988). Paleoenvironments from the Koobi Fora Formation, Turkana Basin, northern Kenya. Ph.D. Dissertation, University of Utah.

Feibel, C. S. (1993). Geological context and the ecology of *Homo erectus* in East Africa. Paper presented at the International *Pithecanthropus* Centennial Conference, Leiden, The Netherlands.

Feibel, C. S. (1994). Freshwater stingrays from the Plio-Pleistocene of the Turkana Basin, Kenya and Ethiopia. *Lethaia*.

Feibel, C. S. & Brown, F. H. (1993). Microstratigraphy and paleoenvironments. In (A. C. Walker & R. E. Leakey, Eds) *The Nariokotome* Homo erectus *Skeleton*, pp. 21–39. Cambridge, MA: Harvard University Press.

Feibel, C. S., Brown, F. H. & McDougall, I. (1989). Stratigraphic context of fossil hominids from the Omo group deposits, northern Turkana Basin, Kenya and Ethiopia. *Am. J. phys. Anthrop.* **78,** 595–622.

Feibel, C. S., Harris, J. M. & Brown, F. H. (1991). Neogene paleoenvironments of the Turkana Basin. In (J. M. Harris, Ed.) *Koobi Fora Research Project, volume 3*, pp. 321–370. Oxford: Clarendon Press.

Foley, R. (1984). Early man and the Red Queen: tropical African community evolution and hominid adaptation. In (R. Foley, Ed.) *Hominid Evolution and Community Ecology: Prehistoric Human Adaptation in Biological Perspective*, pp. 85–110. New York: Academic Press.

Foley, R. (1987). *Another Unique Species: Patterns in Human Evolutionary Ecology*. New York: Longman and John Wiley & Sons.

Haileab, B. & Feibel, C. S. (1993). Tephra from Fejej, Ethiopia. *J. hum. Evol.* **25,** 515–517.

Harris, J. M., Brown, F. H. & Leakey, M. G. (1988). Geology and paleontology of Plio-Pleistocene localities west of Lake Turkana, Kenya. *Contrib. Sci.* **399,** 1–128.

Harris, J. W. K. (1978). The Karari Industry: Its Place in African Prehistory. Ph.D. Dissertation, University of California, Berkeley.

Harris, J. W. K. & Capaldo, S. D. (1993). The earliest stone tools: their implications for an understanding of the activities and behaviour of late Pliocene hominids. In (A. Berthelet & J. Chavaillon, Eds) *The Use of Tools by Human and Non-Human Primates*, pp. 196–220. Oxford: Clarendon Press.

de Heinzelin, J. (1983). *The Omo Group*. Musée Royal de l'Afrique Centrale, Tervuren, Belgique. Annales, Série in 8°, Sciences Géologiques, No. 85.

Holloway, R. L. (1982). *Homo erectus* brain endocasts: Volumetric and morphological observations, with some comments on cerebral asymmetries. *Congres Internationale de Paleontologie Humaine, 1er Congres*, 355–369.

Howell, F. C. (1976). Overview of the Pliocene and Earlier Pleistocene of the Lower Omo Basin, Southern Ethiopia. In (G. L. Isaac & E. R. McCown, Eds) *Human Origins*, pp. 227–268. Menlo Park: W. A. Benjamin.

Howell, F. C., Haeserts, P. & de Heinzelin, J. (1987). Depositional environments, archeological occurrences and hominids from Members E and F of the Shungura Formation (Omo basin, Ethiopia). *J. hum. Evol.* **16,** 665–700.

Isaac, G. L. (1978). Food sharing and human evolution: Archaeological evidence from the Plio-Pleistocene of East Africa. *J. Anthrop. Res.* **34,** 311–325.

Isaac, G. L. & Harris, J. W. K. (1978). Archaeology. In (M. G. Leakey & R. E. F. Leakey, Eds) *Koobi Fora Research Project, Volume 1*, pp. 64–85. Oxford: Clarendon Press.

Isaac, G. L., Harris, J. W. K. & Marshall, F. (1981). Small is informative: the application of the study of mini-sites and least effort criteria in the interpretation of the Early Pleistocene archaeological record at Koobi Fora, Kenya. In (J. D. Clark & G. Isaac, Eds) *Las Industrias mas Antiquas*, pp. 101–119. Mexico City: X Congresso, Union Internacional de Ciencias Prehistoricas y Protohistoricas.

Kaufulu, Z. (1983). Formation and Preservation of some Earlier Stone Age Sites at Koobi Fora, Northern Kenya. Ph.D. Dissertation, University of California, Berkeley.

Kaufulu, Z. (1987). Formation and preservation of some earlier Stone Age sites at Koobi Fora, Northern Kenya. *S. Afr. Archaeol. Bull.* **42,** 23–33.

Kibunjia, M. (1990). Pliocene stone tool technology west of Lake Turkana, Kenya. *Crosscurrents* **4,** 16–26.

Kibunjia, M., Roche, H., Brown, F. & Leakey, R. E. F. (1992). Pliocene and Pleistocene archaeological sites west of Lake Turkana, Kenya. *J. hum. Evol.* **23,** 431–438.

Leakey, R. E. F. (1974). Further evidence of Lower Pleistocene hominids from East Rudolf, North Kenya, 1973. *Nature* **248,** 653–656.

Leakey, R. E. F. & Walker, A. (1985). Further hominids from the Plio-Pleistocene of Koobi Fora, Kenya. *Am. J. phys. Anthrop.* **67,** 135–163.

Leakey, R. E. F. & Walker, A. (Eds) (1993). *The Nariokotome* Homo erectus *Skeleton*. Cambridge, MA: Harvard University Press.

Ludwig, B. V. (1992). An experimental approach to developing methodologies for testing the meaning of technological variation in Early Pleistocene archaeological assemblages from East Africa. Unpublished M.A. Thesis, Rutgers University.

Ludwig, B. V. & Harris, J. W. K. (1994). Hominid skill and hand preference: Their effects on Plio-Pleistocene lithic assemblage variability. Paper to be presented at the 3rd World Archaeological Congress, New Delhi.

Martin, R. D. (1983). Human Brain Evolution in an Ecological Context. 52nd James Arthur Lecture on the Evolution of the Human Brain. American Museum of Natural History, New York.

McDougall, I. (1985). K-Ar and ^{40}Ar/^{39}Ar dating of the hominid-bearing Plio-Pleistocene sequence at Koobi Fora, Lake Turkana, northern Kenya. *Geol. Soc. Am. Bull.* **96,** 159–175.

McHenry, H. M. (1988). New estimates of body weight in early hominids and their significance to encephalization and megadontia in "Robust" Australopithecines. In (F. E. Grine, Ed.) *Evolutionary History of the "Robust" Australopithecines*, pp. 133–148. New York: Aldine de Gruyter.

McHenry, H. M. (1991). Femoral lengths and stature in Plio-Pleistocene hominids. *Am. J. phys. Anthrop.* **85,** 149–158.

McHenry, H. M. (1994). Behavioral ecological implications of early hominid body size. *J. hum. Evol.* **27**, 77–87.

Merrick, H. V. (1976). Recent archaeological research in the Plio-Pleistocene deposits of the Lower Omo Valley, southwestern Ethiopia. In (G. L. Isaac & E. R. McCown, Eds) *Human Origins*, pp. 461–481. Menlo Park: W. A. Benjamin.

Rightmire, G. P. (1990). *The Evolution of* Homo erectus: *Comparative Anatomical Studies of an Extinct Hominid Species.* New York: Cambridge University Press.

Roche, H. (1989). Technological evolution in the early hominids. *OSSA* **14**, 97–98.

Roche, H. & Kibunjia, M. (1994). Les sites archéologiques Plio-Pléistocènes de la Formation de Nachukui, West Turkana, Kenya. *C. R. Acad. Sci. Paris* **318**.

Rogers, M. J. (in prep.). A Landscape Archaeological Study at East Turkana, Kenya. Ph.D. Dissertation, Rutgers University.

Rogers, M. J. (in press). Landscape Archaeology at East Turkana, Kenya. Proceedings of the International Congress in Honour of Dr Mary Douglass Leakey's Outstanding Contribution in Palaeoanthropology, Arusha, Tanzania, August 8–14, 1993.

Rogers, M. J. & Harris, J. W. K. (1992). Recent Investigations into Off-Site Archaeology at East Turkana, Kenya. *Nyame Akuma* **38**, 41–47.

Rogers, M. J. & Sorkowitz, E. (1992). Paleoenvironments and hominid subsistence: new approaches integrating faunal and lithic archaeological remains in the Early Stone Age of East Africa. *Crosscurrents* **5**, 39–58.

Ruff, C. B. (1991). Climate and body shape in hominid evolution. *J. hum. Evol.* **21**, 81–105.

Ruff, C. B. (1993). Climatic adaptation and hominid evolution: the thermoregulatory imperative. *Evol. Anthrop.* **2**, 53–60.

Ruff, C. B. & Walker, A. (1993). Body size and body shape. In (A. C. Walker & R. E. Leakey, Eds) *The Nariokotome* Homo erectus *Skeleton*, pp. 234–265. Cambridge, MA: Harvard University Press.

Shipman, P. & Harris, J. M. (1988). Habitat preference and paleoecology of *Australopithecus boisei* in Eastern Africa. In (F. Grine, Ed.) *Evolutionary History of the "Robust" Australopithecines*, pp. 343–382. New York: Aldine de Gruyter.

Stern, N. (1991). The Scatter-Between-the-Patches: A Study of Early Hominid Land Use Patterns in the Turkana Basin, Kenya. Ph.D. Dissertation, Harvard University.

Toth, N. (1982). The Stone Technologies of Early Hominids at Koobi Fora, Kenya: An Experimental Approach. Ph.D. Dissertation, University of California, Berkeley.

Toth, N. (1985). The Oldowan reassessed: a close look at early stone artifacts. *J. Archaeol. Sci.* **12**, 101–120.

Walker, A. (1993). The origin of the genus *Homo*. In (D. T. Rasmussen, Ed.) *The Origin and Evolution of Humans and Humanness*, pp. 29–47. Boston: Jones and Bartlett.

Walker, A. & Ruff, C. B. (1993). The reconstruction of the pelvis. In (A. C. Walker & R. E. Leakey, Eds) *The Nariokotome* Homo erectus *Skeleton*, pp. 221–233. Cambridge, MA: Harvard University Press.

Wheeler, P. E. (1991). The thermoregulatory advantages of hominid bipedalism in open equatorial environments: The contribution of increased convective heat loss and cutaneous evaporative cooling. *J. hum. Evol.* **21**, 107–115.

Wheeler, P. E. (1992). The thermoregulatory advantages of large body size for hominids foraging in savannah environments. *J. hum. Evol.* **23**, 351–362.

Wheeler, P. E. (1993). The influence of stature and body form on hominid energy and water budgets: A comparison of *Australopithecus* and early *Homo* physiques. *J. hum. Evol.* **24**, 13–28.

White, T. (1988). The comparative biology of "robust" *Australopithecus*: Clues from context. In (F. Grine, Ed.) *Evolutionary History of the "Robust" Australopithecines*, pp. 449–483. New York: Aldine de Gruyter.

Winterhalder, B. (1980). Hominid paleoecology: The competitive exclusion principle and determinants of niche relationships. *Yrbk phys. Anthrop.* **23**, 43–63.

Winterhalder, B. (1981). Hominid paleoecology and competitive exclusion: Limits to similarity, niche differentiation, and the effects of cultural behavior. *Yrbk phys. Anthrop.* **24**, 101–121.

Wood, B. A. (1991). *Koobi Fora Research Project, Volume 4: Hominid Cranial Remains.* Oxford: Clarendon Press.

Wood, B. A. (1992). Origin and evolution of the genus *Homo*. *Nature* **355**, 783–790.

Mzalendo Kibunjia
Division of Archaeology,
National Museums of Kenya,
PO Box 40658, Nairobi, Kenya
and
Department of Anthropology,
Rutgers University, New Brunswick,
New Jersey, 08903 U.S.A.

Received 8 April 1993
Revision received 4 March 1994
and accepted 4 March 1994

Keywords: Lake Turkana basin,
Pliocene, Archaeology,
Technology, Nachukui Formation,
Lokalalei site.

Pliocene archaeological occurrences in the Lake Turkana basin

Archaeological evidence from the Lake Turkana basin, as well as from several other localities in eastern and central Africa, shows that stone tool manufacture and use occurred at least by the later part of the Pliocene, about 2·4 million years (Ma). However, little is known from the archaeological record about the technological characteristics of Pliocene material culture, and related aspects of hominid behavior, such as habitat use and preference and subsistence of the tool makers. Expanded excavations at the West Turkana site of Lokalalei, dated to about 2·35 Ma indicate that hominids making artifacts at this site had little success in striking off whole flakes from the parent core forms. The large number of scars left on these forms consists of step, hinge, and small flakes (>2 cm) which may not have been very successful in cutting or slicing, despite the fact that the raw material utilized was a medium-grained volcanic lava with observable conchoidal fracture mechanics. The fauna is characterized by mainly size 1 or 2 bovids, suggesting early access if scavenged or hunted. This lithic technology patterning, which is shared by other assemblages in the basin, suggests that the Oldowan is not the earliest stone tool technology industry. Rather, the Oldowan represents a point in a continuum from simpler Pliocene technology characterized by little understanding of stone fracture mechanics to greater technological complexity and appreciation of fracture mechanics in the Pleistocene.

Journal of Human Evolution (1994) **27,** 159–171

Introduction

Research undertaken over the last three decades in the sedimentary basins of Olduvai Gorge (Leakey, 1971) and Koobi Fora to the east of Lake Turkana (Isaac, 1984) have laid the foundation of what is known about the archaeology of human origins during the Early Pleistocene. Little, however, is known from the archaeological record about the technological characteristics of hominid material and other related aspects of behavior, such as habitat use and subsistence prior to 2·0 million years (Ma). This gap of knowledge exists because of the limited outcrops dated to this time period. Such deposits have been investigated in Ethiopia at the Omo (Chavaillon, 1976; Merrick, 1976), Hadar (Roche & Tiercelin, 1980; Harris, 1983; Harris & Semaw, 1989), and in eastern Zaire at Senga (Harris *et al.*, 1987; Harris *et al.*, 1990) (Table 1). The ongoing archaeological studies to the west of Lake Turkana (Kibunjia *et al.*, 1992) have complemented the extensive archaeological research done to the east at Koobi Fora (Isaac & Harris, 1978), and to the north at the Omo (de Heinzelin, 1983; Howell *et al.*, 1987). These studies have established that during the Late Pliocene, around 2·4 Ma, contemporaneous groups of stone-tool-making hominids occupied the Turkana Basin to the north and to the west of the lake.

This paper summarizes the artifact characteristics of the West Turkana site of Lokalalei (GaJh 5) and discusses geochronological and archaeological features of known archaeological sites of the Turkana Basin dated to the late Pliocene, particularly those that are post-KBS age (1·88 ± 0·02 Ma, McDougall, 1985) which is an important geochronological marker in the Turkana Basin (Figure 1).

0047–2484/94/010159+13 $08.00/0

Table 1   **Archaeological sites dated at least 2·0 million years old in Eastern and Central Africa**

Country	Region	Sites	Geological member
Ethiopia	Hadar[1]	Kada Gona 2-3-4	Kada Hadar
		West Gona 1 Locality 1	Kada Hadar
		West Gona 1 Locality 2	Kada Hadar
		West Gona 2	Kada Hadar
		West Gona 3	Kada Hadar
	Omo[2]	FtJi 1 Locality 204	Member F
		FtJi 2 Locality 396	Member F
		FtJi 5 Adj. Loc. 4	Member F
		Omo 123	Member F
		Omo 57	Member F
Kenya	West Turkana[3]	Lokalalei	Kalochoro
Zaire	Upper Semliki[4]	Senga 5a	Lusso Beds
		Kanyatsi 2	Lusso Beds

[1]Roche & Tiercelin (1980), Harris (1983), Semaw (1990).
[2]Merrick (1976), Merrick & Merrick (1986), Chavaillon (1976), Howell *et al.* (1987).
[3]Kibunjia *et al.* (1992).
[4]Harris *et al.* (1992), Harris *et al.* (1990).

The Lake Turkana basin

The Kenya (Gregory) Rift Valley System in which the Lake Turkana basin exists was established by volcanism and by rifting movements that began during the Late Miocene about 23 Ma and continued intermittently thereafter (Baker *et al.*, 1976). This long history of volcanism, rifting, erosion and deposition of volcanic sediments accumulated relatively large Miocene sedimentary deposits as well as extensive Plio-Pleistocene and Holocene deposits in lake basins within the Rift Valley. Of all the earth movements associated with the formation of the Rift Valley, the volcanic eruptions, particularly the tephra layers (ash), have proven to be of great importance in that they permit precise time calibration of hominid evolution in East Africa. They also allow reconstruction of the geological and sedimentation history of such basins. For example, detailed geological mapping and chemical finger-printing of aerially exposed tuffs at the Shungura in the north, Koobi Fora to the east and West Turkana to the west of the lake have established excellent stratigraphic and paleogeographic records within the basin that have permitted intra- and extra-basinal geological, paleontological and archaeological comparisons (Brown & Feibel, 1988; Feibel *et al.*, 1989; Haileab & Brown, 1991). It is now clear that the geological and paleogeographical history of the Turkana Basin is an integrated whole (Feibel *et al.*, 1991).

Paleogeography

The earliest Plio-Pleistocene sedimentation in the Turkana Basin apparently began about 4 million years ago (McDougall, 1985; Brown & Feibel, 1988, Feibel *et al.*, 1991). Paleo-geographical reconstruction of the Turkana basin indicated that during the last 4 million years it was dominated by two hydrographic systems: a series of lakes each at different time periods, and a river system interpreted as the Paleo-Omo which dominated and flowed through the basin from the north and exited east into the Indian Ocean (Brown & Feibel, 1988; Feibel *et al.*, 1991). After 2·0 Ma the basin was dominated consecutively by different lakes. Even then

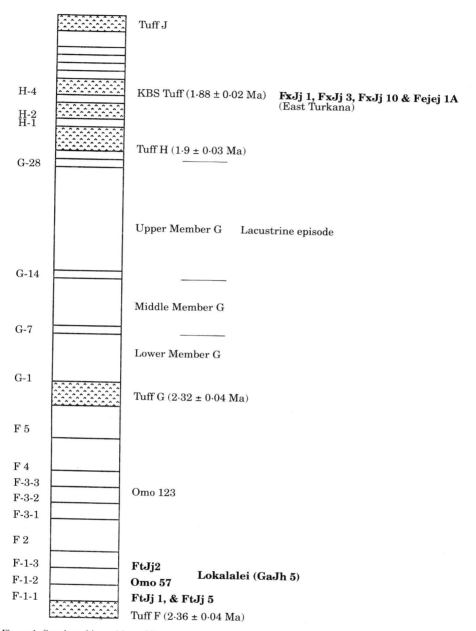

Figure 1. Stratigraphic position of Pliocene archaeological sites (in bold) in the Lake Turkana basin (adapted from Feibel *et al.*, 1989).

the river continued to flow into the lake system and exited the basin to the east until the damming of the exit by the uplifting of Mt Kulal to the southeast. The appearance, expansion, shrinking and drying up of these lakes depended on the water supply and tectonic movements within and outside the basin. Brown & Feibel (1988) argue that these two systems, the

Paleo-Omo and the several lakes, which were controlled by local tectonic movements, probably determined the composition of the plant and animal ecology in the basin as well as the nature of sedimentation. Thus, in contrast to the deteriorating environment elsewhere in the world during the worldwide climatic change that began around 2·5–2·4 Ma (Shackleton & Opdyke, 1977; Vrba, 1988; Vrba et al., 1989), the Lake Turkana basin like other basins in the Rift Valley may have acted as a refugium of plants and animal species.

The West Turkana Region

For over twenty years, paleoanthropological research in the Lake Turkana basin of northern Kenya has yielded abundant data about the emergence of early humans in Africa. Most of the research was targeted to the east of the lake at Koobi Fora, Kenya, and to the north, at Shungura in Ethiopia. Very little was known about the west side of the lake until 1984 when a paleoanthropological research team from the National Museums of Kenya began geological mapping and paleontological collection of fauna (Harris et al., 1988a, 1988b). "West Turkana" lies on the western shores of Lake Turkana north of the point where the Turkwell River turns east from its northwesterly flow to drain into Lake Turkana (Figure 2). The basin margin to the northwest is bounded by the Labur and Murua Rith ranges. The Plio-Pleistocene sedimentary deposits of West Turkana are aerially exposed in an area of about 700 km^2 and reach a thickness of about 730 m. The sequence has been named the Nachukui Formation and is divided into eight members. In order of antiquity they are: Lonyumun, Kataboi, Lomekwi, Lokalalei, Kalochoro, Kaitio, Natoo and Nariokotome (Feibel et al., 1989). Archaeological traces in West Turkana are currently known from the last four members only.

The Lokalalei Site (GaJh 5)

Geology and dating

The Pliocene archaeological assemblage in West Turkana is known from the site of Lokalalei (GaJh 5) which was test-excavated in 1987 (Kibunjia et al., 1992) and extensively excavated in 1991 (Kibunjia, in prep.). The Lokalalei site is situated in an erosional amphitheater along the northern branch of the Lokalalei dry channel (laga). In the Nachukui Formation the site is stratified into the base of the Kalochoro Member. The Kalochoro Member is defined as the geological section that lies between the Kalochoro Tuff (equivalent with Tuff F in the Shungura Formation) and the KBS Tuff [equivalent with Tuff H2 (Harris et al., 1988a,b)]. At the Shungura, the Kalochoro Member is correlative of Member F (ages 2·36 to 2·32 Ma), Member G (ages 2·32 to 1·9 Ma) and lowermost member H [(H-H2) ages 1·9 to 1·88 Ma]. The site is 9 m above the Lokalalei Tuff (equivalent with Tuff F-1) and thus has an age slightly younger than 2·36 ± 0·04 Ma (Kibunjia et al., 1992; Feibel et al., 1989).

At the Koobi Fora Formation to the east, the Kalochoro Member would be contemporaneous with the upper Burgi Member. However, Kalochoro is divisible into lower, middle and upper sections, based on the dominant lithologic units (Harris et al., 1988a,b). The lower part was laid down by a large meandering river, the Proto Omo, while the middle and the upper sections are related to the lacustrine episode that began around 2·0 Ma (Feibel et al., 1989; Feibel et al., 1991). In the Koobi Fora Formation, the upper Burgi Member was deposited during this lacustrine episode, while the lower Burgi Member is older, except for a small section in Area 207. The lower section of the Kalochoro Member, into which the Lokalalei site

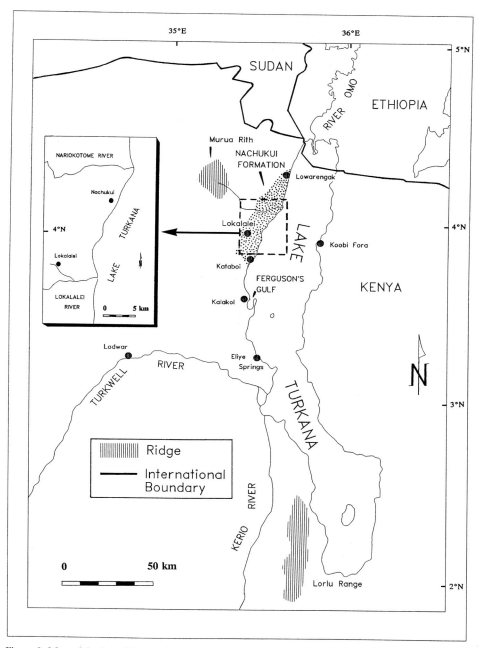

Figure 2. Map of the Lake Turkana basin showing the location of the Nachukui Formation and Lokalalei Site (adapted from M. Siegel 1990, Rutgers Cartography).

(GaJh 5) is stratified, is represented largely by a hiatus between the upper and lower Burgi Members (Feibel, 1988:67–68). This is further supported by the observation that the fauna

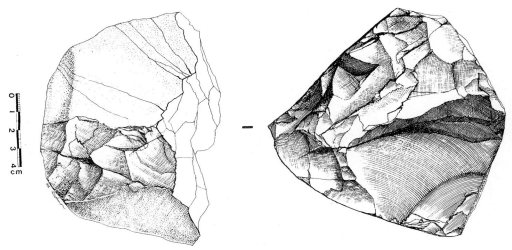

Figure 3. A flaked core from Lokalalei (GaJh 5) Site, West Turkana.

Table 2   Composition of lithic assemblages from Lokalalei site (GaJh5)

Excavated surface	1987 excavated 14 m²		1991 excavated 53 m²		1991 surface	
	n	%	*n*	%	*n*	%
Cores	7	24·1	43	11·1	3	6·1
Whole flakes	14	48·3	59	15·2	6	12·3
Broken flakes and fragments	4	13·8	228	58·8	40	81·6
Fragments (<1 cm)	4	13·8	46	11·9	0	0
Pounded pieces	0	0	12	3·0	0	0
Total	29	100	388	100	49	100

from Kalochoro Member appears to predate the upper portion of Burgi Member (Harris *et al.*, 1988*b*).

Archaeological finds

The 1991 excavation program was intended to retrieve a large sample of artifacts and fauna for information on the technology and paleoecology of the early hominids at this site. The uncovering of a large area has indicated that artifacts at Lokalalei (GaJh 5) are enclosed in silt claystones deposited on top of a gravelly layer. The contact between the gravelly layer and the silty claystones is sharp and indeed no archaeological material was found beyond the contact. All artifacts are fresh with no evidence of abrasion. The cores are relatively large (mean 99·9 mm ± 31·9) and heavy (Figure 3). Their spatial association with small chips (<1 mm) of rock fragments, combined with the geological context indicate that the assemblage was minimally disturbed by fluvial action. Large scale excavations at this site clearly indicate that Lokalalei is a low density site (6·2 artifacts per m², Table 2) in comparison to archaeological sites dated to the Early Pleistocene (Schick, 1987:791, Table 2). The flaked material at the site consists of a variety of medium-grained lava with observable conchoidal fracture, and the source was local. Unlike the rounded cobbles utilized to make Oldowan-type cores at Olduvai Gorge and Koobi Fora, which experimentation has shown require considerable skill to strike the first flake (Toth, 1982, 1985; Leakey, 1971), the cores at this site are of prismatic shape. This means they would have been easier to flake because of the presence of several opportune striking platforms on the blanks. Yet they are not intensively flaked. This has made it possible to determine the shape of the initial cobbles. Flake scar counts of >10 mm on the cores range between 1 and 12. About 80% of the flaking scars on these cores are characterized by step fractures and only a few instances of complete flake removals were observed (Figure 4). Cores appear to have been abandoned after several attempts of flaking if most of the products obtained were the step/hinge flakes. The few number of whole flakes ($N=59$, mean $49·9$ mm $\pm$ $15·2$) and the small ratio of broken flakes and fragments to that of the cores (mean of $6·3$ fragments per core) is a clear indication of this phenomenon. Some of the cores, however, are well flaked but the pattern was not repeated in many other cores. The well flaked cores serve to show that factors other than raw material account for the poor technology.

The 1987 test-excavations recovered 415 faunal specimens, while that of 1991 recovered >3000 faunal specimens. Only the 1987 faunal collection has been studied for taphonomic history. Of the identified bones, $22·7\%$ consist of isolated teeth and tooth fragments, 31% are axial, mainly rib fragments, and $46·3\%$ are appendicular, mainly long bone shaft fragments. Bone surface identification marks were difficult to observe in part due to the obscuring matrix on many of the specimens. In the entire assemblage, only two specimens, both recovered *in situ*, have what may be stone tool cut-marks. These are a thoracic vertebra of a size 2 bovid and a near-epiphyseal long bone shaft fragment from a size 2 mammal. The possible cut-mark on the latter specimen is located on the cortical surface, away from the epiphyseal end. In addition, a long bone shaft fragment of a size 1 or 2 mammal has a possible percussion groove and a carnivore tooth pit. The bone fragment itself resembles an "anvil wedge", a characteristic product of hammerstone breakage of long bones (Blumenschine & Selvaggio, 1988). Direct carnivore activity is indicated by carnivore tooth marks on two specimens found *in situ*, and on six surface specimens. While the 1987 Lokalalei faunal assemblage does not provide definitive evidence for hominid utilization, the completed studies on the large 1991 sample should provide clearer evidence to the nature of hominid activities at the Lokalalei site.

Discussion

Archaeological survey and expanded excavations at the site of Lokalalei (GaJh 5), in West Turkana were intended to explore the questions of the threshold of technology, hominid use of ecological resources around the site and the nature of flaking technology during the Late Pliocene.

Threshold of flaked stone technology
When flaked stone emerged as a novel adaptive threshold among early hominids in the basin, and indeed in the archaeological record, is far from clear. Discussion in the archaeological literature on the beginning of flaked stone tool manufacture and use gives a date of between $2·5$ and $3·0$ Ma as the possible beginning of this significant human adaptation (Harris, 1983; Isaac, 1984; Toth & Schick, 1986). In the Turkana Basin the earliest verified and securely dated archaeological sites cluster around $2·4$ Ma (Howell *et al.*, 1987; Kibunjia *et al.*, 1992), although unsubstantiated surface claims have been made for the presence of artifacts at the Shungura Formation in Tuff C, dated to $2·85$ Ma (de Heinzelin, 1983:53). A huge gap exists between these occurrences and the next securely dated occurrences in the basin, around $1·88$ Ma. Archaeological Survey of deposits in the Nachukui Formation dated around $2·5$ Ma

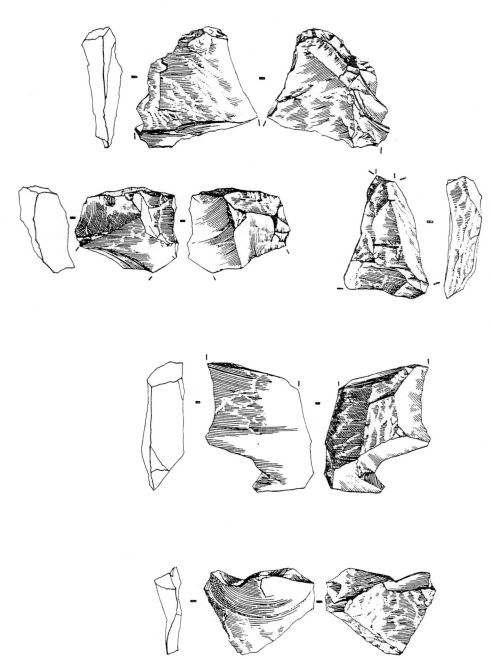

Figure 4. Whole flakes (detached pieces) from Lokalalei Site.

and 2·0 Ma have not yielded any artifacts or bone surface modification marks related to hominid activities. Five archaeological sites are known from Member F of the Shungura Formation. These are Omo 57, 123, FtJj 2, and FtJj 5 (Howell *et al.*, 1987). The Member F is dated to between 2·34 (Tuff F) and 2·32 (Tuff G) million years (Feibel *et al.*, 1989).

Two problematic archaeological assemblages, Omo 71 and Omo 84, have also been reported from Member E, which is dated to between 2·48 (start of Matuyama Reversed Chron.) and 2·36 Ma. The Omo 71 assemblage is a surface collection consisting of one bifacially flaked side chopper made on quartz, and several surface and *in situ* vertebrate fossil bones. A 25 m^2 test-excavation at this site did not recover any *in situ* artifact assemblages, but several vertebrate bones were recovered (Chavaillon, 1970). As a result, Howell *et al.* conclude that "Although the artifactual nature of the specimen is scarcely in doubt, its stratigraphic provenience and hence its age must remain in question" (1987:675). Omo 84, on the other hand, consists of both surface and *in situ* artifact and faunal remains (Chavaillon, 1975). In Howell *et al.*, (1987:675), the site is described as being located in an area consisting of "relatively isolated outcrops, that is considerably disturbed by SW–NE trending faults". For that matter two of the geologists working in the area, J. de Heinzelin and J. Haesaerts argue that "without detailed geological study the exact stratigraphic situation of Omo 84 remains uncertain" (Howell *et al.*, 1987:675, footnote 1). The dating of the Lokalalei (GaJh 5) site to Member F of the Shungura Formation adds to the increasing inventory of archaeological sites dated to 2·3/2·4 million years.

Some researchers in the African continent have taken an environmental approach and argued that the extinction, speciation and culture of hominids was a probable consequence of climatic change (Tobias, 1986; Vrba, 1988, 1992). This hypothesis, however, is not widely accepted in paleoanthropology. Harris calls the suggestion "intriguing" (Harris *et al.*, 1990:265) but falls short of endorsing it fully by suggesting that:

"some populations of early hominids, *presumably in part* as a response to changing environmental conditions, were extending their resource base to include meat in their diet" (Harris 1983:24, emphasis added).

Other researchers, (Delson, 1985; Hill, 1987; White, 1985, 1988), although acknowledging its consistency in explaining evolutionary events of several mammalian species (bovids, suids, equids, cercopithecoids, rodents, etc.) at different time periods have, however, urged caution in applying it to hominid evolution, arguing that correlation is not causation. White, for example, predicts that:

" . . . As more environmental 'episodes' are established, of course, the more probable it becomes that one of them will correlate with some hominid evolutionary 'event', and the less likely it becomes that correlation is evidence of causation" (1988:465).

Hill (1987:588; 1992), citing sampling problems and influence of local geological factors, argues against coupling of global climatic changes and hominid evolutionary changes, and contends that we need to sample more continuously dated sites and/or geological formations. Clearly in spite of the long depositional record of the Turkana Basin (4 million years), the six earliest securely dated archaeological sites in the basin appear only around 2·3 and 2·4 Ma, that is, after the setting of global climatic cooling. Finding archaeological sites dated earlier than 2·4 Ma in the Lake Turkana basin still remains a challenge.

Hominid ecology

Clearly the Lokalalei (GaJh 5) faunal assemblage has not provided clear cut evidence of hominid utilization of animal resources during the late Pliocene. It is doubtful, however, if in such early assemblages we will be able to find hominid bone modification marks at the levels evident at Olduvai Gorge and Koobi Fora (Bunn, 1982). These earlier sites may be different in terms of the intensity of activities performed at these locations where artifacts are found associated with bones. Sites may have been used seasonally and briefly. The near absence or low incidence of hominid bone surface modification marks in the Lokalalei (GaJh 5) faunal assemblage is not unusual compared to other Pliocene archaeological assemblages in the Lake Turkana basin and outside the basin. For example, no vertebrate fossils were found associated with stone artifacts from Omo 123 (Chavaillon, 1976:572) and FtJj 2 (Merrick & Merrick, 1976:575) at the Shungura. At the FtJi 1 site, elephant, hippopotamus and bovid remains were found in what is considered a derived context (Merrick & Merrick, 1976:579), while FtJj 5 contained fresh artifacts with rolled vertebrae fossils (Howell *et al.*, 1987:680; Merrick, 1976).

Outside the basin, no vertebrate fossils were found associated with stone artifacts at the site of Kada Gona at Hadar, Ethiopia (Roche & Tiercelin, 1980); at the site of West Gona, the faunal remains are reported as fragmentary (Harris, 1983:15). A recent exception is the faunal assemblage recovered from Senga 5a in Eastern Zaire which is dated by faunal correlation to 2·0–2·32 Ma (Boaz *et al.*, 1992:511–512). About 230 faunal specimens were recovered from the surface, and 4400 faunal specimens were found *in situ*. With the exception of two cut-marks on a tortoise plastron, one rodent gnaw-mark and two carnivore tooth marks, the rest of the faunal specimens show no bone surface modification marks. Evidently the late Pliocene assemblages are different from Pleistocene faunal assemblages, where evidence of hominid bone surface modification marks such as cut-marks and percussion marks are present in large percentages (Bunn & Kroll, 1986; Potts, 1988).

Technological organization

Experimental knappers (Crabtree, 1972; Callahan, 1979; Johnson, 1979; Young & Bonnichsen, 1985; Cotterel & Kamminga, 1987) have shown that successful flaking, whether to shape a core or to obtain whole flake(s), is dependent on the following crucial variables. (a) The raw material has to be isotropic i.e. a rock with the same properties in all directions. (b) During production the blow must be directed at an appropriate angle for the particular place of impact on the surface of the rocks. High angles in general tend to create micro fractures or pitting on the platform of the cobbles being struck as a result of several unsuccessful attempts to crack the rock. (c) The striking blow must have sufficient force to enable impacted energy to carry through the body of rock. Low energy more often than not tends to produce small flakes, and step and hinge flakes. High energy creates shattering, snapping, etc. (d) Adequate mind, eye and muscular coordination must exist to maneuver and manipulate the core because the act of flaking is a constant routine of cybernation. In prehistoric assemblages only the first of these variables is directly observable. Other variables can only be inferred. Surface observation of the dominant variety of lava utilized at Lokalalei and careful study of the flake termination scars on the core surfaces do not indicate any material flaws such as voids, cracks or lines of weakness which would have led to production errors. It is thus plausible that the other three variables may account for the crude and poor technology exhibited by the Lokalalei assemblage.

Comparisons of the lithic assemblages from the late Pliocene archaeological sites in the Lake Turkana basin may at first appear inappropriate for two reasons. First is the difference in raw material, and second is the size of cobbles that were available to prehistoric knappers in the regions of the Turkana Basin which are discussed here. At West Turkana the dominant raw material was volcanic lava, which was available in large sizes. At the Shungura, quartz was the dominant raw material available. It is difficult to determine the sizes of the blanks available because the cores are said to be small and completely exhausted. The shapes of the cores are described as being discoidal and polyhedric and the fractures on them are steep and angular (Chavaillon, 1976). Quartz tends to smash on impact, however about a half a million years later across the Basin at the site of Fejej 1A, dated to 1·88 Ma (Asfaw *et al.*, 1991) and also at Olduvai Gorge (Leakey, 1971) hominids were able to produce on quartz, either deliberately or as by-products of flake production, Oldowan-type cores similar to those made on lava. I interpret the poor technology exhibited by the Omo assemblages and shared by the Lokalalei and Kada Gona assemblages a result of hominids being near the threshold of stone technology.

Similarities of the Kada Gona and the Omo assemblages are shared with that of the Lokalalei (GaJh 5) site in the sense that most of the cores do not have several scars of step, hinge and small flakes. This has led Roche (1989) and Kibunjia *et al.* (1992) to conclude that the artifacts prior to 2·0 Ma are of poor technology, therefore the Oldowan does not represent the earliest stone tool technology but a point in a continuum. I have suggested using the Omo Industrial Complex to denote industries with these common characteristics: (1) they are dated prior to 2·0 Ma, (2) have low densities of artifacts, (3) the fauna demonstrates low counts of bone surface modification marks, and (4) cores which show less elaboration in terms of retouch and flaking skill than Oldowan-type of cores (Kibunjia, 1990; Kibunjia, in prep.). The West Turkana assemblages which I have named the "Nachukui Industry" (Kibunjia, in prep.) and the Omo assemblages which have been named the "Shungura facies" (Chavaillon, 1976) would be industries in that Complex.

Acknowledgements

Funding was provided by the Leakey Foundation grant and United States, National Science Foundation Dissertation Improvement Grant (BNS 9106186) to Jack Harris of Rutgers University, Total Exploration, Total Oil Products (EA) and Spie Batgniolés. I wish to thank Hélène Roche, Richard Leakey and Meave Leakey for invaluable field and logistical support. Laboratory facilities and some logistical support was provided by the National Museums of Kenya. The excavation crew was led by Bernard Ng'eneo of the Hominid Gang to whom I am grateful. I wish to thank Cregg Madrigal (Rutgers University) for taphonomic studies of fauna and John Kimengich (National Museums of Kenya) for taxonomic identifications.

References

Asfaw, B., Beyenne, Y., Semaw, S., Suwa, G., White, T. & WoldeGabriel, G. (1991). Fejej: a new paleoanthropological research area in Ethiopia. *J. hum. Evol.* **21,** 137–143.

Baker, B. H., Mohr, P.A. & Williams, L. A. J. (1976). Geology of the Eastern Rift System of Africa. *Geol. Soc. Am., Special Paper* **136,** 1–67

Blumenschine, R. J. & Selvaggio, M. M. (1988). Percussion marks on bone surfaces as a new diagnostic of hominid behavior. *Nature* **333,** 763–765.

Boaz, N. T., Bernor, R. L., Brooks, A. S., Cooke, H. B. S., de Heinzelin, J., Dechamps, R., Delson, E., Gentry, A. W., Harris, J. W. K., Meylan, P., Pavlakis, P. P., Sanders, W. J., Stewart, K. M., Verniers, J., Williamson, P. G. & Winkler, A. J. (1992). A new evaluation of the Late Pliocene Lusso Beds, Upper Semliki Valley, Zaire. *J. hum. Evol.* **22,** 505–517.

Brown, F. H. & Feibel, C. S. (1988). "Robust" hominids and Plio-Pleistocene paleogeography of the Turkana Basin, Kenya and Ethiopia. In (F. E. Grine, Ed.) *Evolutionary History of the "Robust" Australopithecines,* pp. 325–341. New York: Aldine de Gruyter.

Bunn, H. T. (1982). Meat-Eating and Human Evolution: studies on the diet and subsistence patterns of Plio-Pleistocene hominids in East Africa. Ph.D. Dissertation, University of California, Berkeley.

Bunn, H. T. & Kroll, E. (1986). Systematic butchery by Plio-Pleistocene Hominids at Olduvai Gorge, Tanzania. *Curr. Anthrop.* **27,** 431–452.

Callahan, D. L. (1979). The basics of biface knapping in the Eastern Fluted Point Tradition: a manual for flintknappers and lithic analysts. *Archaeol. East. North Am.* **7,** 1–180.

Chavaillon, J. (1970). Découverte d'un niveau Oldowayen dans la basse vallée de l'Omo (Ethiopie). *C.R. Seances Mensuelles, Bull. Soc. Prehist. France.* **67,** 7–11.

Chavaillon, J. (1975). Le site paléolithique ancien l'Omo 84 (Ethiopie). *Documents Pour Servir à l'Histoire des civilisations Ethiopiennes* **6,** 9–18. Paris: C.N.R.S.

Chavaillon, J. (1976). Evidence for the technical practices of early Pleistocene hominids. In (Y. Coppens, F. C. Howell, G. L. Isaac & R. E. F. Leakey, Eds) *Earliest Man and Environments in the East Rudolf Basin,* pp. 421–431. Chicago: University of Chicago Press.

Cotterel, B. & Kamminga, J. (1987). The formation of flakes. *Am. Antiquity* **52,** 675–708.

Crabtree, Don E. (1972). An introduction to Flintworking. *Occasional Papers of the Idaho State University* **28.** Pocatello.

de Heinzelin, J. (Ed.) (1983). *The Omo Group. Stratigraphic and Related Earth Sciences Studies in the Lower Omo Basin, southern Ethiopia.* Musee Royal de l'Afrique Centrale, Tervuren, Belgique, Annals Sciences Geologiques, no. 85.

Delson, Eric (1985). Cercopithecid biochronology of the African Plio-Pleistocene: correlation among eastern and southern hominid-bearing localities. *Cour. Forsch. Inst. Senckenberg* **69,** 199–218.

Feibel, C. S. (1988). Paleoenvironments of the Koobi Fora Formation, Turkana basin, northern Kenya. Ph.D. Dissertation, University of Utah.

Feibel, C. S., Brown, F. H. & McDougall, I. (1989). Stratigraphic context of hominids from the Omo Group deposits: northern Turkana basin, Kenya and Ethiopia. *Am. J. phys. Anthrop.* **78,** 595–622.

Feibel, C. S., Harris, J. M. & Brown, F. H. (1991). Neogene paleoenvironments of the Turkana Basin. In (J. M. Harris, Ed.) *Koobi Fora Research Project, Volume 3, Stratigraphy, Artiodactyls and Paleoenvironments,* pp. 321–346. Oxford: Clarendon Press.

Haileab, B. & Brown, F. H. (1991). Turkana basin—Middle Awash valley correlations and the ages of the Sagantole and Hadar Formations. *J. hum. Evol.* **22,** 453–468.

Harris, J. M., Brown, F. H., Leakey, M. G., Walker, A. C. & Leakey, R. E. (1988a). Plio-Pleistocene hominid bearing sites from west of Lake Turkana, Kenya. *Science* **239,** 27–33.

Harris, J. M., Brown, F. H. & Leakey, M. G. (1988b). Stratigraphy and paleontology of Pliocene-Pleistocene localities West of Lake Turkana. *Nat. Hist. Mus. Los Angeles County. Contrib. Sci.* **399,** 1–128.

Harris, J. W. K. (1983). Cultural beginnings: Plio-Pleistocene archaeological occurrences from the Afar, Ethiopia. *Afr. Archaeol. Rev.* **1,** 3–31.

Harris, J. W. K. & Semaw, S. (1989). Further Archaeological studies at Gona River, Ethiopia. *Nyame Akuma* **31,** 19–21.

Harris, J. W. K., Williamson, P. G., Morris, Paul J., de Heinzelin, J., Verniers, J., Helgren, D., Bellomo, R. V., Laden, G., Spang, T. W., Stewart, K. & Tappen, M. J. (1990). Archaeology of Lusso Beds. *Virginia Mus. Nat. Hist. Mem.* **1,** 237–272.

Harris, J. W. K., Williamson, P. G., Verniers, J., Tappen, M. J., Stewart, K., de Heinzelin, J., Boaz, N. T. and Bellomo, R. V. (1987). Late Pliocene hominid occupation in Central Africa: the setting, context, and character of the Senga 5A site, Zaire. *J. hum. Evol.* **16,** 701–728.

Hill A. (1987). Causes of perceived faunal change in the later Neogene of East Africa. *J. hum. Evol.* **17,** 583–596.

Howell, F. C., Haesaertes, P. & de Heinzelin, J. (1987). Depositional environments, archaeological occurrences and hominids from Members E and F of Shungura Formation (Omo Basin, Ethiopia). *J. hum. Evol.* **16,** 643–664.

Isaac, G. Ll. (1984). The archaeology of human origins. In (F. Wendorf & A. E. Close, Eds) *Advances in World Archaeology, Vol. 3,* pp. 1–87. New York: Academic Press.

Isaac, G. Ll. & Harris, J. W. K. (1978). Archaeology. In (M. G. Leakey & R. E. F. Leakey, Eds) *Koobi Fora Research Project,* vol. 1, pp. 64–85. Oxford: Clarendon Press.

Johnson, J. K. (1979). Archaic biface manfacture: production failures, a chronicle of the misbegotten. *Lithic Technology* **8,** 25–35.

Kibunjia, M. (1990). Pliocene stone tool technology West of Lake Turkana, Kenya. *Crosscurrents* **4,** 16–26.

Kibunjia, M. (in prep.) Archaeological Investigations at Lokalalei Late Pliocene site, to the West of Lake Turkana, Kenya. Ph.D Dissertation, Rutgers University.

Kibunjia, M., Roche, R., Brown, F. H. & Leakey, R. E. (1992). Pliocene and Pleistocene archaeological sites of Lake Turkana, Kenya. *J. hum. Evol.* **23,** 432–438.

Leakey, M. D. (1971). *Olduvai Gorge Volume 3.* Cambridge: Cambridge University Press.

McDougall, I. (1985). K-Ar and ^{40}Ar/^{39}Ar dating of the hominid bearing Pliocene-Pleistocene sequence at Koobi Fora, Lake Turkana, northern Kenya. *Bull. Geol. Soc. Am.* **96,** 159–175.

Merrick, H. V. (1976). Recent archaeological research in the Plio-Pleistocene deposits of the Omo Valley, southwestern Ethiopia. In (G. Ll. Isaac & E. R. McCown, Eds) *Human Origins; Louis Leakey and the East African Evidence,* pp. 461–482. Menlo Park, Ca.: W. A. Benjamin.

Merrick, H. V. & Merrick, J. P. S. (1976). Archaeological occurrences of earlier Pleistocene Age, from the Shungura Formation. In (Y. Coppens, F. C. Howell, G. Ll. Isaac & R. E. F. Leakey, Eds) *Earliest Man and Environments in the Lake Rudolf Basin,* pp. 574–584. Chicago: University of Chicago Press.

Potts, R. (1988). *Hominid Activities at Olduvai Gorge.* New York: Aldine.

Roche, H. (1989). Technological evolution in the early hominids. *OSSA, Int. J. Skeletal Res.* **14,** 97–98.

Roche, H. & Tiercelin, J. J. (1980). Industries lithiques de la formation Plio-Pleistocene d'Hadar Ethiopia (campaignè 1976). In (R. E. F. Leakey & B. Ogot, Eds) *Proc. Eighth Panafr. Congr. Prehist. Quat. Stud., Nairobi 1977*, pp. 194–199. Nairobi: TILLMIAP.

Schick, K. D. (1987). Modelling the formation of Early Stone Age artifact concentrations. *J. hum. Evol.* **16,** 789–807.

Shackleton, N. J. & Opdyke, N. D. (1977). Oxygen isotope and paleomagnetic evidence for early northern hemisphere glaciation. *Nature* **270,** 216–219.

Tobias, P. V. (1986). Delineation and dating of some major phases in hominidization since the middle Miocene. *South Afr. J. Sci.* **82,** 91–93.

Toth, N. (1982). The stone technologies of early hominids at Koobi Fora, Kenya: Experimental approach. Unpublished Ph.D. Dissertation, University of California, Berkeley.

Toth, N. (1985). The Oldowan reassessed: a closer look at early stone artifacts. *J. Archaeol. Sci.* **12,** 101–120.

Toth, N. & Schick, K. (1986). The first million years: the archaeology of protohuman culture. *Adv. Archaeol. Method Theory* **9,** 1–96.

Vrba, E. S. (1988). Late Pliocene climatic events and hominid evolution. In (F. Grine, Ed.) *Evolutionary History of the "Robust" Australopithecines*, pp. 405–426. New York: Aldine de Gruyter.

Vrba, E. S. (1992). Mammals as a key to evolutionary theory. *J. Mammal.* **73,** 1–28.

Vrba, E. A., Denton, G. H. & Prentice, M. L. (1989). Climatic influence on early hominid behavior. *OSSA, Int. J. Skeletal Res.* **14,** 127–156.

White, T. D. (1985). African suid evolution: the last six million years. *South Afr. J. Sci.* **81,** 271.

White, T. D. (1988). The comparative biology of "Robust" Australopithecus: clues from the context. In (F. Grine, Ed.) *Evolutionary History of the "Robust" Australopithecines*, pp. 449–483. New York: Aldine de Gruyter.

Young, D. E. & Bonnichsen, R. (1985). Cognition, behavior and material culture. In (M. G. Plew, J. C. Woods & M. G. Pavesic, Eds) *Stone Tool Analyses*, pp. 91–131. Albuquerque: University of New Mexico Press.

Randy V. Bellomo
Department of Anthropology, University of South Florida, 4202 East Fowler Avenue, SOC 107, Tampa, Florida 33620-8100, U.S.A.

Received 27 September 1993
Revision received 16 February 1994 and accepted 16 February 1994

Keywords: Koobi Fora, FxJj 20 Main, hominid-controlled fire, fire-related behavioral activities, nearest-neighbor analysis, Hodder and Okell's *A*, local density analysis, pure locational clustering, stone artifact class frequencies.

Methods of determining early hominid behavioral activities associated with the controlled use of fire at FxJj 20 Main, Koobi Fora, Kenya

A methodological approach which can discriminate between archaeological evidence of fire resulting from natural processes and archaeological evidence of fire resulting from human activities has recently been used to identify evidence of hominid-controlled fire at FxJj 20 Main, an early Pleistocene archaeological site near Koobi Fora, Kenya. The evidence of fire at FxJj 20 Main consists of highly localized, fully oxidized sediment features found near the base of the archaeological horizon. Similar features are preserved at the nearby sites of FxJj 20 East and FxJj 20 AB, but evidence of hominid-controlled fire has not yet been confirmed at these sites. The use and control of fire is regarded as a major technological breakthrough which would have provided hominids with a number of adaptive advantages that have important evolutionary implications. The virtual lack of thermally-altered stone artifacts and bones, the localized configurations of the oxidized features, and the spatial patterning of the artifacts in relation to the oxidized features, indicate that the early hominids did not use fire for the purpose of hunting, cooking, preserving food, intentional plant selection, vegetation clearing, or improving the flaking characteristics of lithic materials. At present, it appears that the early hominids at FxJj 20 Main used controlled fire primarily as a source of protection against predators, as a source of light, and/or as a source of heat.

Journal of Human Evolution (1994) **27,** 173–195

Introduction

In recent years unequivocal evidence of hominid-controlled fire has been reported from the site of FxJj 20 Main, Koobi Fora, Kenya (Figure 1), dated at 1·6 Mya (Bellomo, 1990; Bellomo & Kean, 1994). The evidence of hominid-controlled fire was confirmed using a methodological approach which can distinguish between traces of fire resulting from natural processes (e.g., grass fires, brush fires, tree stump fires, and forest fires) and traces of fire resulting from human activities (e.g., multiple-burn campfires, hearths, ovens, or similar kinds of humanly-produced fires) based on macroscopic, magnetic, and archaeomagnetic techniques of analysis (Bellomo, 1990, 1991, 1993). Discussions regarding the sources of fire in antiquity are beyond the scope of this paper, but have been presented in Bellomo (1990), Clark & Harris (1985), Harrison (1954), and Oakley (1955, 1956, 1961, 1970). It is most likely that fire was captured and used long before hominids possessed the technological capabilities to make it (e.g., Stewart, 1958:116).

This paper will focus on methods of determining the behavioral activities associated with the use of controlled fire at FxJj 20 Main. The methods include (1) analysis and classification of fire types and associated activities, (2) examinations of stone artifacts and bones for evidence of thermal alteration, (3) analyses of the spatial patterning of the stone artifacts in relation to the oxidized features, and (4) analyses of stone artifact frequencies by type, and by size and depth to assess whether or not hominid or non-hominid agencies could have been responsible for the observed spatial patterning at the site.

Fire types and associated activities

The use of fire provided hominids with a number of adaptive advantages (Figure 2), and its use is associated with different activities (see Bellomo, 1990; Clark & Harris, 1985). Based on

0047–2484/94/010173+23 $08.00/0

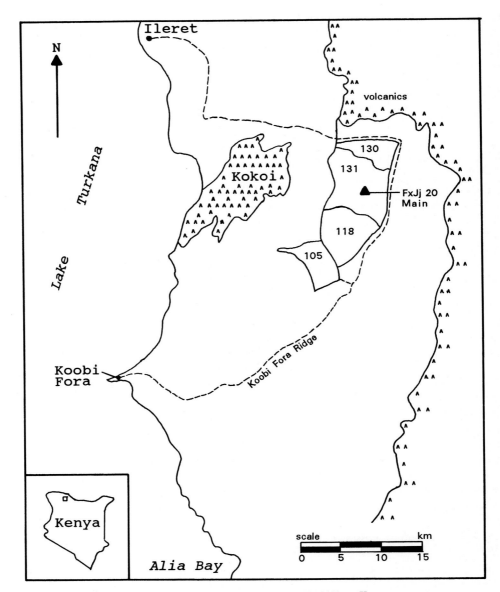

Figure 1. The location of FxJj 20 Main, Koobi Fora, Kenya.

considerations of task requirements and overall efficiency, these activities can be grouped into three fire type categories (Table 1), including (1) campfires, (2) tree stump fires, and (3) grass and brush fires. Grass and brush fires are set to clear vegetated areas efficiently, to control pest populations, to hunt animals, and for purposes of intentional plant selection (Stewart, 1958). Campfires are lit and maintained to provide an efficient source of heat for personal comfort, cooking, preserving food, and for tempering the tips of wooden tools or improving the flaking

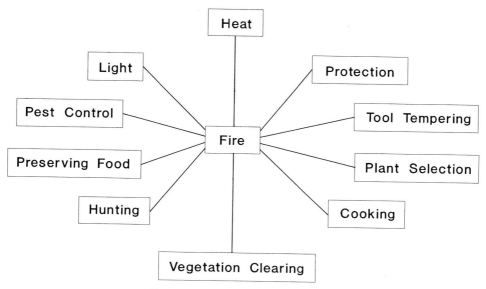

Figure 2. The adaptive advantages of fire.

Table 1   **Fire type categories and associated activities**

	Fire type categories		
Activity*	Campfires	Stump fires	Grass/brush fires
Source of heat	X	X	
Source of light	X	X	
Protection	X		
Tool tempering	X		
Cooking	X	X	
Preserving food	X		
Vegetation clearing			X
Pest control			X
Plant selection			X
Hunting			X

*Excluding metal smelting and pottery firing.

characteristics of certain stone materials, as well as to provide a source of light, and protection from large carnivores and other dangerous animals (Clark & Harris, 1985; Oakley, 1955, 1970; Pfeiffer, 1971). Although burning stumps can be used as a source of light, heat, and for cooking, it is unlikely that stumps were primarily ignited for these purposes during the early Pleistocene because burning stumps produce a soft glow, and low relative temperatures (see Bellomo, 1990, 1993). In addition, experiments undertaken by the author have shown that stumps are more difficult to ignite and maintain than campfires. However, stumps ignited by natural processes were probably used opportunistically as a source of fire, and as a temporary focal point while travelling across the landscape.

A previously developed methodological approach (Bellomo, 1990, 1993) has demonstrated that samples and data from multiple-burn campfire experiments and known archaeological hearths and ovens produce results which are mutually exclusive from the results obtained by samples and data from grass fire, tree stump fire, and forest fire contexts. Experimental and observational data indicate that campfires and hearths produce highly oxidized features which are basin-shaped in profile, with fully oxidized sediments extending to depths of 15 cm below ground surface. In contrast, burned tree stumps leave a hole in the substratum, and oxidation of sediments is rare except for the top 1 or 2 cm around the circumference of forest fire stump holes. Grass fires leave no visible traces of oxidation, because they produce relatively low temperatures and move rapidly across the landscape. In addition, samples from multiple-burn campfires and known archaeological hearth and oven features exhibit dramatic increases in magnetic susceptibility, highly stable magnetic directions and intensities, dramatic changes in magnetic mineralogy, and characteristic linear paleointensity distributions which are distinguishable from the magnitudes displayed by unfired control samples, samples from isolated tree stump fire and forest fire contexts, and grass fire samples (Bellomo, 1990, 1993). Thus, each of these three fire type categories produce mutually exclusive feature configurations and sample responses, which can be identified and distinguished in the archaeological record provided that the sites are found in primary context.

This methodological approach (Bellomo, 1990, 1993) has been used to demonstrate that one of two oxidized features at FxJj 20 Main was produced by either a multiple-burn campfire or a campfire which was kept burning for more than three or four days (see Bellomo, 1990; Bellomo & Kean, 1994). Although unequivocal evidence of fire was not confirmed for the second oxidized feature because some of the methodological results did not quite meet the minimum criteria necessary for identifying unequivocal evidence of humanly-controlled fire (see Bellomo, 1993; Bellomo & Kean, 1994), the characteristics of the features and sample responses suggest that the second feature was probably also produced by a campfire. The identification of a campfire feature, characterized by a highly localized configuration of fully oxidized sediments to depths of at least 5 cm, indicates that hominid fire use at FxJj 20 Main was not associated with vegetation clearing, pest control, intentional plant selection, or hunting (see Table 1). Although campfire use is known to be associated with at least six behavioral activities, archaeological evidence from FxJj 20 Main provides important clues regarding the specific use of fire at that site. The archaeological evidence is presented in the following sections.

Archaeological evidence of fire related activities

Condition of stone artifacts and bones

Over 2500 stone artifacts and nearly 3000 bone specimens were recovered from excavated proveniences during the original site investigations at FxJj 20 Main (Harris, 1978). All of the stone and bone artifacts recovered from excavation units located within 2 m of the oxidized features were examined in 1986 and 1988 at the National Museum of Kenya in Nairobi. These examinations were undertaken to determine if evidence of thermal alteration could be identified using published criteria and direct comparisons with samples recovered from controlled fire experiments undertaken by the author.

Thermally altered stones are commonly recognized by changes in color, differential luster, potlidding, cracking, and crazing (e.g., Purdy & Brooks, 1971; Mandeville, 1971). Thermally altered bones are often recognized by changes in color, changes in structure, and degree of calcination (e.g., Knight, 1985; Shipman *et al.*, 1984).

Only three of the 335 stone artifacts examined from the site of FxJj 20 Main were believed to exhibit evidence of thermal alteration. These specimens included two chert artifacts with reddish and yellowish spot discolorations (Artifact numbers 1461a and 20311), and one basalt artifact (number 1255) which appears to contain evidence of potlidding. It is also important to recognize that over 90% of the stone artifacts recovered from FxJj 20 Main were made of volcanic lava (see Harris, 1978). In 1986, I observed lava raw materials from Koobi Fora explode within 2 minutes after being placed in an experimental campfire, causing hot rock projectiles to be hurled from the fire. Each of those lava fragments displayed evidence of thermal alteration manifest in the form of numerous yellow spot discolorations. Since most of the stone artifacts from FxJj 20 Main were made of lava, and evidence of thermal alteration of lithic artifacts is extremely marginal, it can be concluded that the use of fire at FxJj 20 Main was not employed for purposes of thermal pre-treatment designed to improve the flaking characteristics of lithic raw materials, or as an alternative method of lithic reduction.

An examination of approximately 250 fossilized bones recovered from FxJj 20 Main revealed that all of the bones generally exhibited white exterior surfaces and bluish-gray to black interior surfaces. These color ranges are comparable to those observed on burned bones recovered from experimental contexts, based on published data and direct comparisons (Shipman *et al.*, 1984; Bellomo, 1990:286). However, the large number of bones which consistently exhibit the same range of colors and color distribution patterns suggest that the colors and patterns displayed on the bones from FxJj 20 Main did not result from thermal processes (e.g., see Bellomo, 1990), but probably formed during the fossilization process. The lack of burned bone at FxJj 20 Main suggests that the use of fire was not associated with cooking or food preservation activities.

Spatial patterning of features and stone artifacts

The locations of the two oxidized features from FxJj 20 Main are shown in Figure 3. Both of these features were investigated, although only the eastern feature (see Figure 3) was confirmed to have been produced by a hominid-controlled campfire (Bellomo, 1990; Bellomo & Kean, 1994). The observed spatial distribution of all of the *in situ* stone artifacts recovered from the site of FxJj 20 Main between 99·81 m and 100·08 m of elevation is shown in Figure 4. However, this plot provides little analytical value because it includes a number of artifacts found at elevations which are beyond reasonable limits to evaluate whether or not any association exists between the stone artifacts and the oxidized features.

To improve the resolution for analytical purposes, two separate stone artifact distribution plots were generated for each feature in an initial attempt to explore whether or not the stone artifacts may have been behaviorally associated with the oxidized features. One plot per feature included all of the *in situ* stone artifacts recovered at the same relative elevation as the top of the feature (Figures 5 and 6), and the second plot per feature included all of the *in situ* stone artifacts recovered from relative elevations ranging between the top surface of the feature down to a depth of 5 cm below the top surface of the feature (Figures 7 and 8). The latter plots were generated because archaeological sites are the product of complex patterns of behavior, and trampling and other processes result in the vertical dispersion and size sorting of artifacts (see Gifford-Gonzalez *et al.*, 1985; Schiffer, 1983; Stevenson, 1985, 1991; Villa, 1982). This 5 cm range of elevations is considered to be a conservative estimate of vertical dispersion (cf. Stockton, 1973).

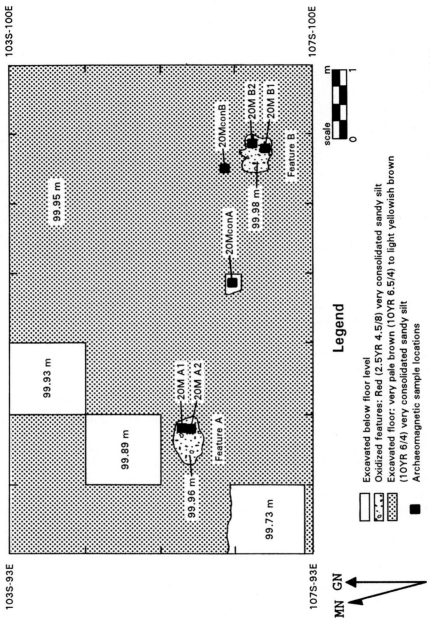

Figure 3. The locations of two of the oxidized features discovered near the base of the archaeological horizon at FxJj 20 Main.

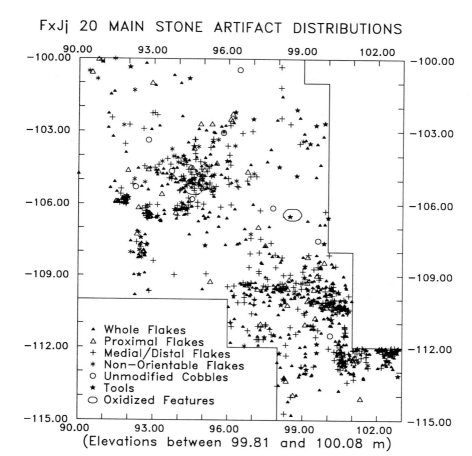

Figure 4. FxJj 20 Main stone artifact distributions between 99·81 and 100·08 m of elevation.

The artifact distribution plots (Figures 5 through 8) show that six different classes of stone artifacts and debitage are included in levels apparently associated with the oxidized features, including unmodified cobbles, whole flakes, proximal flakes, medial/distal flakes, non-orientable fragments, and tools. Examination of the plots suggests that the spatial distributions of the different stone artifact categories relative to the oxidized features (Figures 5 through 8) may have resulted from unintensive lithic reduction activities (e.g., tool production and/or maintenance) which were performed near the oxidized features (see Harris, 1978:377). Unfortunately, any spatial associations which may have existed between the artifacts and features cannot be determined through examinations of the artifact distribution plots alone. In order to demonstrate that the stone artifacts and oxidized features were spatially and behaviorally associated, rigorous quantitative analyses are required since the site records a palimpsest of occupations through time, and numerous hominid activities (e.g., tool manufacture, site maintenance and cleaning, tool sharpening and re-use) and post-depositional processes (e.g., water transport, burrowing of animals, soil movement, and trampling) are known to affect the distribution of artifacts within archaeological contexts (see

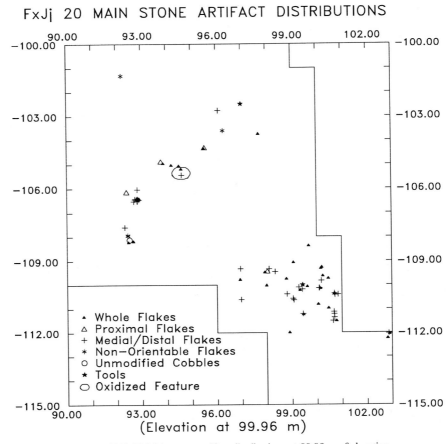

Figure 5. FxJj 20 Main stone artifact distributions at 99·96 m of elevation.

Gifford-Gonzalez *et al.*, 1985; Schiffer, 1983; Stevenson, 1985, 1991; Villa, 1982). In addition, geological data indicate that FxJj 20 Main was located on a floodplain in close proximity to a river channel (Harris, 1978:153–158; Kaufulu, 1983:152–162).

To explore whether or not the observed spatial distributions may have resulted from hominid activities or post-depositional processes, artifact size distribution plots (Figures 9 and 10) were generated to examine if evidence of size sorting and/or vertical dispersion could be identified. Three different artifact size categories (based on maximum dimensions) were utilized for this purpose. Figure 9 shows that the majority of the artifacts which were recovered *in situ* are in the smallest size category (i.e., less than 3 cm in maximum dimension). Although the number of artifacts greater than 6 cm in maximum dimension is low, the relative locations of these larger artifacts could be an indication that intentional size sorting was periodically practiced to keep the central site area somewhat clear so that intrasite movement was facilitated. However, since the largest size categories are found vertically dispersed throughout the oxidized feature horizon (Figure 10), it is also possible that separate episodes of site occupation are represented between 99·91 and 99·98 m of elevation.

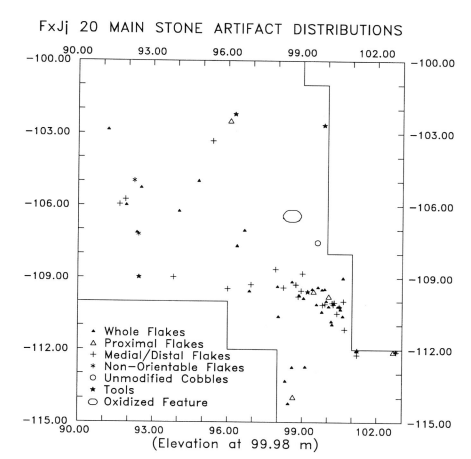

Figure 6. FxJj 20 Main stone artifact distributions at 99·98 m of elevation.

While the plots of the observed spatial distributions of *in situ* stone artifacts relative to the oxidized features at FxJj 20 Main suggest that the artifacts and features may be behaviorally associated, it is necessary to employ more objective methods to assess whether or not any spatial association exists between the different artifact classes and the oxidized features, because the observed spatial distributions may be the product of numerous post-depositional processes and/or subsequent hominid reoccupations at the site locality. To identify whether or not the spatial distributions observed may have resulted from hominid-related activities, the stone artifact data from FxJj 20 Main were analysed using four indices of spatial patterning. For all methods employed, only those stone artifacts which were recovered *in situ* from depths between 99·93 and 99·98 m of elevation were included. This range of elevations was chosen to include all artifacts recovered between the top surface of the uppermost oxidized feature down to an arbitrary depth of 5 cm below the top surface of the uppermost oxidized feature (see Figure 3). This strategy was employed in an attempt to minimize the effects of any possible admixture between archaeological events which may have occurred at different times. This range of elevations also includes both of the oxidized features depicted in Figure 3.

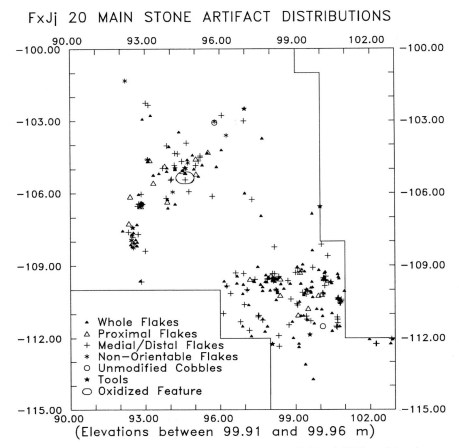

Figure 7. FxJj 20 Main stone artifact distributions between 99·91 and 99·96 m of elevation.

Nearest neighbor analysis

Nearest-neighbor analysis has been shown to be a sensitive indicator of non-random spatial clustering (Clark & Evans, 1954; Whallon, 1974). Within a population, the nearest-neighbor of a point is another point which lies closer to it than to any other point. The nearest-neighbor of each point is then calculated for the entire population. The nearest-neighbor coefficient (*R*) is the ratio of the average distance between all nearest-neighbors observed in a population divided by the average distance which would be expected if the same number of points were randomly distributed over that same area. A value of *R* between 0 and 1·0 indicates a clustered distribution of points, a value of *R* greater than 1·0 indicates a more even distribution of points than would be expected at random, and a value of *R* near 1·0 is indicative of a random distribution of points.

Nearest-neighbor analyses of archaeological data, however, must be undertaken with caution since boundary problems (e.g., the absolute boundaries of a site or region are not always known) and the limited range of data required for the computations (i.e., total area, number of points, and the distribution of distances between each point and its nearest neighbor) present special difficulties for archaeological data (Hodder & Orton, 1976; Pinder

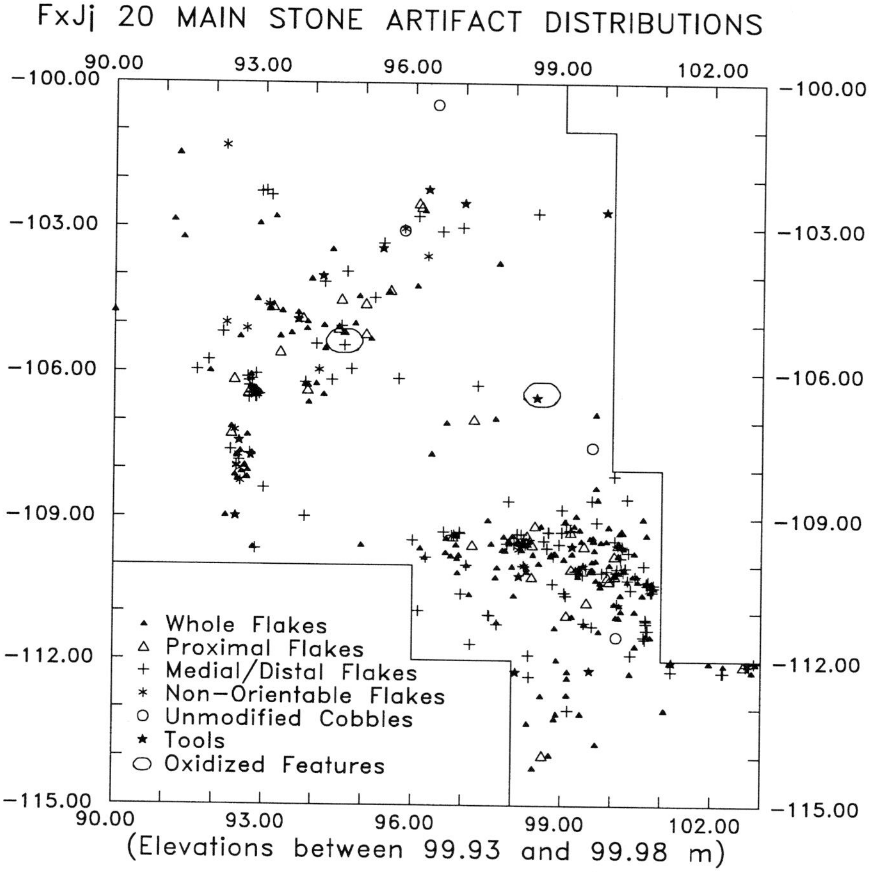

Figure 8. FxJj 20 Main stone artifact distributions between 99·93 and 99·98 m of elevation.

et al., 1979). Despite these potential problems and difficulties, nearest-neighbor analyses can be used cautiously as a measure of relative patterning of different classes of artifacts (Hivernel & Hodder, 1984:100). For example, it has been demonstrated that the ratio of the nearest-neighbor coefficients calculated for two different tool types will be the same regardless of the area used in the nearest-neighbor calculations, providing that all points are included in both analyses (e.g., Kintigh, 1990). Thus, the ratios of nearest-neighbor coefficients can be compared to determine if one artifact class is more or less clustered than another class because this method avoids the difficulties known to be associated with archaeological data.

Since only the lithic artifact data from FxJj 20 Main were available for analysis, and lithic reduction activities produce four discrete categories of debitage in addition to any tool forms which are produced, a between-class nearest-neighbor analysis was undertaken to determine the degree of spatial clustering between the different classes of stone artifacts. The results of this analysis could then be used to determine if the observed spatial distributions resulted from hominid activities or non-hominid agencies. For example, hominid activities involving lithic reduction would be expected to produce a spatial distribution in which all four debitage categories were found clustered together, providing that the site was discovered in primary

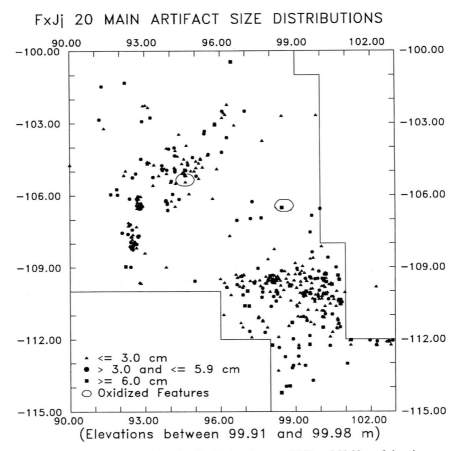

Figure 9. FxJj 20 Main artifact size distributions between 99·91 and 99·98 m of elevation.

context. This expectation is based on the fact that lithic reduction activities are known to produce all four classes of debitage (e.g., Newcomer, 1971). A lack of clustering between members of the different debitage classes would suggest that the observed spatial distributions probably resulted from post-depositional processes.

The ratios of the between-class nearest-neighbor coefficients are presented in Table 2. The coefficients have been standardized by taking the ratio of the row variable to the column variable. When interpreting between-class nearest-neighbor ratios, values near 1·0 are indicative of a randomly intermingled distribution of points between two classes, values less than 1·0 are indicative of a spatially aggregated distribution of points between two classes, and values greater than 1·0 are indicative of a spatially segregated distribution of points between two classes (Kintigh, 1990). Table 2 shows that all four debitage categories (whole flakes, proximal flakes, medial/distal flakes, and non-orientable fragments) are spatially aggregated, since the nearest-neighbor ratios between any debitage class and each of the other three debitage classes is significantly less than 1·0.

Table 2 also facilitates comparisons between the different stone artifact classes. For example, the nearest-neighbor ratio of whole flakes to tools is 0·63, while the ratio of tools to whole

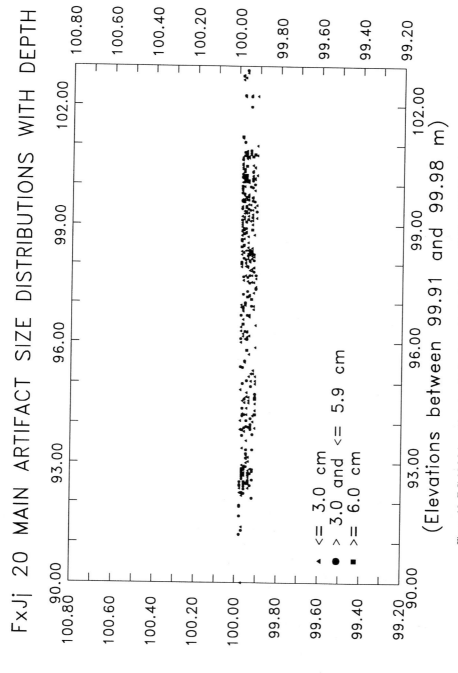

Figure 10. FxJj 20 Main artifact size distributions with depth between 99·91 and 99·98 m of elevation.

Table 2 **Ratios of between-class nearest-neighbor coefficients calculated for stone artifacts recovered from FxJj 20 Main between 99·93 and 99·98 m of elevation**

	n	Whole flakes	Proximal flakes	Medial/distal flakes	Non-orientable fragments	Unmodified cobbles	Tools
Whole flakes	198	0·62	0·56	0·62	0·88	0·76	0·63
Proximal flakes	35	0·37	0·63	0·55	0·73	0·77	0·52
Medial/distal flakes	140	0·51	0·60	0·52	0·86	0·72	0·66
Non-orientable fragments	16	0·59	0·71	0·59	0·65	0·68	0·62
Unmodified cobbles	5	1·92	1·30	1·76	1·09	1·35	0·74
Tools	17	1·14	0·94	0·92	0·89	0·64	0·81

flakes is 1·14. This indicates that the observed distances from whole flakes to tools is a little over half of what would be expected given the overall number of tools, but the distance from tools to whole flakes is what would be expected given the number of whole flakes. In regard to the observed spatial distributions, this indicates that all whole flakes have nearby tools, but not all tools have nearby whole flakes. As shown in Table 2, similar ratios exist between the other three debitage classes and tools. These results can be understood by comparing lithic reduction activities with specific tool using activities: a tool may be expected to occur near whole flakes and other debitage categories following production but before it was used for its intended purpose, while tools picked up and used for specific activities would be expected to occur away from the location of manufacture.

The between-class nearest-neighbor data suggest that the spatial distribution of lithic artifacts observed at FxJj 20 Main is consistent with patterns which would be expected following both lithic reduction activities (Callahan, 1980; Newcomer & Sieveking, 1980) and tool-using activities. Thus, the nearest-neighbor data seem to support the interpretation that the observed spatial distribution resulted from hominid activities rather than from non-hominid (post-depositional) agencies.

Hodder and Okell's A analysis
Hodder & Okell (1978) recognized that the first nearest-neighbor distances of artifacts may not be good indicators of culturally interpretable clustering. To compensate for the fundamental inadequacies of nearest-neighbor statistics, Hodder & Okell (1978) developed a measure of spatial association between two artifact classes which uses both first nearest-neighbor distances and distance data from each point to every other point in the distribution regardless of class. Their index, known as Hodder and Okell's A, is computed from the intertype distances observed in a sample. An intertype distance is defined as the mean linear distance between every point in one artifact class and every point in a second artifact class.

Hodder and Okell's A is a coefficient of spatial association or spatial disassociation between artifact classes. A value of A less than 1·0 indicates that two artifact classes are spatially segregated, while a value of A near 1·0 indicates that two artifact classes are spatially intermingled. Hodder and Okell's A values for stone artifact classes recovered from FxJj 20 Main are presented in Table 3. The data in Table 3 indicate that all stone artifact classes are spatially intermingled, that is, all classes tend to occur and not occur in the same places, with the exception of the non-orientable fragments. Non-orientable fragments appear to be more spatially segregated from whole flakes, proximal flakes, and medial/distal flakes than any of the other stone artifact classes. One possible explanation for this might be that whole, proximal,

Table 3   Hodder and Okell's *A* calculated for stone artifacts recovered from FxJj 20 Main between 99·93 and 99·98 m of elevation

	n	Whole flakes	Proximal flakes	Medial/distal flakes	Non-orientable fragments	Unmodified cobbles	Tools
Whole flakes	198	1·00	0·99	1·00	0·65	0·92	0·95
Proximal flakes	35	0·99	1·00	0·98	0·74	1·03	1·02
Medial/distal flakes	140	1·00	0·98	1·00	0·62	0·92	0·94
Non-orientable fragments	16	0·65	0·74	0·62	1·00	0·96	0·85
Unmodified cobbles	5	0·92	1·03	0·92	0·96	1·00	1·20
Tools	17	0·95	1·02	0·94	0·85	1·20	1·00

and medial/distal flakes may have been picked up and used as expedient tools within the immediate site area, while non-orientable fragments were left *in situ* because they offered little utility for activities requiring the use of lithic objects as tools. It is also possible that the lower relative degree of spatial association between non-orientable fragments and the other three debitage categories is due to the small number of non-orientable fragments in the sample. Thus, the Hodder & Okell's *A* data seem to support the interpretation that the observed spatial distributions resulted from hominid activities associated with a combination of lithic reduction activities and use of expedient tools within the immediate site area. However, the Hodder and Okell's *A* results could also be an indication that the sample represents a palimpsest of multiple occupations or has been somewhat disturbed by post-depositional processes.

Local density analysis

Local density analysis is a measure of artifact class association which considers interpoint distances within a fixed radius of each point (Johnson, 1976, 1984). The local density coefficient reflects the mean density of points of one type (e.g., whole flakes) in the neighborhood of points of another type (e.g., proximal flakes), divided by the global density of points of the first type (e.g., whole flakes). To calculate the local density coefficients, each point is considered in turn. Thus, the local density of proximal flakes relative to a single whole flake location is equal to the number of proximal flakes within a fixed radius of the single whole flake divided by the area of the neighborhood. The global density of proximal flakes is equal to the total number of proximal flakes divided by the area of the site. A local density coefficient value near 1·0 indicates that there is no aggregation or segregation between two data classes at that particular scale of analysis (i.e., the fixed radius). In other words, the mean actual and expected densities are approximately the same. A local density coefficient value greater than 1·0 indicates that two classes are spatially associated, while a local density coefficient value less than 1·0 indicates that two classes are spatially segregated.

One advantage of local density analysis is that it provides information about the patterning of artifact classes at different scales (i.e., the different radii). An examination of the distributions of intratype local density coefficients for each stone artifact class recovered from FxJj 20 Main reveals that whole flakes, proximal flakes, medial/distal flakes, and non-orientable fragments are spatially associated, as indicated by the higher relative local density coefficient values regardless of neighborhood radius (Figure 11). In addition, the spatial association is greatest for radii of 1 m or less.

The local density coefficients calculated for a neighborhood radius of 0·5 m is presented in Table 4, while the local density coefficients calculated for a neighborhood radius of 1·0 m is

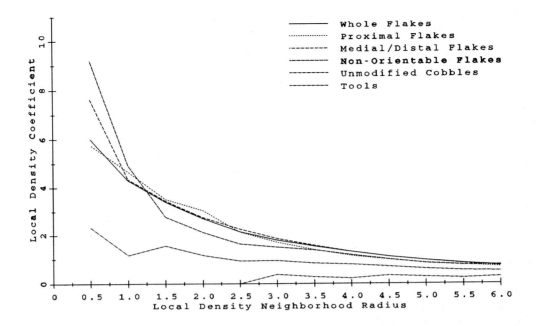

FxJj 20 MAIN BETWEEN 99.93 AND 99.98 m

Figure 11. FxJj 20 Main intratype local density plot for stone artifacts recovered between 99·93 and 99·98 m of elevation.

Table 4 Local density coefficients calculated for stone artifacts recovered from FxJj 20 Main between 99·93 and 99·98 m of elevation using a neighborhood radius of 0·5 m

	n	Whole flakes	Proximal flakes	Medial/distal flakes	Non-orientable fragments	Unmodified cobbles	Tools
Whole flakes	198	6·00	6·55	5·90	5·20	2·38	4·04
Proximal flakes	35	6·55	5·76	6·65	3·60	0·00	4·24
Medial/distal flakes	140	5·90	6·65	7·63	3·98	1·44	2·68
Non-orientable fragments	16	5·20	3·60	3·98	9·19	10·50	4·94
Unmodified cobbles	5	2·38	0·00	1·44	10·50	0·00	1·98
Tools	17	4·04	4·24	2·68	4·94	1·98	2·33

shown in Table 5. The local density coefficient values provided in Tables 4 and 5 indicate that all of the different artifact classes are spatially associated, although whole flakes, proximal flakes and medial/distal flakes are more strongly associated than any other classes. In addition, the degree of spatial association between almost every pair of stone artifact classes is significantly greater than what would be expected for a random distribution. For example, Table 4 shows that the local density coefficient for whole flakes and proximal flakes is 6·55, which indicates that there are more than six and a half times as many proximal flakes in the neighborhood of whole flakes than would be expected if the members of both classes were randomly distributed.

Table 5   Local density coefficients calculated for stone artifacts recovered from FxJj 20 Main between 99·93 and 99·98 m of elevation using a neighborhood radius of 1·0 m

	n	Whole flakes	Proximal flakes	Medial/distal flakes	Non-orientable fragments	Unmodified cobbles	Tools
Whole flakes	198	4·30	4·72	4·36	2·51	1·27	2·67
Proximal flakes	35	4·72	4·66	4·37	2·48	0·96	3·11
Medial/distal flakes	140	4·36	4·37	4·34	2·04	1·20	2·38
Non-orientable fragments	16	2·51	2·48	2·04	4·92	3·68	2·94
Unmodified cobbles	5	1·27	0·96	1·20	3·68	0·00	2·97
Tools	17	2·67	3·11	2·38	2·94	2·97	1·16

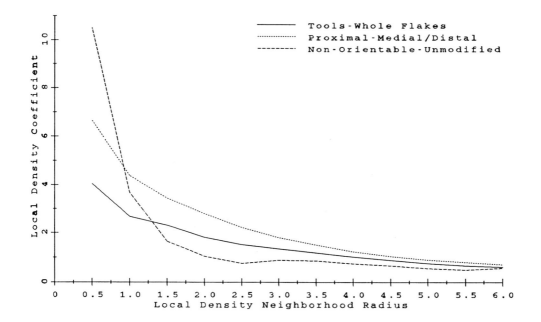

Figure 12. FxJj 20 Main intertype local density plot for stone artifacts recovered between 99·93 and 99·98 m of elevation.

An examination of the distributions of intertype local density coefficients for pairs of stone artifact classes from FxJj 20 Main (Figure 12) reveals that the non-orientable fragments–unmodified cobbles coefficient is significantly greater than either the tools–whole flakes or the proximal flakes–medial/distal flakes coefficients at the smallest radius value. However, at a radius of 1·5 m and beyond, the non-orientable fragments–unmodified cobbles coefficient value indicates that the spatial association between this pair of artifact classes is much less than for the other two stone artifact class combinations.

The local density data indicate that the greatest spatial associations exist between whole flakes, proximal flakes, and medial/distal flakes. These results seem to support the interpretation that the observed spatial distribution resulted from hominid activities rather than from

Table 6 **Pure locational clustering of stone artifacts recovered from FxJj 20 Main between 99·93 and 99·98 m of elevation showing the cluster composition for a 12 cluster solution. Percentages shown are for individual clusters**

Cluster	n	RMS	Whole flakes	Proximal flakes	Medial/distal flakes	Non-orientable fragments	Unmodified cobbles	Tools
1	56	0·48	28/50%	6/11%	18/32%	2/4%	0/0%	2/4%
2	33	0·67	16/48%	1/3%	7/21%	5/15%	1/3%	3/9%
3	40	0·89	20/50%	7/18%	9/23%	2/5%	0/0%	2/5%
4	18	1·33	3/17%	2/11%	5/28%	2/11%	2/11%	4/22%
5	71	0·68	29/41%	5/7%	36/51%	0/0%	1/1%	0/0%
6	10	1·39	6/60%	0/0%	3/30%	1/10%	0/0%	0/0%
7	20	0·93	14/70%	1/5%	3/15%	0/0%	0/0%	2/10%
8	8	1·25	4/50%	1/13%	2/25%	0/0%	0/0%	1/13%
9	15	0·71	6/40%	1/7%	6/40%	1/7%	0/0%	1/7%
10	44	0·69	18/41%	4/9%	19/43%	3/7%	0/0%	0/0%
11	32	0·86	16/50%	2/6%	14/44%	0/0%	0/0%	0/0%
12	64	0·68	38/59%	5/8%	18/28%	0/0%	1/2%	2/3%

non-hominid agencies. It is likely that the lower relative degree of spatial association between non-orientable fragments and the other three debitage categories is due to the small number of non-orientable fragments in the sample.

Pure locational clustering analysis

Pure locational clustering is a rigorous procedure designed to identify spatial clusters with their component points (Kintigh & Ammerman, 1982; Kintigh, 1990). The method performs a *k*-means non-hierarchical divisive cluster analysis which attempts to minimize the intracluster variances while maximizing the intercluster variances (Kintigh & Ammerman, 1982). Pure locational clustering allocates every point (based on *x* and *y* coordinate data) into one of a number of possible clusters, up to the maximum number of clusters selected prior to analysis. Points are allocated to the most appropriate clusters to insure that the sum-squared error (SSE) is minimized. The SSE is the sum of the squared distances from each point to the center of the cluster to which it has been assigned.

The procedure works best if the maximum number of clusters specified exceeds the number of clusters expected to be of interest. The analysis provides the best solutions by minimizing the SSE values for each cluster level up to the maximum number specified. Following the analysis, the nature and degree of spatial patterning can be assessed by comparing the SSE plot of the observed data with the SSE plots of *k*-means analyses of randomized data (Kintigh & Ammerman, 1982). Data are significantly clustered at a given cluster level if the SSE value is below the SSE values provided by the randomized data analyses. In addition, the best clustering levels are indicated by inflections in the SSE curve where the absolute value of the slope decreases (Kintigh, 1990).

The SSE plot of the observed and randomized data are presented in Figure 13. Examination of Figure 13 reveals that the actual distribution of points from FxJj 20 Main is clustered when compared with the randomized data, and that inflections occur at two, four, six, and twelve cluster levels. The tabulation of the twelve cluster solution is presented in Table 6. The division of the 411 points into twelve clusters is shown in Figure 14. In regard to Figure 14, the cluster number to which a point is allocated is displayed at the point's *x* and *y*

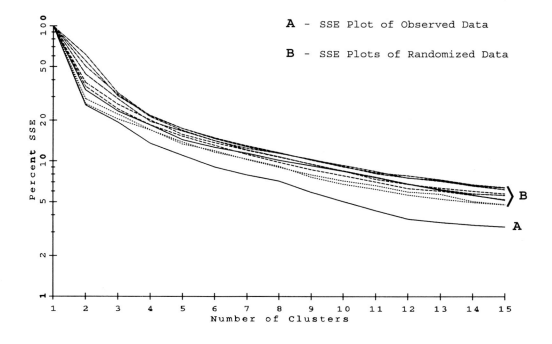

A - SSE Plot of Observed Data

B - SSE Plots of Randomized Data

FxJj 20 MAIN BETWEEN 99.93 AND 99.98 m

Figure 13. FxJj 20 Main pure locational clustering SSE plots for observed data (solid line, A) recovered between 99·93 and 99·98 m of elevation and randomized data (broken lines, B).

coordinate location, and each cluster centroid is depicted by a cross. A cluster radius or RMS, determined by the root of the mean of the squared distances from each point to its cluster center, is drawn around each cluster. Comparison of Figures 3 and 14 reveals that the western oxidized feature at FxJj 20 Main is contained within the RMS of cluster 3, while the eastern oxidized feature is located just outside the boundary of the RMS of cluster 8. According to the data presented in Table 6, 95% of the stone artifacts allocated within cluster 3 are classified as debitage, including 20 whole flakes, seven proximal flakes, nine medial/distal flakes, and two non-orientable fragments. In addition, two stone artifacts within cluster 3 are classified as tools. Cluster 8 is comprised of eight stone artifacts, consisting of four whole flakes, one proximal flake, two medial distal flakes, and one tool.

The pure locational clustering data for a 12 cluster solution show an interesting relationship between the oxidized feature locations and two of the twelve clusters. The data suggest that some lithic reduction activities occurred in the immediate vicinity of the western oxidized feature (see Figures 3 and 14), and further implies that the feature and the activity were associated. However, the data also suggest that other activities involving the use of stone materials were conducted a short distance away from the eastern oxidized feature (see Figures 3 and 14), as well as in other areas away from the oxidized features.

Stone artifact frequencies by type, and by size and depth
To assess whether or not the stone artifact class distributions observed in the FxJj 20 Main sample are indicative of specific hominid activities or post-depositional processes, the *in situ*

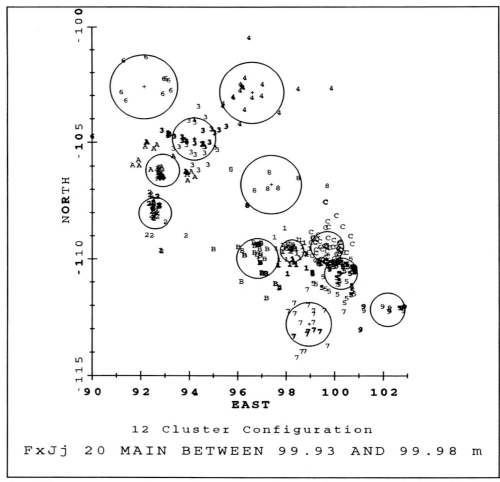

Figure 14. FxJj 20 Main pure locational clustering plot of the 12 cluster configuration for stone artifacts recovered between 99·93 and 99·98 m of elevation. Numbers (1–12) and letters (a–d) occur at actual data point locations and indicate the cluster to which a point is assigned as a result of analysis. Circles are RMS cluster radii.

lithic data were analysed using the approach developed by Sullivan & Rozen (1985), to determine what technological activities were undertaken based on the relative frequencies of stone artifact and debitage categories (Table 7). Their interpretations are based on comparisons with experimentally produced assemblages and assemblages recovered from New World archaeological sites, so their approach may not be entirely applicable for Karari Industry assemblages. However, from a purely technological perspective, lithic reduction activities result in the production of four debitage classes (e.g., Callahan, 1980; Newcomer, 1971; Newcomer & Sieveking, 1980), so the approach of Sullivan & Rozen (1985) should be valid regardless of the age of the assemblage or its associated technological level.

Table 7 **Percentages of stone artifacts recovered from FxJj 20 Main by category and elevation (after Sullivan & Rozen, 1985)**

Artifact category	Elevation			
	99·96 m	99·98 m	99·91–99·96 m	99·93–99·98 m
Whole flakes	43·2	48·0	48·1	48·8
Proximal flakes	5·4	6·7	7·1	8·6
Medial/distal flakes	43·2	34·7	37·1	34·5
Non-orientable fragments	5·4	3·9	4·7	3·9
Tools	2·8	6·7	3·0	4·2

Table 8 **Frequencies of stone artifacts recovered from FxJj 20 Main by size and elevation**

Maximum dimension	Elevation			
	99·97–99·98 m	99·95–99·96 m	99·93–99·94 m	99·91–99·92 m
≤1·0 cm	5	4	3	0
>1·0 and ≤2·0 cm	43	37	32	11
>2·0 and ≤3·0 cm	50	47	27	25
>3·0 and ≤4·0 cm	38	30	23	11
>4·0 and ≤5·0 cm	15	10	8	5
>5·0 and ≤6·0 cm	9	6	5	2
>6·0 and ≤7·0 cm	4	3	3	2
>7·0 and ≤8·0 cm	5	1	1	1
>8·0 and ≤9·0 cm	0	3	0	0

Based on Sullivan & Rozen (1985) the relative percentages of the FxJj 20 Main artifact types (Table 7) recovered in apparent association with the oxidized features suggest that the hominids were involved in core reduction and tool manufacturing activities at the site. These data support the interpretation that the observed stone artifact distributions resulted from hominid activities rather than from significant post-depositional disturbances. This interpretation is consistent with the observed spatial distributions for the stone artifact type categories presented in Figures 5 through 8, and with the results of the various indices of spatial patterning discussed in the previous sections.

Lithic size data were also analysed in relation to depth to determine if evidence of size sorting or site maintenance could be identified. Table 8 presents the relative frequencies of artifacts by size categories and elevation. As shown in Table 8, the majority of artifacts are less than or equal to 4 cm in maximum dimension, suggesting that larger artifacts may have been intentionally removed (transported?) to facilitate intrasite movement, or were further reduced in size. The vertical dispersion and size sorting of artifacts are implied, but can not be confirmed at the present time. Further detailed analyses of the FxJj 20 Main data, involving three-dimensional distribution data along with size and weight variables, would be required to address this complex issue. Such analyses are beyond the scope of this study, but should be undertaken in the future to gain a better understanding of all of the processes which contributed to the formation of the observed spatial distributions at FxJj 20 Main.

Conclusions

Archaeological data from FxJj 20 Main were analysed to determine the possible uses of controlled fire by early hominids. Since the evidence of fire at the site resulted from controlled campfires (Bellomo, 1990; Bellomo & Kean, 1994), hominid fire use could not have been associated with hunting, plant selection, vegetation clearing, or pest control. The lack of thermally altered stone and bone artifacts suggests that fire use was not associated with activities designed to improve the flaking characteristics of the stone raw materials, for cooking, or for preserving food.

Four indices of spatial patterning, including nearest-neighbor, Hodder & Okell's *A*, local density analysis, and pure locational clustering, as well as comparisons of artifact class frequencies were employed to determine if the observed spatial distributions at FxJj 20 Main resulted from hominid activities or post-depositional processes. The combined data suggest that hominid lithic manufacturing and maintenance activities were undertaken both in the immediate vicinity of the western oxidized feature (see Figure 3) and in peripheral site areas which were still in close relative proximity to the oxidized features.

The spatial patterning of the stone artifacts relative to the fire features suggests that the campfires provided a central focus of activities, including the production and maintenance of stone tools and the consumption of food. It is most likely that the early hominids at FxJj 20 Main primarily used fire as a source of protection from predators, a source of light, and/or as a source of heat. Further, more detailed studies involving three-dimensional point location data, as well as point size and weight data, are recommended for both stone artifacts and bones recovered from FxJj 20 Main to gain a better understanding of all of the processes responsible for producing the observed spatial distributions at the site.

Acknowledgements

This research is an extension of previous work made possible through grants from the National Science Foundation, the L. S. B. Leakey Foundation, the Boise Fund, the Nato Advanced Study Institute, and Sigma Xi, The Scientific Research Society. I would also like to express my thanks to several individuals who helped make this research possible and/or provided valuable assistance, including J. W. K. Harris, Richard Leakey, Diana Holt, Kathlyn Stewart, and Greg Laden, as well as members of the 1988 Koobi Fora Field School and the Semliki Research Expedition.

References

Bellomo, R. V. (1990). Methods for documenting unequivocal evidence of humanly controlled fire at Early Pleistocene archaeological sites in Africa: The role of actualistic studies. Ph.D. Dissertation, University of Wisconsin-Milwaukee.

Bellomo, R. V. (1991). Identifying traces of natural and humanly-controlled fire in the archaeological record: the role of actualistic studies. *Archaeol. Montana* **32,** 75–93.

Bellomo, R. V. (1993). A methodological approach for identifying archaeological evidence of fire resulting from human activities. *J. Archaeol. Sci.* **20,** 525–555.

Bellomo, R. V. & Kean, W. F. (1994). Evidence of hominid-controlled fire at the FxJj 20 Site Complex, Karari Escarpment, Koobi Fora, Kenya. Accepted for publication in (G. Ll. Isaac & Barbara Isaac, Eds) *Koobi Fora Research Project Monograph Series Volume 3: Archaeology.* Oxford: Clarendon Press.

Callahan, E. (1980). Spatial organization of the work areas of three contemporary flintknappers. *Bull. Archaeol. Soc. Virginia* **35,** 101–108.

Clark, P. J. & Evans, F. C. (1954). Distance to nearest neighbor as a measure of spatial relationships in populations. *Ecology* **35,** 445–453.

Clark, J. D. & Harris, J. W. K. (1985). Fire and its roles in early hominid lifeways. *Afr. Archaeol. Rev.* **3**, 3–27.

Gifford-Gonzalez, D. P., Damrosch, D. B., Damrosch, D. R., Pryor, J. & Thunen, R. L. (1985). The third dimension in site structure: an experiment in trampling and vertical dispersal. *Am. Antiquity* **50**, 803–818.

Harris, J. W. K. (1978). The Karari industry: Its place in East African prehistory. Ph.D. Thesis, University of California-Berkeley.

Harrison, H. S. (1954). Fire-making, fuel, and lightning. In (C. Singer, E. J. Holmyard & A. R. Hall, Eds) *A History of Technology, Volume 1*, pp. 216–237. Oxford: Clarendon Press.

Hivernel, F. & Hodder, I. (1984). Analysis of artifact distribution at Ngenyn (Kenya): depositional and post depositional effects. In (H. J. Hietala, Ed.) *Intrasite Spatial Analysis in Archaeology*, pp. 97–115. Cambridge: Cambridge University Press.

Hodder, I. & Okell, E. (1978). A new method for assessing the association between distributions of points in archaeology. In (I. Hodder, Ed.) *Simulation Studies in Archaeology*, pp. 97–107. Cambridge: Cambridge University Press.

Hodder, I. & Orton, C. (1976). *Spatial Analysis in Archaeology*. Cambridge: Cambridge University Press.

Johnson, I. (1976). Contribution méthodologique à l'étude de la répartition des vestiges dans des niveaux archaéologiques. Thesis for obtaining a Diplome des Études Superieurs, Université de Bordeaux I.

Johnson, I. (1984). Cell frequency recording and analysis of artifact distributions. In (H. J. Hietala, Ed.) *Intrasite Spatial Analysis in Archaeology*, pp. 75–96. Cambridge: Cambridge University Press.

Kaufulu, Z. M. (1983). The geological context of some early archaeological sites in Kenya, Malawi and Tanzania: microstratigraphy, site formation and interpretation. Ph.D. Dissertation, University of California, Berkeley.

Kintigh, K. W. (1990). Intrasite spatial analysis: a commentary on major methods. In (A. Voorrips, Ed.) *Mathematics and Information Science in Archaeology*, pp. 165–200. Bonn: Holos-Verlag.

Kintigh, K. W. & Ammerman, A. J. (1982). Heuristic approaches to spatial analysis in archaeology. *Am. Antiquity* **47**, 31–63.

Knight, J. A. (1985). Differential preservation of calcined bone at the Hirundo site, Alton, Maine. MS Thesis, University of Maine at Orono.

Mandeville, M. (1971). The baked and the half baked: a consideration of the thermal pretreatment of chert. MS Thesis, University of Missouri, Columbia.

Newcomer, M. H. (1971). Some quantitative experiments in handaxe manufacture. *World Archaeol.* **3**, 85–93.

Newcomer, M. H. & Sieveking, G. de G. (1980). Experimental flake scatter patterns: a new interpretive technique. *J. Field Archaeol.* **7**, 345–352.

Oakley, K. P. (1955). Fire as a Palaeolithic tool and weapon. *Proc. Prehist. Soc. 1955* **21**, 36–48.

Oakley, K. P. (1956). The earliest fire-makers. *Antiquity* **30**, 102–107.

Oakley, K. P. (1961). Possible origins of the use of fire. *Man* **61**, 244.

Oakley, K. P. (1970). On man's use of fire, with comments on tool-making and hunting. In (S. L. Washburn, Ed.) *Social Life of Early Man*, pp. 176–193. Chicago: Aldine.

Pfeiffer, J. (1971). When *Homo erectus* tamed fire, he tamed himself. In (H. R. Bleibtreu & J. F. Downs, Eds) *Human Variation: Readings in Physical Anthropology*, pp. 193–203. California: Glencoe Press.

Pinder, D., Shimada, I. & Gregory, D. (1979). The nearest-neighbor statistic: archaeological application and new developments. *Am. Antiquity* **44**, 430–445.

Purdy, B. A. & Brooks, H. K. (1971). Thermal alteration of silica minerals: an archaeological approach. *Science* **173**, 322–325.

Schiffer, M. B. (1983). Toward the identification of formation processes. *Am. Antiquity* **48**, 675–706.

Shipman, P., Foster, G. & Schoeninger, M. (1984). Burnt bones and teeth: an experimental study of color, morphology, crystal structure and shrinkage. *J. Archaeol. Sci.* **11**, 307–325.

Stevenson, M. G. (1985). The formation of artifact assemblages at workshop/habitation sites: models from Peace Point in Northern Alberta. *Am. Antiquity* **50**, 63–81.

Stevenson, M. G. (1991). Beyond the formation of hearth-associated artifact assemblages. In (E. M. Kroll & T. D. Price, Eds) *The Interpretation of Archaeological Spatial Patterning*, pp. 269–299. New York: Plenum Press.

Stewart, O. C. (1958). Fire as the first great force employed by man. In (W. L. Thomas, Jr, Ed.) *Man's Role in Changing the Face of the Earth*, pp. 115–133. Chicago: University of Chicago Press.

Stockton, E. D. (1973). Shaw's Creek Shelter: human displacement of artifacts and its significance. *Mankind* **9**, 112–117.

Sullivan, A. P., III & Rozen, K. C. (1985). Debitage analysis and archaeological interpretation. *Am. Antiquity* **50**, 755–779.

Villa, P. (1982). Conjoinable pieces and site formation processes. *Am. Antiquity* **47**, 276–290.

Whallon, R. (1974). Spatial analysis of occupation floors II: the application of nearest-neighbor analysis. *Am. Antiquity* **39**, 16–34.

Robert J.
Blumenschine
John A. Cavallo &
Salvatore D. Capaldo
*Department of Anthropology, Rutgers
University, New Brunswick, NJ 08903
U.S.A.*

Received 27 September 1993
Revision received 16 February
1994 and accepted 17 February
1994

Keywords: Paleoanthropology,
zooarchaeology, hominid
behavioral ecology, ecological
taphonomy.

Competition for carcasses and early hominid behavioral ecology: A case study and conceptual framework

The behavioral-ecological perspective provides paleoanthropology with the opportunity to uncover the course and causes of human behavioral evolution. At this time, however, there is no established conceptual framework for conducting behavioral-ecological studies of individuals of extinct hominid species. We provide such a framework by integrating two mutually reinforcing lines of inference currently pursued separately by paleoanthropologists: middle range research (defined broadly here to incorporate inferences based on archaeological, functional morphological, and paleoenvironmental data) and behavioral-ecological modelling. We argue that a testable and detailed behavioral ecology of prehistoric hominids must be grounded in such ecologically-oriented middle range research. We provide a zooarchaeological case study that recasts the current debates over carcass acquisition by early hominids and the socio-economic function of early archaeological sites within the framework provided by our behavioral-ecological approach. Specifically, we use the results of field studies and experiments to examine the ecological conditions and behavioral strategies that may have increased or decreased competition between hominids and carnivores for carcass foods. The case study generates testable hypotheses regarding the ecological causes and behavioral consequences of various foraging strategies that have been advanced for early hominids.

Journal of Human Evolution (1994) **27**, 197–213

The challenge of behavioral ecology in paleoanthropology

Viewing the hominid fossil and archaeological records through the perspective of behavioral ecology has the potential for providing a detailed understanding of the course and causes of human behavioral evolution. Behavioral ecology views an individual organism's behavior as one of a series of alternative strategies that maximizes its fitness under prevailing ecological conditions. In addition to using ecological factors to explain interspecific differences in behavior (e.g., the classic work of Crook, 1964; Crook & Gartlan, 1966; Emlen & Oring, 1977; Jarman, 1974; Lack, 1968), behavioral ecology allows for behavioral flexibility on the individual level within constraints imposed by phylogeny, physiology, ecology (Crook, 1989; Krebs & Davies, 1993), and, significantly for hominids, technology and other cultural traits. As such, the perspective recognizes that the ecological underpinnings of a behavioral strategy are themselves temporally and spatially variable, being defined by an interplay of numerous physical and biotic components of the environment. Isolating the ecological factors that affect specific hominid behaviors is therefore tantamount to identifying the selective agents that shaped hominid behavioral evolution.

As appealing as the concept of behavioral ecology is for paleoanthropology, there is no established procedure for conducting behavioral-ecological analyses of individuals of extinct species, including pre-*sapiens Homo* and *Australopithecus*. As a result, paleoanthropologists have no behavioral scenarios for early hominids that are both ecologically sound *and* firmly anchored in a detailed explication of the fossil evidence for hominid activities and their paleoenvironmental context.

Paleoanthropologists confront several challenges to understanding prehistoric hominid behavior in ecological terms. First, the archaeological and human paleontological evidence for

0047–2484/94/010197+17 $08.00/0

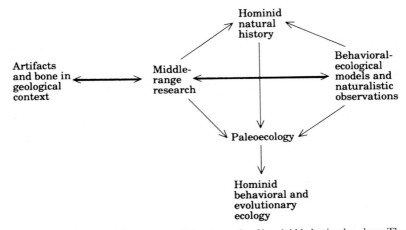

Figure 1. A schematic framework for conceptualizing the study of hominid behavioral ecology. The vertical axis represents the extinct hominid counterpart of behavioral-ecological studies of modern organisms. The horizontal axis represents taphonomic and naturalistic studies of modern processes and organisms applied to hominid fossils, archaeological remains, and paleontological and geological evidence for paleoenvironments. Arrows indicate the direction of inference within and between the axes. The relative thickness of the arrows reflects the importance we place on the middle range (defined broadly; see text), actualistic link between fossil evidence and behavioral ecological models, which is often de-emphasized in current behavioral-ecological modelling of fossil hominids.

hominid activities, and geological evidence for paleoenvironments, must be decoded accurately and in detail. Second, these reconstructions of activities and environments must be placed in their respective behavioral and paleoecological contexts. Further, this contextualized information on activities and environments must be merged according to models and principles of behavioral ecology that have been developed through the study of modern organisms. A conceptual framework for pursuing hominid behavioral ecology is needed. One is offered here and presented schematically in Figure 1.

The behavioral ecology of prehistoric hominids, or any extinct organism, should be based on the same components as the study of the behavioral ecology of modern organisms (vertical axis in Figure 1). As with modern organisms, accounts of activity repertoires, or what we refer to as "natural history", are essential building blocks of hominid behavioral ecology. However, since natural history emphasizes the description of behavior, even those natural history accounts that correlate behavior with aspects of environmental setting do not satisfy the mandate of behavioral ecology, which is to explain behavior in ecological terms. Many accounts of prehistoric hominid behavior, while using the jargon of behavioral ecology, are in fact natural history packaged in a loose paleoenvironmental context (e.g., geographic setting, rainfall, flora and fauna present) that fall short of specifying the paleoecological interactions (i.e., selective agents) that elicit a behavioral response in hominids. The paleoecological interactions may involve an aspect of the physical environment such as temperature or rainfall, or they may occur in the biotic realm in a form such as the distribution and abundance of food, the nature of competitive interactions, or the intensity of predator pressure. Finally, the validity of any ecological explanation rests on the extent to which it matches the predictions of existing behavioral-ecological models (e.g., those concerning diet breadth, ranging behavior, reproductive strategies), which in turn arise from consideration of general neo-Darwinian theory.

As challenging as it is to construct meaningful behavioral ecologies for modern organisms, the limited fossil evidence for activities and paleoenvironments demands that paleoanthropologists use an additional line of inference to construct prehistoric behavioral ecologies (horizontal axis of Figure 1). Unlike modern organisms, whose behavior and ecological relations can be observed directly, paleoanthropologists must work with artifacts, hominid fossils and those of other organisms, and a variety of geological data that are physical proxies for prehistoric activities and their environmental context. Further, the sparseness and fragmentary nature of the proxies render evidence for behavioral repertoires and environmental settings incomplete. As a result, even the construction of natural history, which is a fairly straightforward matter of empirical concern for modern ethologists, becomes inferential and subject to uncertainty when performed on extinct organisms (see Gifford-Gonzalez, 1991 for a useful and explicit hierarchical scheme for depicting the chain of inference from archaeological object to the generation of probability statements about what we refer to as natural history). Yet as Isaac (1968) pointed out, this fossil evidence is the most direct source of information on hominid behavioral strategies, and therefore the only factual basis for prehistoric hominid behavior ecology.

Inferences about hominid natural history and paleoecology, while necessarily based on fossil evidence, are guided by two types of studies of the modern world. First, the proxy nature of the behavioral and paleoecological records requires that middle range research play a central role in reconstructing hominid behavioral ecology (Figure 1). Here, we use the term middle-range research to characterize studies that link theory and data within a much broader range of disciplines than archaeology, which was Binford's (e.g., 1968, 1981) sole concern (see, for example, Raab & Goodyear, 1984, for the early usage of middle range theory in sociology). We are specifically concerned with that part of middle range research that documents the linkage between a modern process relevant to the formation of the archaeological, paleontological, or geological records, and the traces it produces. Hence, middle range research as depicted in Figure 1 incorporates not only the Binfordian components of taphonomy and actualistic studies of archaeological site formation, but also biomechanical and functional morphological studies that engender behavioral inferences from hominid fossils. Middle range research also includes research on the vast array of geochemical, sedimentological, paleontological, and palynological remains in the geological record for which the general methodology of inferring paleoenvironments forms a precedent for middle range research in its more limited archaeological useage (see, for example, Bradley, 1985 for a paleoclimatological formulation of middle range research). Second, naturalistic and experimental observations of the behavior of modern organisms must be used in conjunction with behavioral-ecological models to both flesh out behavioral strategies that could be decoded only partially from the sparse and incomplete fossil remains, and to conceive alternative ones.

In our view, the middle range and behavioral-ecological inputs from the modern world are interdependent and are ideally tapped simultaneously. On the one hand, the accuracy of behavioral-ecological reconstructions depends on the correct calibration of the fossil proxies for activities and paleoenvironments. On the other hand, the richness and realism of the reconstruction depends on the extent to which the modern behavioral models and observations have been understood and applied. In the absence of such models, the fossil record alone is unlikely to provide us with many details on, for example, hominid social organizational responses to different ecological conditions.

In our framework for reconstructing hominid behavioral ecology (Figure 1), middle range research forms a two-way inferential link between fossil evidence and behavioral-ecological models. Behavioral-ecological models developed from naturalistic observations and experimentation are used to deduce behavioral implications for the fossil record. Before being applied, the implications need to be translated into the "language" of the fossil record by middle range research. Conversely, it is the effectiveness by which middle range studies can be used to induce activities and aspects of paleoenvironments from fossil evidence that controls the type of behavioral-ecological model that can be applied in a testable manner, and the form that the model takes. For example, currently understood traces of paleodiets limit the archaeological application of optimal diet breadth models to the animal food component of the diet (Madsen, 1993), and, among parts of a given class of carcasses, only to marrow-bearing larger mammal long bones (cf. Blumenschine & Madrigal, 1993). In this way, we view middle range research as playing a pivotal role in operationalizing the behavioral ecological perspective in paleoanthropology.

In sum, our scheme for conceptualizing hominid behavioral ecology emphasizes the intersection of two mutually dependent lines of inference, one behavioral-ecological and the other taphonomic. Each is essential, for a testable hominid behavioral ecology cannot be developed without balanced attention to both. While paleoanthropologists have made impressive progress in developing the two lines of inference, the field of prehistoric hominid behavioral ecology is embryonic at best because these efforts have not been integrated.

One group of paleoanthropologists, the taphonomists or middle range researchers in the broadest sense, have been increasingly successful in documenting the nature of particular hominid activities. This research requires, as Isaac (1967:191) phrased it, "a fine exegesis of the excavated documents", where archaeological remains and hominid bones are interpreted through models derived from taphonomic research (experimental or naturalistic), and biomechanical and functional anatomical analyses, respectively. The result of this effort has been to document the existence of relatively simple, individual activities, such as stone tool manufacturing, transport and curation strategies (e.g., Schick, 1987; Toth, 1987), carcass butchery (e.g., Bunn & Kroll, 1986), aspects of hominid diet (e.g., Kay & Grine, 1988; Grine, 1981; Peters, 1981; Sillen, 1992; Ungar & Grine, 1991), and modes of hominid locomotion (e.g., Susman & Stern, 1982; Trinkhaus, 1986). Aspects of hominid social behavior, such as group size and composition, day range length, and mating systems are still largely beyond the resolution of middle range research. Regardless, the current achievements can be labelled "natural historic", for they simply describe hominid activities and biomechanical abilities, and, perhaps, also place them in paleoenvironmental contexts. This is a legitimate and important current goal of paleoanthropology (Cachel & Blumenschine, n.d.). However, such a descriptive, natural history approach does not enable us to explain behavioral variability in ecological terms (see also Gifford-Gonzalez, 1991). Most middle-range based efforts to view hominid activities within a wider behavioral and paleoenvironmental context lack the ecological realism that can only be appreciated from long-term naturalistic observations of living systems.

A second group of paleoanthropologists, who might be labelled the general theorists, have been applying ecological theory and principles in an attempt to explain general patterns of human evolution. Within current early hominid studies, this group is best represented by Foley (e.g., 1987, 1989; Foley & Lee, 1989; see also Bettinger, 1991; Earle & Christensen, 1980; Gamble, 1986; Ghiglieri, 1987; Gaulin, 1979; Jochim, 1979; Winterhalder, 1981; Wrangham, 1987). This approach replaces those of the late 1950s and 1960s, where the boldest reconstructions of early hominid lifestyles were produced not by those recovering and describing the fossil evidence, but rather by primatologists (e.g., Washburn & DeVore, 1961), zoologists (e.g., Morris, 1967), social scientists (e.g., Tiger & Fox, 1966), and a host of others working exclusively with modern organisms. The current efforts, while still emphasizing modern organisms, are distinguished from earlier ones in being explicitly ecological, such that the veracity of the reconstruction is based not on some formal property shared by the preferred modern analogue and the early hominid in question, but rather on general ecological principles relating aspects of the environment to behavioral responses in a wide range of species in different habitats (see, e.g., Andrews, 1983 & Dunbar, 1992 for examples from non-hominid fossil primates). These applications of modern ecological theory to hominids have been important for introducing modern principles from evolutionary and behavioral ecology to paleoanthropology, for addressing aspects of adaptation that are at best minimally visible in the fossil record (e.g., social behavior), and for enhancing the realism of scenarios of hominid lifestyles. However, many are tied too loosely to the available fossil evidence for activities and paleoenvironment due to insufficient attention to middle range research. The best current theoretical applications rely on gross proxies of hominid biology and behavior such as body size, and broad estimates of basic environmental parameters such as rainfall and temperature, such that many details of the fossil/geological evidence become subordinated to the models' predictions, serving at best as only a very general test of predictions. If modelling efforts were designed to include detailed middle range studies of natural history and paleoecology, the fossil evidence might attain its preferred place as the final arbiter of the models' appropriateness. Not only would the richness of behavioral reconstructions be enhanced, but we would be positioned to enjoy the ultimate benefit of adopting the behavioral-ecological perspective, namely, isolating in a verifiable way the selective agents shaping hominid behavioral strategies.

Paleoanthropologists can only realize the benefits of the behavioral-ecological perspective if their behavioral reconstructions are grounded in ecologically-oriented middle range research (*sensu lato*). In the remainder of this paper, this formulation of prehistoric hominid behavioral ecology is exemplified by examining larger mammal carcass utilization by hominids. We restrict ourselves to the manner in which zooarchaeological evidence for carcass utilization can be interpreted using behavioral-ecological models that are based on our experimental and naturalistic observations on the availability and use of carcasses by a variety of free-ranging carnivores in northern Tanzania. The critical ecological variable linking zooarchaeological data to carcass exploitation strategies is competition, a universal ecological phenomenon central to evolutionary and behavioral ecology (Krebs & Davies, 1993; Pianka, 1978). Competition is also one of the few ecological variables that have been incorporated widely into discussions of hominid evolution (e.g., Ambrose, 1986; Cachel, 1975; Foley, 1984, 1987; Schaffer, 1968; Winterhalder, 1981; Wolpoff, 1971; see also Grine, 1986). Here, we additionally show how the concept of competition can be used by paleoanthropologists to link the two taphonomic and behavioral-ecological components of our framework into a more effective tool for understanding many aspects of paleoecology and extinct hominid behavior that are currently poorly resolved by fossil evidence.

Towards a hominid behavioral ecology

We suggest that studying archaeological bone remains through the perspective of competition for carcasses can lead to testable reconstructions of part of the subsistence component of hominid behavioral ecology. We argue that bones are barometers of the level of competition between consumers of carcass tissues. The particular mix of skeletal parts, species, ages, the types and incidence of modification, and the spatial configuration of specimens in bone assemblages are informative about the level of competition, the identity of the competitors, and the strategic responses of carcass consumers to other competitors. It is difficult to conceive of another ecological variable that is more directly manifested in as many processes that affect bone during its relatively brief time as a food resource, and, later, in as many bone attributes that can be recorded by a zooarchaeologist. For the zooarchaeologist trying to reconstruct hominid adaptation, competition is the ecological filter that translates carcass exploitation behaviors into bony remains. Competition is a key process that warrants the perspective we refer to as "ecological taphonomy" (Blumenschine, 1988a), the manner in which ecological processes interact with the mechanical and energetic properties of bone to determine the taphonomic trajectory of a carcass' skeleton during its resource life.

In behavioral-ecological terms, competition for carcasses is any interaction between two individuals or groups of the same or different species that reduces access to a carcass, or reduces the nutritional benefit accrued from it. Competition, either direct (interference) or indirect (exploitative), can restrict, or elicit modifications in, the strategies used to obtain carcasses. Given the extreme brevity of the resource life of larger mammal carcasses in tropical African settings, and their relative rarity compared to most plant foods, competition for carcasses is likely to be intense relative to that experienced over other food resources. Many fundamental aspects of the niche and behavioral repertoire of a carcass consumer should be shaped by considerations that offset competition. These would include strategies such as the timing of access to carcasses by consumers and their efficiency and thoroughness of carcass consumption, all of which are ultimately based on such ecological variables as seasonality in the abundance and distribution of carcasses, and carnivore:herbivore biomass ratios.

The taphonomy of competition

Ecosystemic processes control the availability of ungulate and other larger mammal carcasses to consumers through the proximate mechanism of competition. Carcass availability is greatest and competition least in contexts where the carcass:consumer ratio (particularly the carcass: spotted hyena ratio) is the highest (Blumenschine, 1989). In the Serengeti region of northern Tanzania, carcass availability, as measured by either the persistence of edible tissues or the edible tissue completeness of carcasses upon discovery, has been shown to be greater in contexts where spotted hyenas, a bone crunching carnivore that can thoroughly consume all but buffalo-sized and larger carcasses, occur in small numbers, or are slow to discover carcasses (Table 1; Blumenschine, 1986a). Hence, carcass availability is greater:

(a) in ecosystems with a low carnivore:ungulate ratio arising from annual migrations of herd animals (e.g., Serengeti *vs.* Ngorongoro Crater);

(b) within the Serengeti during the dry season in the northern refuge of the migratory herds, and during the long rainy season on the short grass plains;

(c) in riparian woodlands versus non-riparian or open riparian settings;

(d) when flesh-eating carnivores (felids) are the initial carcass consumer;

(e) when carcasses are located in trees (as in cached leopard kills) versus on the ground;

(f) probably during mid-day, when carnivore activity is low, rather than at dawn or at dusk; and,

(g) when multiple carcasses are simultaneously available, as in mass drownings, mass death due to starvation, and multiple kills.

These ecological correlates of reduced competition for carcasses have distinct bony signatures, many of which we have documented with a variety of experimental and naturalistic observations that can be labelled as ecological taphonomy. These studies have either

Table 1 Ecological, zooarchaeological, and behavioral correlates of competition for larger mammal carcasses, as documented in the Serengeti ecosystem of northern Tanzania. See text for citations

	Low competition	High competition
Ecosystemic variable (N. Tanzania)		
Habitat	Riparian woodlands	Non-riparian *or* non-wooded
Season and place (Serengeti only)	Dry in north, wet on plains	Wet in north? Dry on plains
Type of competing carnivore	Flesh-eater	Bone-cruncher
Carnivore:ungulate ratio	Low (Serengeti)	High (Ngorongoro)
Stratum	Arboreal	Terrestrial
Time of day	Mid-day	Dawn/dusk
Cause of death	Mass drowning	Predation by bone-crunching carnivore
Zooarchaeological traces		
Skeletal parts	Little differential destruction	Denser bones survive preferentially
Portions	High epiphysis:shaft fragment ratio	Low epiphysis:shaft fragment ratio
Species	Low carnivore:ungulate ratio; dominated by single species as in glut; more small species	High carnivore:ungulate ratio; high species diversity?; more large species
Age (mortality)	High neonate:adult ratio	Low neonate:adult ratio
Fragmentation	Low NISP:MNE or MNI	High NISP:MNE or MNI
Surface marks	Low incidence; single agent	High incidence; multiple agents
Behavioral strategy		
Timing of access	Planned search?; unrestricted access, including early in resource life	Opportunistic search?; access restricted to end of resource life
Processing/ consumption	Energy limited?; complete consumption	Time limited?; consumption of highest net yield available
Transport	Process and consume where found	Transport easily removed parts to safe/concealed locale
Defense/pilfering	No strategies needed	Cooperative and/or artificial techniques
Predation risk from competitors	Lower	Higher

demonstrated or suggested causal links between ecology and taphonomic consequences through a variety of mechanical and energetic mechanisms. Many of these observations have been reported, but here they are contextualized within the single framework provided by competition (Table 1).

Skeletal part and portion profiles have been shown to be sensitive to the level of competition for carcasses and carcass parts, respectively. In contexts of high competition, carcasses and individual skeletal parts are consumed relatively completely, so that skeletal part survival is biased toward the densest bones, the densest portions of individual skeletal parts, and/or to the lowest yielding bones [these subsets of the skeleton are generally the same, given the covariation between a part's food content and its density (Lyman, 1985)]. This has been shown by comparing the composition of landscape bone assemblages found in Ngorongoro Crater (highly competitive) with those from various habitats within the Serengeti (Blumenschine, 1989), and by comparing the amount of disturbance to experimental bone assemblages in different habitats of the Serengeti (Blumenschine, 1988*b*; Capaldo, n.d.). Conversely, the simultaneous availability of numerous carcasses creates glut conditions, where the local carnivore population is saturated with food. Occurring as a result of mass drownings (Capaldo & Peters, 1994), mass kills (Kruuk, 1972:195–196) or high rates of mortality due to drought conditions (Behrensmeyer & Boaz, 1980; Blumenschine, 1986*a*; Haynes, 1982, 1988; Shipman, 1975), glut results in the very incomplete use of carcasses, generally only the most nutritious parts. For the Serengeti region, these ecosystem- and habitat-based patterns in skeletal part profiles have been linked by direct observations to the relative rapidity and completeness of consumption of fresh carcasses by carnivores (Blumenschine, 1986*a*,*b*).

The relationship between bone density and survivability has been known since Brain's (1967, 1969) pioneering work, but only now is it becoming apparent that heavily ravaged assemblages (i.e., those over which intense competition occurred) will come to be represented by a similar, heavily depleted inventory of parts virtually *regardless* of initial composition (Binford 1981:224–225; similar results can be seen in data provided by Brain, 1981; Crader, 1983, and Yellen, 1977). This renders as suspect the common use of skeletal part profiles to infer aspects of an assemblage's resource-life history, including the mode of carcass acquisition (Binford, 1981; Blumenschine, 1986*b*; Bunn & Kroll, 1986; Potts, 1983), the identity of bone assemblage accumulators (e.g., Klein, 1980), and whether the assemblage has been transported (Bunn & Kroll, 1986; Bunn *et al.*, 1988; O'Connell *et al.*, 1988; Perkins & Daly, 1968). Before such behavioral diagnoses can be made, we must determine the degree to which carnivore destruction and/or dispersal depleted the inventory of postcranial axial elements, epiphyseal portions of long bones, or non-tooth portions of crania and mandibles (Marean *et al.*, 1992). In more heavily ravaged assemblages, it may be that skeletal part and portion profiles are securely informative only about the intensity of competitive interactions over nutrients remaining *within* bones (marrow, grease, brain, etc.).

Potts (1984) has used species profiles, particularly the ratio of mammalian carnivore to ungulate individuals, to infer the degree of competition between carnivores and hominids for carcass parts at Early Pleistocene sites from Olduvai Gorge. His criterion is based on Klein's (1975) demonstration that carnivore den assemblages typically preserve a higher proportion of carnivore individuals than those in other contexts. It is unclear, however, whether the ratio is sensitive to competition because the behavioral and ecological mechanisms that produce the ratio have not been specified or tested.

Another aspect of species representation, species diversity, may be linked to levels of competition for the flesh, marrow, and grease foods in carcasses. Very low species diversity in an assemblage, for example, is commonly linked to mass procurement episodes, including mass kills and the aforementioned drownings and drought conditions which can provide numerous individuals of the same species to scavengers. However, higher species diversity is not necessarily produced in more competitive contexts.

The size and age diversity of an ungulate bone assemblage, particularly the proportion of small species and young individuals, are sensitive to levels of competition for carcasses at procurement and processing sites. Many researchers have documented the lower resistance of bones of smaller animals to destructive forces, and the completeness by which carnivores can consume these carcasses. Controlling for differences in the size structure of animal communities and age structure of particular species, bone assemblages biased towards larger species or adult individuals can therefore be seen to have accumulated in a more competitive context.

More recently, it is starting to become clear that bone modification can be informative about the intensity and object of competition among carnivores for carcasses, and the identity of the competitors. This seems to be the case for fracture features such as notches (Capaldo & Blumenschine, 1994), and especially for marks on bone surfaces (Blumenschine, 1988*b*; Blumenschine & Marean, 1992). It has been suggested for some time that the mere presence of carnivore tooth-marked bone in a hominid butchered assemblage is indicative of competition between the two. However, the simplicity of the criterion has led to opposed interpretations of the effect of the competitors on one another. At one extreme, Binford's interpretation of gnawed bone at early Olduvai sites (e.g., 1981; 1986) can be read to mean that carnivores outcompeted hominids for all but the most marginal carcass foods. Bunn (Bunn *et al.*, 1980; Bunn & Kroll, 1986), at the other end, places hominids as the primary beneficiaries of carcass foods, with carnivores inflicting most of the gnaw damage only later. Bunn & Ezzo's (1993) recent advocacy of confrontational scavenging suggests that, in our terms, Bunn would see hominids as not only superior exploitative competitors, but also as potent interference competitors for the carcasses they acquired. To make such determinations about interspecific competition, however, we need information on not simply the types, but also on the incidence and location of surface marks and notches inflicted by a particular agent. Experiments have shown, for instance, that the incidence of percussion-marked and tooth-marked long bone fragments can resolve the sequence by which hominids and carnivores gained access to marrow and grease (Blumenschine, 1988*b*, 1994; see Oliver, 1994 for similar analyses). Further, in combination with epiphyseal:shaft fragment ratios, the incidence of tooth-marked long bone fragments may be able to distinguish whether carnivore ravaging of a hominid-butchered assemblage occurred in a setting characterized by high or low competition (Blumenschine, 1988*b*; Blumenschine & Marean, 1992).

Competitively-based behavioral strategies of hominids

The above summary of ecological taphonomy is informative mainly about the manner in which carnivores pattern bone in response to different levels of competition. These patterns of carcass availability and associated bony traces can be argued to form a baseline from which hominid interactions with carcasses and other consumers arise. We note that the baseline can include no preliminary carnivore input, where hominids enjoy uncontested access to complete carcasses.

The intensity of competitive interactions can be used to define alternative strategies employed by hominids in the search for, and procurement, transport (if any), handling, and consumption of carcasses (Table 1). In this way, we can reformulate in behavioral-ecological terms and integrate two current issues in Plio-Pleistocene paleoanthropology which we will view as sequentially enacted strategies for offsetting competition with carnivores: methods of carcass procurement by hominids (focussing on various types of scavenging), and the socio-economic function of archaeological sites (routed foraging, refuging, central place foraging, and "stone caching", see Figure 2). In focussing on the behavioral-ecological issue of strategies of carcass utilization, we assume a foundation of information more routinely generated by middle range explications of the fossil and geological records. This includes natural historic evidence for hominid involvement with carcasses (e.g., butchery marks) and paleoenvironmental data on vegetation structure and the structure of mammalian carnivore and herbivore guilds (e.g., plant and animal fossils).

We argue that hominids should adopt behavioral strategies to acquire and process carcasses that minimize competition from carnivores. Competition reduces the benefit accrued to hominids from carcass utilization because it can:

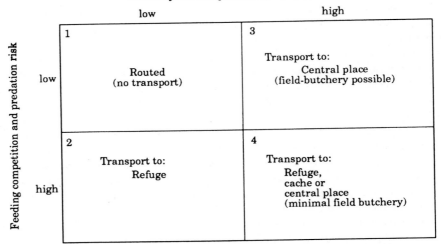

Figure 2. Behavioral-ecological model of the foraging strategies of early flaked-stone-tool-using hominids as they are contingent upon circumstances encountered at a carcass acquisition site (after Blumenschine, 1991). Circumstances are defined by intensity of intra- and interspecific competition, level of predation risk, and carcass yield in relation to processing equipment needs (stone knives and hammers). Carcass utilization strategies are labelled according to the foraging mode responsible for the creation of archaeological sites at which carcass remains are found, as hypothesized by Binford (1981; routed foraging), Isaac (1983; central place foraging and the concept of refuging; see Blumenschine, 1991); and Potts (1984; stone caching). Each of the four contexts of carcass acquisition (number in each quadrant) is defined by specific scavenging opportunities, which include:

Quadrant 1: tree-stored leopard kills of small (size 1) subadult ungulates;
Quadrant 2: open-habitat cheetah or jackal kills of size 1 ungulates;
Quadrant 3: tree-stored leopard kills of adult size 1 and larger ungulates; larger felid (lion, saber-tooth) kills abandoned on the ground; drownings, if in wooded habitat;
Quadrant 4: natural deaths or abandoned kills of medium-sized and larger ungulates in open habitats; usurping social carnivore kills.

(a) increase search time by rendering carcass location less predictable, and decrease the overall abundance of carcasses;

(b) increase procurement and possession costs by requiring coordinated and active pilfering or defense strategies;

(c) decrease procurable yields due to the more complete consumption of any carcasses found;

(d) decrease the efficiency and/or completeness of carcass processing by forcing hominids to incur the cost of transporting carcasses to less competitive locales, and/or by forcing hominids to process carcasses in a time-limited mode whereby only the most quickly processed units are removed for consumption; and

(e) increase predation risk when competition is direct (interference) and involves the larger felids, hyaenids, or canids. Even competition from vultures increases this risk because large carnivores are attracted to carcasses by vulture movements.

Our naturalistic observations have shown that the predictability, yield, and risk associated with opportunities to scavenge larger mammal carcasses are extremely variable. The best opportunities occur in the low competition contexts, including remains of kills abandoned on the ground or in trees in riparian woodland, especially in the dry season in the Seronera area of the Serengeti (Blumenschine & Cavallo, 1992; Cavallo, n.d.; Cavallo & Blumenschine, 1989). Today, lion kills and tree-stored leopard kills in these settings are predictably located. The proximity of trees, which can serve as refuge sites and as shields from discovery of carcasses by vultures, renders these opportunities low risk. As well, the generally high yield of abandoned kills, particularly tree-stored leopard kills, suggests that the more risky strategy of confrontational scavenging may not be the optimal one, nor even necessary to explain the composition of known Plio-Pleistocene archaeological bone assemblages (*contra* Bunn & Ezzo, 1993).

Other high-yielding strategies include scavenging from natural deaths of very large animals, and mass deaths resulting from drowning or drought. As bountiful as these opportunities can be, they are rendered more opportunistic by their relative rarity and unpredictability. As well, unless a sufficient number of deaths occurred simultaneously or sequentially so as to swamp local scavengers with food, the open-country setting of many of these carcasses would have also exposed hominids to dangerous competitors. We infer that all of the above scavenging opportunities probably existed in Pleistocene savanna-woodlands along with the additional low competition opportunity possibly provided by abandoned sabertooth cat kills in the same habitats (Blumenschine, 1987; Marean, 1989).

Once a carcass is located and procured, processing and consumption can occur on the spot, or the whole carcass or portions of it can be transported to a different locale for consumption (Table 1 and Figure 2). In a model where competition is the only constraint on maximizing the net yield of carcass consumption, hominids acquiring carcasses in low competition contexts should simply process and consume the carcass at the discovery point (routed foraging). Abandoned felid kills found in riparian woodland contexts would seem to qualify particularly well for this type of treatment. Any carcass acquired in a higher competition context, such as natural deaths or felid kills found in open settings, should be removed for processing to wooded areas where the carcass would be concealed by trees and hence any vultures that would attract potent competitors (Cavallo, n.d.). This strategy can be labelled "refuging" (cf. Isaac, 1983; Blumenschine, 1991; see Figure 2).

These considerations of the model suggest that the marginal scavenging strategy and associated routed foraging mode advocated by Binford (e.g., 1981) for hominids were ecologically incompatible. His postulation that hominids exploited carcasses consumed thoroughly by carnivores suggests that scavenging occurred in highly competitive contexts that additionally posed considerable predation risk and other disadvantages associated with high competition that were itemized earlier. Yet routed foraging is a strategy modelled here to occur when levels of competition and predation risk are too low to motivate the carcass transport associated with other foraging modes (Figure 2 and below). Heavily ravaged carcasses are unlikely to have been encountered frequently and exploited by hominids practicing routed foraging.

Predation risk and stone tool processing needs in relation to carcass yield are two other factors impinging on carcass transport decisions (Figure 2). The need for carcass processing equipment is a constraint motivating carcass transport regardless of the level of competition at the carcass acquisition site. The equipment constraint underlies Potts' (1984) suggestion that hominids should transport carcasses to "caches" of previously stockpiled stone. However, because Potts views stone caches as foci of dangerously competitive interactions between hominids and carnivores (Figure 2), it is clear that this strategy should only be practiced if low-competition contexts for carcass processing did not exist. Indeed, we model avoidance of predation risk as another motivation for transporting carcass parts, but here to a safe refuge for further processing. Since dangerous predators form a subset of potential competitors for carcasses, the characteristics of the transport site are the same regardless of whether the motivation to transport is to avoid predators or competitors, or to meet processing equipment needs. Finally, a carcass providing yields in excess of what the foraging party can consume in a single sitting might encourage transport to a protected and central place, where the surplus could provision other group members (Isaac, e.g., 1978) or alternatively be cached for future use (Cavallo, n.d.).

Summary and future directions

We have attempted to show that foraging modes practiced by hominids can be viewed as alternative behavioral strategies used to exploit a variety of specific scavenging opportunities. The scavenging opportunities most profitably exploited, in turn, are defined by a series of ecological variables influencing the abundance, spatio-temporal distribution, and, therefore, the degree of competition for carcasses. A behavioral-ecological perspective requires us to view hominids as organisms potentially capable of employing multiple strategies to acquire and process carcasses in correspondence to prevailing ecological conditions, as reflected in our case study by levels of competition for carcass nutrients.

According to the carcass utilization model, routed foraging should occur only in contexts of low competition, where the predation risk is too low to necessitate carcass transport. In these settings, hominids would find relatively intact carcasses, not the heavily ravaged, nutritionally marginal ones inferred by Binford. Likewise, our model predicts that hominids should transport carcasses for processing to dangerous stone caches only if low-competition processing locales are unavailable. The absence of safe refuges at fossil localities such as Olduvai Gorge, however, seems unlikely given the commonness of groves of climbable trees in modern savanna-woodland mosaics, particularly those in riparian settings. These and other predictions of the carcass utilization model are testable because we have expressed them in terms of competition, an ecological process that we have also shown to have specific and detectable zooarchaeological signatures.

Our zooarchaeological case study is a step toward implementing our conceptual framework for reconstructing the behavioral ecology of extinct hominids, which is fundamentally based on an ecologically-oriented, middle range assessment of the fossil record. Viewing paleo-anthropological issues such as carcass utilization strategies in behavioral-ecological terms is desirable, but at least four areas of related research need to be pursued to fully develop this zooarchaeological case study alone.

First, following the methodology of wildlife ecologists, we advocate that archaeologists conduct long term, continuous observations of competitive feeding interactions among carnivores, and related aspects of carnivore behavior. At the same time, we need to study the movements and other habits of prey species including anthropoid primates (e.g., Boesch, 1991; Boesch & Boesch, 1989; Busse, 1980; Isbell *et al.*, 1990), whose response to predation risk from carnivores may illustrate strategies adopted by hominids to avoid predation when exploiting food sources, including scavengeable carcasses. The best appreciation of carnivore and prey behavior would be gained through comparative interpopulational field studies, where ecosystemic differences would be expected to elicit a greater diversity of behavioral responses explicable in terms of habitat or other ecological factors visible in the fossil record. Other benefits of studying these kinds of ecosystemic differences include the chance to assess the universality of environmental cues used to locate carcasses (e.g., circling and stooping vultures), the strategies used to prevent competitors from picking up on the cues (e.g., tree-storage and tree-cover), and the relationship between the location of carcasses, non-carcass foods (e.g., insects, fruit), and other affordances (e.g., potable water, refuge trees) that might reinforce the attraction of hominid consumers to particular feeding opportunities (Peters & Blumenschine, n.d.). In this fashion, we are invoking the rationale of ethnoarchaeology: as with ethnographic databases compiled by ethnographers, wildlife studies conducted by zoologists rarely provide archaeologists with the exact types and form of data that can be applied to the fossil record.

We also need a much more diverse and detailed understanding of ecological taphonomy. Most fundamentally, we need to investigate whether the bony correlates of competition for carcasses based on experimental and naturalistic observations in the Serengeti pertain to other savanna-woodland ecosystems (cf. Behrensmeyer, 1993). As well, experimental simulated sites need to be expanded in scope in the directions initiated by Capaldo (1990, n.d.) to include all skeletal parts, a greater size and species diversity of bones, a greater variety of butchery methods, and information on the spatial displacement of remains. The extent to which these taphonomic variables are affected by competition must be investigated not only by placing simulated sites in contexts where competition is known to vary on other grounds, but also by directly observing the number and identity of scavenging individuals who disturb the sites. We also need to define in more detail the ecological taphonomy of additional scavenging opportunities including drowning (Capaldo & Peters, 1994), drought (e.g., Blumenschine, 1986a), and those provided by extinct species (e.g., Marean, 1989). Finally, we need to develop new and more sensitive osteological signatures of carcass utilization, such as Selvaggio's (1994) efforts to recognize taxon-specific tooth-marking by carnivores.

We can improve the zooarchaeological contribution to the behavioral ecology of hominid carcass utilization if the following data are systematically recorded and reported for many more assemblages than the current handful. According to the ecological taphonomy results we have obtained to date, we require information on (a) the location and incidence of carnivore tooth marks and hominid butchery marks; (b) more comprehensive element and portion profiles; and (c) epiphyseal:shaft fragment ratios for long bones. While these data are routinely recorded by some zooarchaeologists, the methods used to derive the data are often incompatible, and even flawed: for example, recognition of tooth marks must not be restricted to conspicuous areas of gross gnawing (Blumenschine & Marean, 1992). Further, minimum numbers profiles of skeletal parts, such as those for long bones, need to be based on shaft fragments *in addition to* epiphyses (Bunn & Kroll, 1986; Capaldo, n.d.; Marean & Spencer, 1991; see Lyman, 1994 for other examples).

Finally, we need to alter methods used to sample the fossil record for archaeological and paleoenvironmental data that are relevant to hominid behavioral ecology. The standard, diachronic approach relied upon by Early Stone Age archaeologists (e.g., Leakey, 1971) is poorly suited to this end because it only samples a minute fraction of the behavioral repertoire and paleoenvironmental setting for any one time period. Rather, we need to sample evidence of hominid behavior and paleoenvironments on regional scales, to the extent that this is permitted by the limited exposure of laterally extensive isochronous units of Plio-Pleistocene age. This is one goal of the landscape archaeology projects recently initiated at several Early Pleistocene localities (Blumenschine & Masao, 1991; Isaac & Harris, 1980; Potts, 1989; Stern, 1993), where data on the regional density distribution, composition, and condition of fossil bones and artifacts can be related to paleogeography, depositional environment, and aspects of soil chemistry (Blumenschine & Masao, 1991). Analysed for attributes shown to be sensitive to ecological variables such as competition, and placed within the framework provided by naturalistic observations of the behavioral strategies of modern species, these fossil data can be integrated in a way that will permit us to understand the behavioral ecology of extinct hominid populations.

Acknowledgements

This paper is based on a manuscript prepared for the symposium entitled, "Early Hominid Behavioral Ecology: New Looks at Old Questions" at the 62nd Annual Meeting of the American Association of Physical Anthropologists, Toronto, Canada, 1993. We thank the symposium organizers, Jim Oliver, Nancy Sikes, and Kathy Stewart, for the invitation to prepare this paper, and Lynne Isbell, Nancy Sikes, Dieter Steklis and three reviewers including Rob Foley for valuable comments on the penultimate draft.

References

Ambrose, S. H. (1986). The earliest archaeological traces: proximate and ultimate causes. Paper presented at the 85th Annual Meeting of the American Anthropological Association, Philadelphia.

Andrews, P. J. (1983). The natural history of *Sivapithecus*. In (R. L. Ciochon & R. S. Corruccini, Eds) *New Interpretations of Ape and Human Ancestry*, pp. 441–463. New York: Plenum.

Behrensmeyer, A. K. (1993). Discussion: noncultural processes. In (J. Hudson, Ed.). *From Bones to Behavior*, pp. 342–348. Southern Illinois University at Carbondale.

Behrensmeyer, A. K. & Boaz, D. E. D. (1980). The recent bones of Amboseli National Park, Kenya, in relation to East African paleoecology. In (A. K. Behrensmeyer & A. P. Hill, Eds) *Fossils in the Making: Vertebrate Taphonomy and Paleoecology*, pp. 72–92. Chicago: University of Chicago Press.

Bettinger, R. L. (1991). *Hunter-Gatherers; Archaeological and Evolutionary Theory*. New York: Plenum Press.

Binford, L. R. (1968). Archaeological perspectives. In (S. R. Binford & L. R. Binford, Eds) *New Perspectives in Archaeology*, pp. 5–33. Chicago: Aldine Press.

Binford, L. R. (1981). *Bones: Ancient Men and Modern Myths*. New York: Academic Press.

Binford, L. R. (1986). Comment to Bunn, H. T. & Kroll, E. M. (1986). Systematic butchery by Plio-Pleistocene hominids at Olduvai Gorge, Tanzania. *Curr. Anthropol.* **27**, 444–446.

Blumenschine, R. J. (1986a). *Early Hominid Scavenging Opportunities: Implications of Carcass Availability in the Serengeti and Ngorongoro Ecosystems*. Oxford: British Archaeological Reports International Series 283.

Blumenschine, R. J. (1986b). Carcass consumption sequences and the archaeological distinction of scavenging and hunting. *J. hum. Evol.* **15**, 639–659.

Blumenschine, R. J. (1987). Characteristics of an early hominid scavenging niche. *Curr. Anthropol.* **28**, 383–407.

Blumenschine, R. J. (1988a). The taphonomic trio: mechanics, energetics, and ecology. Paper presented at the 53rd Annual Meeting of the Society for American Archaeology, Phoenix.

Blumenschine, R. J. (1988b). An experimental model of the timing of hominid and carnivore influence on archaeological bone assemblages. *J. Archaeol. Sci.* **15**, 483–502.

Blumenschine, R. J. (1989). A landscape taphonomic model of the scale of prehistoric scavenging opportunities. *J. hum. Evol.* **18**, 345–371.

Blumenschine, R. J. (1991). Hominid carnivory and foraging strategies, and the socio-economic function of early archaeological sites. *Phil. Trans. R. Soc. Lond. B* **334**, 211–221.

Blumenschine, R. J. (1994). Percussion marks, tooth marks, and experimental determinations of the timing of hominid and carnivore access to long bones at FLK *Zinjanthropus*, Olduvai Gorge, Tanzania. *J. hum. Evol.* (in press).

Blumenschine, R. J. & Cavallo, J. A. (1992). Scavenging and human evolution. *Sci. Am.* **267**(4), 90–96.

Blumenschine, R. J. & Madrigal, T. C. (1993). Variability in long bone marrow yields of East African ungulates and its zooarchaeological implications. *J. Archaeol. Sci.* **20**, 555–587.

Blumenschine, R. J. & Marean, C. W. (1992). A carnivore's view of archaeological bone assemblages. In (J. Hudson, Ed.) *From Bones to Behavior*, pp. 273–300. Southern Illinois University at Carbondale.

Blumenschine, R. J. & Masao, F. T. (1991). Living sites at Olduvai Gorge, Tanzania? Preliminary landscape archaeology results in the basal Bed II lake margin zone. *J. hum. Evol.* **21**, 451–462.

Boesch, C. (1991). The effects of leopard predation on grouping patterns in forest chimpanzees. *Behavior* **117**, 220–241.

Boesch, C. & Boesch, H. (1989). Hunting behavior of wild chimpanzees in the Tai National Park. *Am. J. phys. Anthropol.* **78,** 547–573.

Bradley, R. S. (1985). *Quaternary Paleoclimatology.* Boston: Allen & Unwin.

Brain, C. K. (1967). Hottentot food remains and their bearing on the interpretation of fossil bone assemblages. *Sci. Papers Namib Desert Res. Stn.* No. **32,** 1–7.

Brain, C. K. (1969). The contribution of Namib Desert Hottentots to an understanding of Australopithecine bone accumulations. *Sci. Papers Namib Desert Res. Stn.* No. **39,** 13–22.

Brain, C. K. (1981). *The Hunters or the Hunted? An Introduction to African Cave Taphonomy.* Chicago: University of Chicago Press.

Bunn, H. T. & Ezzo, J. A. (1993). Hunting and scavenging by Plio-Pleistocene hominids: Nutritional constraints, archaeological patterns, and behavioural implications. *J. Archaeol. Sci.* **20,** 365–389.

Bunn, H. T. & Kroll, E. M. (1986). Systematic butchery by Plio-Pleistocene hominids at Olduvai Gorge, Tanzania. *Curr. Anthrop.* **27,** 431–452.

Bunn, H. T., Bartram, L. E. & Kroll, E. M. (1988). Variability in bone assemblage formation from Hadza hunting, scavenging, and carcass processing. *J. Anthropol. Archaeol.* **7,** 412–457.

Bunn, H. T., Harris, J. W. K., Isaac, G. Ll., Kaufulu, Z., Kroll, E. M., Schick, K., Toth, N. & Behrensmeyer, A. K. (1980). FxJj50: an early Pleistocene site in northern Kenya. *World Archaeol.* **12,** 109–136.

Busse, C. (1980). Leopard and lion predation upon chacma baboons living in the Moremi Wildlife Reserve, Botswana. *Botswana Notes and Records* **27,** 15–21.

Cachel, S. (1975). A new view of speciation in *Australopithecus.* In (R. H. Tuttle, Ed.) *Paleoanthropology: Morphology and Paleoecology,* pp. 183–201. The Hague: Mouton Press.

Cachel, S. & Blumenschine, R. J. (n.d.). The natural history strategy in paleoanthropology. Manuscript on file, Dept. Anthropology, Rutgers University.

Capaldo, S. D. (1990). Differential treatment of axial and appendicular elements by scavenging carnivores at simulated archaeological sites. Paper presented at the Biennial Conference of the Society for Africanist Archaeologists, Gainesville, Florida.

Capaldo, S. D. (n.d). Inferring hominid and carnivore behavior from dual-patterned archaeofaunal assemblages. Ph.D. Dissertation, Rutgers University.

Capaldo, S. D. & Blumenschine, R. J. (1994). A quantitative diagnosis of notches made by hammerstone percussion and carnivore gnawing on bovid long bones. *Am. Antiquity* (in press).

Capaldo, S. D. & Peters, C. R. (1994). Skeletal inventories from wildebeest drownings at lakes Masek and Ndutu in the Serengeti ecosystem of Tanzania. *J. Archaeol. Sci.* (in press).

Cavallo, J. A. (n.d.). Back to bases: A reconsideration of central place foraging by early East African hominids. Manuscript on file, Dept. Anthropology, Rutgers University.

Cavallo, J. A. & Blumenschine, R. J. (1989). Tree-stored leopard kills: expanding the hominid scavenging niche. *J. hum. Evol.* **18,** 393–399.

Crader, D. C. (1983). Recent single-carcass bone scatters and the problem of "butchery" sites in the archaeological record. In (J. Clutton-Brock & C. Grigson, Eds) *Animals and Archaeology. 1. Hunters and their Prey,* pp. 107–141. Oxford: British Archaeological Reports International Series 163.

Crook, J. H. (1964). The evolution of social organisation and visual communication in the weaver birds (Ploceinae). *Behaviour* Suppl. **10,** 1–178.

Crook, J. H. (1989). Socioecological paradigms, evolution and history: perspectives for the 1990s. In (V. Standen & R. A. Foley, Eds) *Comparative Socioecology,* pp. 1–36. Oxford: Blackwell Scientific Publications.

Crook, J. H. & Gartlan, J. S. (1966). Evolution of primate societies. *Nature* **210,** 1200–1203.

Dunbar, R. I. M. (1992). Behavioral ecology of the extinct papionines. *J. hum. Evol.* **22,** 407–421.

Earle, T. K. & Christensen, A. L. (1980). *Modeling Change in Prehistoric Subsistence Economies.* New York: Academic Press.

Emlen, S. T. & Oring, L. W. (1977). Ecology, sexual selection and the evolution of mating systems. *Science* **197,** 215–223.

Foley, R. A. (1984). Early man and the red queen: Tropical African community evolution and hominid adaptation. In (R. A. Foley, Ed.) *Hominid Evolution and Community Ecology: Prehistoric Human Adaptation in Biological Perspective,* pp. 1–24. New York: Academic Press.

Foley, R. A. (1987). *Another Unique Species: Patterns in Human Evolutionary Ecology.* New York: John Wiley & Sons, Inc.

Foley, R. A. (1989). The evolution of hominid social behaviour. In (V. Standen & R. A. Foley, Eds) *Comparative Socioecology,* pp. 473–494. Oxford: Blackwell Scientific Publications.

Foley, R. A. & Lee, P. C. (1989). Finite social space, evolutionary pathways, and reconstructing hominid behavior. *Science* **243,** 901–906.

Gamble, C. S. (1986). *The Palaeolithic Settlement of Europe.* Cambridge: Cambridge University Press.

Gaulin, S. J. C. (1979). A Jarman/Bell model of primate feeding niches. *Hum. Ecol.* **7,** 1–20.

Ghiglieri, M. P. (1987). Sociobiology of the great apes and the hominid ancestor. *J. hum. Evol.* **16,** 319–357.

Gifford-Gonzalez, D. (1991). Bones are not enough: Analogues, knowledge, and interpretive strategies in zooarchaeology. *J. Anthropol. Archaeol.* **10,** 215–254.

Grine, F. E. (1981). Trophic differences between "Gracile" and "Robust" Australopithecines: A scanning electron microscope analysis of occlusal events. *S. Afr. J. Sci.* **77,** 203–230.

Grine, F. E. (1986). Ecological causality and the pattern of Plio-Pleistocene hominid evolution in Africa. *S. Afr. J. Sci.* **82,** 87–89.

Haynes, G. (1982). Prey bones and predators: Potential ecologic information from analysis of bone sites. *Ossa* **7,** 75–97.

Haynes, G. (1988). Mass deaths and serial predation: Comparative taphonomic studies of modern large mammal death sites. *J. Archaeol. Sci.* **15,** 219–235.

Isaac, G. Ll. (1967). Towards the interpretation of occupation debris: Some experiments and observations. *Kroeber Anthropol. Soc. Papers* **37,** 30–57.

Isaac, G. Ll. (1968). Traces of Pleistocene hunters: An East African example. In (R. B. Lee & I. DeVore, Eds) *Man the Hunter.* Chicago: Aldine Publishing Company.

Isaac, G. Ll. (1978). Food-sharing and human evolution: Archaeological evidence from the Plio-Pleistocene of East Africa. *J. Anthropol. Res.* **34,** 311–325.

Isaac, G. Ll. (1983). Bones in contention: Competing explanations for the juxtaposition of Early Pleistocene artifacts and faunal remains. In (J. Clutton-Brock & C. Grigson, Eds) *Animals and Archaeology. 1. Hunters and their Prey*, pp. 3–19. Oxford: British Archaeological Reports International Series 163.

Isaac, G. Ll. & Harris, J. W. K. (1980). A method for determining the characteristics of artefacts between the sites in the Upper Member of the Koobi Fora Formation, East Lake Turkana. In (R. E. Leakey & B. A. Ogot, Eds) *Proceedings of the 8th Panafrican Congress of Prehistory and Quaternary Studies*, pp. 19–22. Nairobi: The International Louis Leakey Memorial Institute for African Prehistory.

Isbell, L. A., Cheney, D. L. & Seyfarth, R. M. (1990). Costs and benefits of home range shifts among vervet monkeys (*Cercopithecus aethiops*) in Amboseli National Park, Kenya. *Behav. Ecol. Sociobiol.* **27,** 351–358.

Jarman, P. J. (1974). The social organization of antelope in relation to their ecology. *Behaviour* **48,** 215–267.

Jochim, M. A. (1979). Breaking down the system: Recent ecological approaches in archaeology. *Adv. Archaeol. Meth. Theory* **2,** 77–117.

Kay, R. F. & Grine, F. E. (1988). Tooth morphology, wear, and diet in *Australopithcus* and *Paranthropus* from Southern Africa. In (F. E. Grine, Ed.) *Evolutionary History of the "Robust" Australopithecines*, pp. 427–447. New York: Aldine de Gruyter.

Klein, R. G. (1975). Paleoanthropological implications of the nonarchaeological bone assemblage from Swartklip I, South-Western Cape Province, South Africa. *Quat. Res.* **5,** 275–288.

Klein, R. G. (1980). The interpretation of mammalian faunas from Stone-Age archaeological sites, with special reference to sites in the Southern Cape Province, South Africa. In (A. K. Behrensmeyer & A. P. Hill, Eds) *Fossils in the Making: Vertebrate Taphonomy and Paleoecology*, pp. 223–246. Chicago: University of Chicago Press.

Krebs, J. R. & Davies, N. B. (1993). *An Introduction to Behavioural Ecology* (3rd Ed.). Oxford: Blackwell Scientific Publications.

Kruuk, H. (1972). *The Spotted Hyena.* Chicago: University of Chicago Press.

Lack, D. (1968). *Ecological Adaptations for Breeding in Birds.* London: Methuen.

Leakey, M. D. (1971). *Olduvai Gorge. 3. Excavations in Beds I and II, 1960–1963.* Cambridge: Cambridge University Press.

Lyman, R. L. (1985). Bone frequencies: Differential transport, *in situ* destruction, and the MGUI. *J. Archaeol. Sci.* **12,** 221–236.

Lyman, R. L. (1994). Quantitative units and terminology in zooarchaeology. *Am. Antiquity* **59,** 36–71.

Madsen, D. B. (1993). Testing diet breadth models: Examining adaptive change in the late Prehistoric Great Basin. *J. Archaeol. Sci.* **20,** 321–329.

Marean, C. W. (1989). Sabertooth cats and their relevance for early hominid diet and evolution. *J. Archaeol. Sci.* **18,** 559–582.

Marean, C. W. & Spencer, L. M. (1991). Impact of carnivore ravaging on zooarchaeological measures of element abundance. *Am. Antiq.* **56,** 645–658.

Marean, C. W., Spencer, L. M., Blumenschine, R. J. & Capaldo, S. D. (1992). Captive hyaena bone choice and destruction, the schlepp effect and Olduvai archaeofaunas. *J. Archaeol. Sci.* **19,** 101–121.

Morris, D. (1967). *The Naked Ape.* New York: Dell.

O'Connell, J. F., Hawkes, K. & Blurton-Jones, N. (1988). Hadza hunting, butchering and bone transport and their archaeological implications. *J. Anthropol. Res.* **44,** 113–161.

Oliver, J. S. (1994). Estimates of hominid and carnivore involvement in the FLK *Zinjanthropus* fossil assemblage: some socioecological implications. *J. hum. Evol.* **27,** 267–294.

Peters, C. R. (1981). Robust vs. gracile early hominid masticatory capabilities: The advantages of the megadonts. In (L. L. Mai, E. Shanklin & R. W. Sussman, Eds) *The Perception of Evolution*, pp. 161–181. Anthropology, UCLA vol. 7.

Peters, C. R. & Blumenschine, R. J. (n.d.). Landscape perspectives on possible land use patterns for Early Pleistocene hominids in the Olduvai Basin. Submitted to *J. hum. Evol.*, June, 1994.

Perkins, D. & Daly, P. (1968). A hunters' village in Neolithic Turkey. *Sci. Am.* **219,** 97–106.

Pianka, E. R. (1978). *Evolutionary Ecology* (2nd Ed.). New York: Harper & Row, Publishers.

Potts, R. (1983). Foraging for faunal resources by early hominids at Olduvai Gorge, Tanzania. In (J. Clutton-Brock & C. Grigson, Eds) *Animals and Archaeology. 1. Hunters and their Prey*, pp. 51–62. Oxford: British Archaeological Reports International Series 163.

Potts, R. (1984). Home bases and early hominids. *Am. Sci.* **72,** 338–347.

Potts, R. (1989). Olorgesailie: New excavations and findings in Early and Middle Pleistocene contexts, southern Kenya rift valley. *J. hum. Evol.* **18,** 477–484.

Raab, L. M. & Goodyear, A. C. (1984). Middle-range theory in archaeology: A critical review of origins and applications. *Am. Antiq.* **49,** 255–268.

Schaffer, W. M. (1968). Character displacement and the evolution of the Hominidae. *Am. Nat.* **102,** 559–571.

Schick, K. D. (1987). Modeling the formation of Early Stone Age artifact concentrations. *J. hum. Evol.* **16,** 789–807.

Selvaggio, M. M. (1994). Carnivore tooth marks and stone tool butchery marks on scavenged bone: Archaeological implications. *J. hum. Evol.* **27,** 215–227.

Shipman, P. (1975). Implications of drought from vertebrate fossil assemblages. *Nature* **257,** 667–668.

Sillen, A. (1992). Strontium-calcium ratios (Sr/Ca) of *Australopithecus robustus* and associated fauna from Swartkrans. *J. hum. Evol.* **23,** 495–516.

Stern, N. (1993). The structure of the Lower Pleistocene archaeological record. *Curr. Anthrop.* **34,** 201–226.

Susman, R. L. & Stern, J. T. Jr (1982). Functional morphology of *Homo habilis*, *Science* **217,** 931–934.

Tiger, L. & Fox, R. (1966). The zoological perspective in social science. *Man* **1,** 75–81.

Toth, N. (1987). Behavioral inferences from early stone artifact assemblages: An experimental model. *J. hum. Evol.* **16,** 763–787.

Trinkhaus, E. (1986). The Neandertals and modern human origins. *Ann. Rev. Anthrop.* **15,** 193–218.

Ungar, P. S. & Grine, F. E. (1991). Incisor size and wear in *Australopithecus africanus* and *Paranthropus robustus*. *J. hum. Evol.* **20,** 313–340.

Washburn, S. L. & DeVore, I. (1961). Social behavior of baboons and early man. In (S. L. Washburn, Ed.) *Social Life of Early Man*, pp. 91–105. Chicago: Aldine.

Winterhalder, B. (1981). Hominid paleoecology and competitive exclusion: Limits to similarity, niche differentiation, and the effects of cultural behavior. *Yrbk Phys. Anthrop.* **24,** 101–121.

Wolpoff, M. H. (1971). Competitive exclusion among lower Pleistocene hominids: The single species hypothesis. *Man* **6,** 601–614.

Wrangham, R. W. (1987). The significance of African apes for reconstructing human social evolution. In (W. G. Kinzey, Ed.) *The Evolution of Human Behavior: Primate Models*. Albany: State University of New York Press.

Yellen, J. E. (1977). *Archaeological Approaches to the Present; Models for Reconstructing the Past*. New York: Academic Press.

Marie M. Selvaggio
Southern Connecticut State University, Department of Sociology and Anthropology, 501 Crescent Street, New Haven, CT 06515, U.S.A.

Received 5 October 1993
Revision received 2 March 1994
and accepted 28 March 1994

Keywords: experimental archaeology, scavenging, carnivore feeding behavior, taphonomy, bone modification.

Carnivore tooth marks and stone tool butchery marks on scavenged bones: archaeological implications

This paper compares naturalistic observations on carnivore feeding behavior and carcass abandonment to results of experimental butchery of long bones collected after carcasses were abandoned by carnivores. Experiments were conducted to assess the incidence of butchery marks and carnivore tooth marks on long bones where butchery followed carnivore defleshing of limbs. One hundred and sixty-eight long bones bearing marrow and occasionally scraps of flesh were collected after carcasses were abandoned by free-ranging East African carnivores including lions, leopards, cheetahs, spotted hyenas and jackals. Flesh scraps and marrow were removed using lithic materials found commonly at Plio-Pleistocene archaeological sites. Minimally, marrow was recovered from all long bones. However, marrow was generally the only nutrient available on limbs from carcasses defleshed by nine or more carnivores. Conversely, both flesh scraps and marrow were frequently available on limbs from carcasses abandoned by solitary carnivores or groups of carnivores comprised of fewer than five animals. Data presented indicate that the incidence of butchery marks and carnivore tooth marks is related to the condition of bones upon carnivore abandonment. Quantitative data are presented on the incidence of long bone specimens bearing at least one carnivore tooth mark, one cut mark or one hammerstone percussion mark. Variability within the sample is examined with respect to the ratios of carnivore tooth marks and butchery marks on bones defleshed by different numbers of carnivores. Results indicate that variability within the sample is related to the number of carnivores involved in defleshing bones. Data presented provide a controlled basis for evaluating similar marks in archaeological bone assemblages.

Journal of Human Evolution (1994) **27,** 215–228

Introduction

The extent to which Plio-Pleistocene hominids procured carcasses through hunting or scavenging has broad socio-ecological implications with respect to their diet, foraging strategies and social adaptations. Various criteria have been advanced for distinguishing hunting from scavenging in archaeological bone assemblages, including size/age profiles (Klein, 1982; Vrba, 1980), skeletal part profiles (Binford, 1984; Blumenschine, 1986; Potts, 1983), and marks on bone surfaces (Binford, 1981; Bunn, 1981, 1983; Shipman, 1983). While there are documented problems with each strategy (e.g., Behrensmeyer *et al.*, 1986; Binford, 1984; Bunn & Blumenschine, 1987), carnivore tooth marks and stone-tool-butchery marks provide the most appropriate criteria, because they represent both carnivore and hominid involvement with archaeological bone assemblages and they are sensitive to the nature and timing of each agent's access to carcasses (Selvaggio n.d.*a*). Recognition of the archaeological traces of early access versus late access (Potts, 1983) to carcasses by Plio-Pleistocene hominids is necessary for the development of plausible models of early hominid behavioral ecology.

Recent experiments designed to identify the effects of carnivore disturbance on butchered bones have established major taphonomic indicators of carnivore bone modification and bone deletion when carnivores are exclusively the final consumers of bone nutrients (Binford *et al.*,

Presented at the American Association of Physical Anthropologists meeting, symposium entitled "*Early Hominid Behavioral Ecology: New Looks at Old Questions*" April, 1993.

1988; Blumenschine, 1988; Blumenschine & Selvaggio, 1991; Marean & Spencer, 1991). However, experimental butchery of the bones prior to carnivore ravaging of assemblages mimics taphonomic patterns where hominids have initial access to flesh and marrow from carcasses and carnivore involvement is limited to extraction of any remaining nutrients. Therefore, cut marks on the bones reflect disarticulation and removal of major muscles, while carnivore tooth marks were inflicted during extraction of bone grease or morsels of marrow from the hammerstone-fragmented refuse. Implicitly, such experiments model early access to carcasses by hominids and secondary access to remaining nutrients by bone-crunching carnivores.

In contrast, research reported here models the reverse scenario where hominid access to carcasses follows defleshing of major muscle masses by carnivores. For experiments reported here, tooth marks represent only consumption of major muscles and disarticulation of the bones by carnivores. Likewise, cut marks represent removal of flesh scraps, skin and tendons and bone disarticulation. The most notable difference between experiments conducted by others (e.g., Blumenschine, 1988; Binford *et al.*, 1988) and those reported here, is that carnivore involvement is restricted to final access to bones in the former and initial access to bones in the latter. As well, few carnivore species are represented in previous research where the effects of bone-crunching carnivores on butchered assemblages was the goal of such studies. Most large East African carnivores are represented in the research reported here including lions, leopards, cheetahs, jackals and spotted hyenas. A similarity shared by previous research and that reported here, is that cut marks were inflicted on whole long bones prior to marrow extraction.

Both cut marks and carnivore tooth marks have been a major focus of research on the timing and sequence of hominid and carnivore involvement with archaeological bone assemblages (e.g., Binford, 1981; Bunn *et al.*, 1980; Bunn, 1981; Bunn & Kroll, 1986; Haynes, 1980; Potts, 1988; Potts & Shipman, 1981; Shipman, 1986). However, such studies have not resulted in a consensus on the relative contributions of hominids and carnivores to Plio-Pleistocene bone assemblages (e.g., Binford, 1981; Binford *et al.*, 1988; Bunn & Kroll, 1986).

Percussion marks provide a new source of evidence for carcass processing that has not been previously integrated into research on hunting and scavenging by early hominids. Percussion marks are inflicted on bones during hammerstone impact and occur as patches of microstriations or as pits or grooves containing dense patches of microstriations (Blumenschine & Selvaggio, 1988, 1991). Given good bone surface condition, percussion marks provide direct evidence for marrow extraction by hominids. Identification of percussion marks and documentation of their comparative frequency relative to carnivore tooth marks and cut marks in archaeological bone assemblages is essential for recognizing whether marrow or flesh was the primary focus of carcass processing by hominids.

Data presented are part of a broad research strategy designed to identify traces of hunting and scavenging by hominids that can be recognized in archaeological bone assemblages. This paper represents the first systematic analysis of stone-tool-butchery marks and carnivore tooth marks on long bones initially defleshed by different numbers of free-ranging carnivores. Quantitative data are presented on the incidence of long bone specimens bearing at least one carnivore tooth mark, one stone tool cut mark and one hammerstone percussion mark. Since individual bone specimens can exhibit more than one type of mark, variability within the sample is examined using the total number of each type of mark in the sample to construct ratios of cut marks to carnivore tooth marks and ratios of percussion marks to cut marks. Mark inflation resulting from hammerstone impact is controlled by comparing the ratios of marks.

The sample is comprised of long bones where marrow was available after carnivore abandonment of carcasses. Bones were not subjected to carnivore ravaging after butchery. Consequently, carnivore tooth marks represent only defleshing and disarticulation of bones by carnivores. Cut marks represent the same processes and are, therefore, sensitive to the condition of the bone upon carnivore abandonment. Results indicate that the relative availability of flesh versus marrow on long bones defleshed and abandoned by different numbers of carnivores varies in a predictable way and leaves recognizable traces on the bones. The relative availability of such nutrients to hominid scavengers can be estimated where archaeological bone assemblages preserve carnivore tooth marks, cut marks and hammerstone percussion marks.

Methods

Research with free-ranging carnivores was conducted for 7 months during 1989–1990 in the Serengeti National Park and Ngorongoro Conservation Area, protected wildlife refuges located in northern Tanzania. These tropical savanna-woodland ecosystems support both migratory and resident populations of large mammalian herbivores and carnivores (Sinclair & Norton-Griffiths, 1979).

Observations were conducted from a four-wheel drive vehicle positioned to afford a good view of carnivore consumption of carcasses, yet minimize disturbance of carnivore activity. While carnivore hunting of prey animals was observed for only 19 of the carcasses in this sample, all long bones in the sample represent observed episodes of limb flesh consumption by carnivores. With the exception of femora, limb flesh is consumed late in the carnivore consumption sequence (Blumenschine, 1985). Therefore, it is not unusual to encounter a carnivore consuming a carcass where limb flesh is still intact. The analysis is restricted to long bones where carnivore defleshing was observed. The focus on long bones is justified due to their high representation in archaeological bone assemblages.

Nutrients were available on carcasses abandoned by carnivores in this sample. Generally, the major muscle masses of all skeletal elements were consumed by carnivores. However, head contents, bone marrow and sometimes scraps of flesh were available after carnivore abandonment of carcasses. Long bones containing marrow and occasionally scraps of flesh were collected when carnivores were approximately 0·5 km from the carcasses.

Lithics found commonly at Plio-Pleistocene archaeological sites, such as basalt and quartzite, were used to disarticulate limbs and remove remaining tissue. Approximately half of the butchery experiments were conducted under my supervision by a Tanzanian butcher who was naïve about the purpose of the experiments. Preliminary analysis indicated that variability within the sample is not related to different butchers.

Butchery of long bones included disarticulation and removal of any remaining flesh scraps, skin or tendons that would impede hammerstone fracture. A conservative butchery strategy was practiced for all experiments in that bones were not scraped either to remove periosteum or minute scraps of flesh. In order to systematize hammerstone fracture of the marrow cavity, the initial impact was inflicted near the proximal end of all long bones. Additional hammerstone impacts were inflicted along the bone's long axis until the marrow cavity was exposed. Long bones of small gazelle-sized prey usually required less than four hammerstone impacts in order to expose the marrow cavity. Prey species larger than gazelles commonly required at least five hammerstone impacts in order to expose the marrow cavity. Marrow was

Table 1 Sample of scavenged long bones by number of carcasses represented, number of nutrient-bearing limbs collected and number of specimens ⩾2 cm in length represented by long bone portion and feeding group sizes of carnivore consumers

Carnivore group size	Number of carcasses	Number of limbs	Long bone portions				Total specimens
			CO	EPIPH	NEF	MSH	
Solitary	15	88	23	97	35	103	258
Small groups	13	72	12	121	22	130	285
Large groups	4	8	0	9	11	21	41
Total	32	168	35	227	68	254	584

Carnivore feeding groups are comprised of solitary consumers, small groups of two to four animals and large groups of nine or 11 animals. CO=complete limbs (not hammerstone broken), EPIPH=epiphyseal fragments, NEF=near-epiphyseal fragments, MSH=midshaft fragments.

removed from bones with a blunt wooden probe. After butchery, bones were cleaned by boiling, air dried and wrapped in tissue for protection during shipment.

Carnivore tooth marks and butchery marks were distinguished by published criteria (Binford, 1981; Bunn, 1981; Blumenschine & Selvaggio, 1988, 1991). Marks were identified using a 16 power hand lens under strong, low incidence light. Each mark was counted once. Where overlapping marks occurred each mark was counted separately. In seven cases, the textured surface of bones from sub-adult animals hindered accurate identification of marks. These specimens were excluded from the sample.

Sample

The sample is comprised of 168 long bones from 32 carcasses represented by 584 specimens (Table 1). With the exception of 35 long bones from seven carcasses, all specimens result from hammerstone impact. The 35 whole long bones allow comparison of data presented here to complete long bones in archaeological bone assemblages. To control for bias in the recovery of very small bone fragments that result from hammerstone impact, the sample is comprised of bone specimens equal to or greater than 2 cm in length.

Long bones were identified with respect to skeletal element and bone portion. Long bone portions are stratified into four categories: (1) complete or whole long bones, which were not hammerstone broken, (2) epiphyseal fragments, defined here as any bone fragment exhibiting an articular surface, (3) near-epiphyseal fragments, identifiable on bovids by cancellous tissue on the medullary surface or by refitting and land marks in other species and, (4) midshafts or diaphyseal fragments.

Preliminary analysis revealed that variability in the incidence of carnivore tooth marks is not related to the size of prey and that variability in the incidence of cut marks and percussion marks is not related to butchery by different individuals (Selvaggio n.d.a). The sample of long bones is stratified by the number of carnivores involved in flesh consumption (Figure 1). This stratification enables comparison of carnivore tooth marks and butchery marks on long bones from carcasses abandoned by different numbers of carnivores. A distribution identified three groups of carnivore consumers: (1) solitary animals, (2) small groups, consisting of two to four animals and, (3) large groups, comprised of nine or 11 animals. Solitary consumers experienced no competition while feeding and frequently abandoned carcasses with minimal disarticulation and with skin, tendons and flesh scraps remaining on limbs. Only a moderate degree of competition, as defined by aggression among animals, was observed for small groups

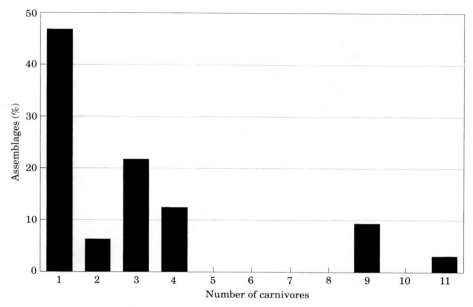

Figure 1. Distribution of the number of carnivores observed to feed on 32 carcasses.

of carnivores. Well-represented among the sample of small feeding parties are closely related animals such as females and their offspring and members of the same lion pride. Carcasses were usually abandoned by small feeding parties with minimal disarticulation and flesh scraps remaining on the bones. In this sample, large groups of consumers are comprised exclusively of members of at least two different carnivore taxa: felids, spotted hyenas and/or jackals. Competition for carcass nutrients was most intense among large groups of carnivores. Carcasses consumed by large carnivore feeding parties were usually abandoned with limb elements disarticulated from carcasses and frequently epiphyses were heavily damaged or completely gnawed away. During this study, scavenging opportunities from large groups of carnivores were infrequent as the sample size indicates (Table 1).

Analysis

Each assemblage is comprised of long bones collected from one carcass. Two analyses are presented: (1) the mean assemblage percent and the standard deviation of assemblages comprised of specimens bearing at least one carnivore tooth mark, one cut mark or one percussion mark and, (2) the ratio of carnivore tooth marks to cut marks and the ratio of percussion marks to cut marks. Variability in the former is examined by comparing the ratio of marks inflicted on bones where different numbers of carnivores were involved in defleshing carcasses. Since individual bone specimens frequently bear more than one type of mark, data on ratios are derived from the total number of carnivore tooth marks, cut marks and percussion marks on specimens in each assemblage.

Results

When all specimens are considered, carnivore tooth-marked specimens are more frequently represented than either cut-marked specimens or percussion-marked specimens (Figure 2 and

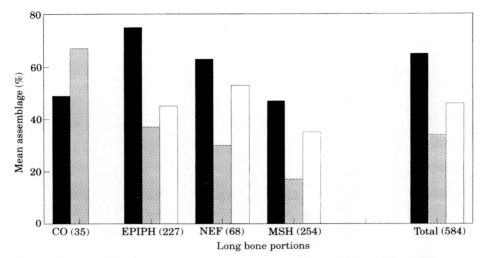

Figure 2. Mean assemblage percent of specimens bearing at least one carnivore tooth mark (■), one cut mark (▨) or one percussion mark (□). CO=complete long bones, EPIPH=epiphyses, NEF=near epiphyseal fragments, MSH=midshafts or diaphyseal fragments. See Table 2 for % ± 1 s.d.

Table 2 **The mean assemblage percent of specimens ≥2 cm in length bearing at least one carnivore tooth mark, one cut mark or, one percussion mark, ± 1 standard deviation**

| | Long bone portions | | | | |
	CO	EPIPH	NEF	MSH	Total
% Tooth-marked	49·4	75·0	63·0	47·0	65·0
± 1 S.D.	(37·0)	(25·0)	(41·0)	(27·0)	(20·0)
% Cut-marked	67·4	37·0	30·0	17·4	34·1
± 1 S.D.	(32·3)	(28·0)	(32·0)	(22·1)	(22·0)
% Percussion-marked	—	45·0	53·0	35·0	46·0
± 1 S.D.	—	(35·0)	(41·0)	(28·0)	(24·0)

Each assemblage is composed of specimens from long bones collected from one carcass ($n=32$). Complete limbs do not bear percussion marks since they were not hammerstone broken. Seven assemblages comprised of complete limbs are excluded from the percussion-marked sample.

Table 2). As well, the mean percent of specimens bearing at least one carnivore tooth mark is greater than the mean percent of specimens bearing at least one percussion mark or one cut mark for epiphyses, near epiphyses and midshafts. Among hammerstone-generated specimens, epiphyseal specimens are more frequently tooth-marked or cut-marked compared to near epiphyseal specimens or midshaft specimens. Among butchery marks, percussion-marked specimens occur more frequently than cut-marked specimens when all hammerstone-generated specimens are compared. Where long bones were not subjected to hammerstone fracture the mean percent of bones bearing cut marks is higher than the mean percent of bones bearing tooth marks.

The large standard deviations for marks on all long bone-portions indicate great variability within the sample. Variability within the sample is examined relative to the number of carnivores involved in defleshing carcasses. The ratio of cut marks to carnivore tooth marks,

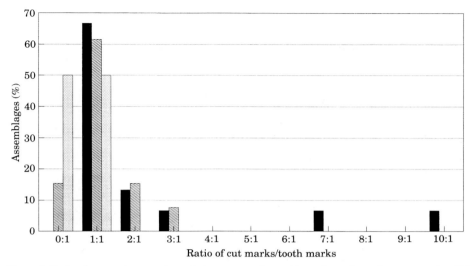

Figure 3. Ratio of cut marks to tooth marks by percent of assemblages for each carnivore group size. (■) solitary (*n*=15); (▨) small (2–4) (*n*=13); (☐) large (9–11) (*n*=4).

Table 3a   Ratios of cut marks to tooth marks by percent of assemblages initially defleshed by different numbers of carnivores

| | Ratio of cut marks/tooth marks | | | | |
Carnivore group size	0:1	1:1	2:1	3:1	≥4:1
Solitary	—	0·68	0·13	0·07	0·12
Small group (2–4)	0·15	0·62	0·15	0·08	—
Large group (9–11)	0·50	0·50	—	—	—

Solitary animals (*n*=15), small carnivore groups (*n*=13), large carnivore groups (*n*=4).

and percussion marks to cut marks, provide an indicator with respect to the number of carnivores involved in defleshing bones, hence the relative availability of flesh on limbs after carnivore abandonment.

High ratios of cut marks to carnivore tooth marks are found on long bones where there was little or no competition among carnivores and where flesh scraps were commonly available after carnivore abandonment of carcasses (Figure 3 and Table 3a). Ratios of cut marks to carnivore tooth marks equal to or greater than 2 to 1 are found solely on bones abandoned by solitary animals and small carnivore feeding parties. The highest ratios of cut marks to carnivore tooth marks, those greater than 3 to 1, are found exclusively on long bones abandoned by solitary consumers. Conversely, long bones abandoned by large groups of carnivores exhibit the lowest ratio of cut marks to carnivore tooth marks. The number of carnivores involved in flesh consumption can be estimated in archaeological bone assemblages given the ratio of cut marks to carnivore tooth marks (Table 3b).

As well, the ratio of percussion marks to cut marks signals the condition of bones upon carnivore abandonment. A distinctively higher ratio of percussion marks to cut marks is

Table 3b   Probability of the number of carnivores involved in flesh consumption given the ratios of cut marks to carnivore tooth marks

Carnivore group size	Ratio of cut marks/tooth marks				
	0:1	1:1	2:1	3:1	≥4:1
Solitary	0·00	0·38	0·46	0·47	1·00
Small group (2–4)	0·23	0·34	0·54	0·53	0·00
Large group (9–11)	0·77	0·28	0·00	0·00	0·00

Probabilities are derived from data presented in Table 3a. The percent indicated in Table 3a for each carnivore group size for a specific ratio of marks is divided by the total percent of that specific ratio category.

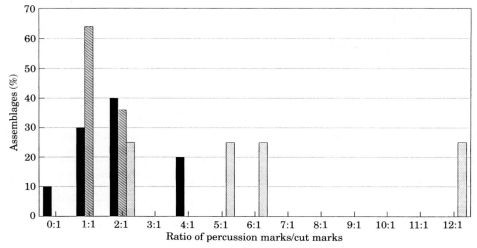

Figure 4. Ratio of percussion marks to cut marks by percent of assemblages for each carnivore group size. Seven assemblages not hammerstone-broken are excluded from this analysis. (■) solitary (*n*=10); (▨) small (2–4) (*n*=11); (☐) large (9–11) (*n*=4).

exhibited on long bones abandoned by large groups of carnivores where marrow was commonly the only remaining nutrient, in comparison to solitary animals or small groups of carnivores where both flesh scraps and marrow were usually present on bones (Figure 4 and Table 4a). Seventy-five percent of long bones from carcasses abandoned by large groups of carnivores exhibit a ratio of percussion marks to cut marks equal to or greater than 5 to 1, while all long bones abandoned by solitary animals or small groups of carnivores exhibit a ratio of less than 5 to 1. Given the ratio of percussion marks to cut marks in archaeological bone assemblages the relative availability of flesh versus marrow to a hominid scavenger can be estimated (Table 4b).

Discussion

The extent to which long bone assemblages bear butchery marks in this sample is related to the condition of the bones upon carnivore abandonment. When all specimens are considered,

Table 4a   Ratios of percussion marks to cut marks by percent of assemblages initially defleshed by different numbers of carnivores

Carnivore group size	Ratio of percussion marks/cut marks					
	0:1	1:1	2:1	3:1	4:1	≥5:1
Solitary	0·10	0·30	0·40	—	0·20	—
Small group (2–4)	—	0·64	0·36	—	—	—
Large group (9–11)	—	—	0·25	—	—	0·75

Seven assemblages not hammerstone broken are excluded from this analysis. Solitary animals ($n=10$), small carnivore groups ($n=11$), large carnivore groups ($n=4$).

Table 4b   Probability of the number of carnivores involved in flesh consumption given the ratio of percussion marks to cut marks. Probabilities are derived from data presented in Table 4a

Carnivore group size	Ratio of percussion marks/cut marks					
	0:1	1:1	2:1	3:1	4:1	≥5:1
Solitary	1·00	0·32	0·40	—	1·00	0·00
Small group (2–4)	0·00	0·68	0·36	—	0·00	0·00
Large group (9–11)	0·00	0·00	0·24	—	0·00	0·00

those bearing at least one carnivore tooth mark are more frequently represented than specimens bearing at least one percussion mark or one cut mark (Figure 2 and Table 2). Among hammerstone-generated long bone portions, the highest percentages of tooth-marked and cut-marked specimens are associated with epiphyseal portions. Epiphyses are sites of articulation and muscle attachments where tissue removal and disarticulation by both butchery and carnivore consumers is highly concentrated. Therefore, it is not unexpected that epiphyseal fragments exhibit the highest mean percentages of both cut-marked and tooth-marked specimens.

Among all hammerstone-generated long bone fragments, midshaft specimens exhibit the lowest mean percent of specimens bearing at least one cut mark or one tooth mark. The low frequency of specimens bearing at least one cut mark or one tooth mark is expected since disarticulation and muscle attachments are not associated with this portion of long bones.

Near epiphyseal fragments exhibit an intermediate percentage of cut-marked and tooth-marked specimens. Near epiphyses are sites of muscle attachments and reflect flesh removal by both carnivores and butchery. In this sample, tooth-marked near epiphyseal fragments are represented more frequently than cut-marked near epiphyseal specimens. In fact, tooth-marked specimens occur almost twice as frequently as cut-marked specimens for all hammerstone-generated long bone portions. The predominance of tooth-marked specimens for all hammerstone-generated fragments is expected since carnivores had initial access to limb flesh.

Among hammerstone-generated specimens, percussion-marked specimens are more frequently represented than cut-marked specimens. Marrow was the major nutrient available after carnivore abandonment of carcasses. Therefore, it is not unexpected that percussion-marked specimens occur as the dominant indicator of butchery in this sample.

The sample of long bones not subjected to hammerstone fracture is comprised solely of bones abandoned by solitary consumers and small groups of carnivores where disarticulation of limbs by carnivores was infrequent and flesh scraps were commonly available after carnivores abandoned carcasses (see Table 1). Disarticulation of limbs with stone flakes as well as the removal of remaining skin, tendons and flesh scraps by butchery explains the high proportion of complete bones bearing at least one cut mark.

With respect to tissue removal, both carnivore tooth marks and cut marks represent defleshing and disarticulation, while marrow extraction is solely represented by percussion marks. Taken together, the proportion of specimens bearing at least one carnivore tooth mark or one butchery mark among long bone portions reflects initial defleshing of major muscle masses by carnivores and removal of remaining tissue by butchery. However, standard deviations for marks on all long bone portions indicate great variability within the sample. Variability among the assemblages is related to the condition of the bones upon carnivore abandonment.

The ratio of cut marks to carnivore tooth marks reflects both the agent of disarticulation and the extent that flesh was available on long bones abandoned by different numbers of carnivores (Figure 3 and Tables 3a and 3b). The high ratio of cut marks to carnivore tooth marks on long bones collected from carcasses abandoned by solitary animals and small groups of carnivores, results from necessary disarticulation of the bones from carcasses by butchery, as well as the greater availability of flesh scraps on long bones compared to those abandoned by large carnivore feeding parties. Conversely, low ratios of cut marks to carnivore tooth marks on long bones abandoned by large numbers of carnivores is explained by intense competition among carnivores for flesh.

Data on the ratios of percussion marks to cut marks support the above analysis in that these ratios reflect the relative availability of marrow and flesh on long bones after carnivore abandonment (Figure 4 and Tables 4a and 4b). Among solitary consumers and small carnivore feeding parties where flesh scraps were commonly available, the ratio of percussion marks to cut marks is low (less than 5 to 1) for all long bones. In contrast, long bones collected from large carnivore feeding parties, where marrow was usually the only nutrient available, exhibit markedly higher ratios of percussion marks to cut marks than those on long bones abandoned by solitary consumers or small feeding parties. Seventy-five percent of long bones collected from carcasses defleshed by nine or more carnivores exhibit ratios of percussion marks to cut marks equal to or greater than 5 to 1.

Data on the ratios of marks in this sample indicate that variability in nutrient yields from carcasses abandoned by different numbers of carnivores is predictable and leaves recognizable traces on bones. As well, from these data it can be inferred that where the reverse scenario of hominid and carnivore access to carcasses exists, ratios of cut marks to carnivore tooth marks should be higher than those reported here and ratios of percussion marks to cut marks should be limited to the lowest ratio categories. Given good bone surface preservation, carnivore tooth marks, cut marks and percussion marks in archaeological bone assemblages can be used to estimate the relative yields of flesh or marrow extracted from long bones by carnivores or butchery.

Archaeological implications

The relationship of carnivore tooth marks and butchery marks on long bones reflects the competitive context of carcass consumption by different numbers of carnivores, hence the extent to which flesh and marrow were available after carnivores abandoned carcasses. Data presented indicates marrow would be the primary nutrient available to hominid scavengers from carcasses defleshed and abandoned by large numbers of carnivores, while both flesh scraps and marrow would be more frequently encountered on carcasses abandoned by solitary animals or small carnivore feeding parties.

Secondary access to long bones by early hominids can be identified in archaeological assemblages where bone surfaces preserve traces of hominid butchery and carnivore gnawing. The effects of post-butchery disturbance by carnivores on patterns described here is minimal if archaeological analysis is restricted to midshaft specimens with good surface condition. Experiments on carnivore disturbance of butchered bones indicate that nutrient-depleted, hammerstone-generated midshaft fragments are ignored by carnivorous scavengers (Binford *et al.*, 1988; Blumenschine, 1988; Marean & Spencer, 1991; pers. obs.). Therefore, midshaft specimens preserve direct physical traces of carnivore and hominid involvement with carcasses prior to post-butchery ravaging by carnivores (Selvaggio, n.d.*a*).

Currently, the presence of both carnivore tooth marks and stone-tool-cut marks in archaeological bone assemblages defies interpretation as to whether hominids or carnivores acted first. Most notable, cut marks and carnivore gnaw damage on bones in the FLK *Zinjanthropus* asemblage have been used to support two different scenarios of hominid and carnivore access to carcasses (e.g., Bunn & Kroll, 1986; Binford *et al.*, 1988). In one scenario, hominids are considered to have had early access to fleshy carcasses either through hunting or usurping fresh kills from carnivores (e.g., Bunn & Kroll, 1986). In this scenario, carnivore tooth marks are thought to represent damage inflicted on bones after all flesh and marrow was removed by hominids. The opposing view is that carnivore gnaw damage on bones in the FLK *Zinjanthropus* assemblage represents initial access to carcasses by carnivores and cut marks represent disarticulation or removal of dry surface tissue (Binford, 1981; Binford *et al.*, 1988).

The assumption that carnivore involvement with bones in the FLK *Zinjanthropus* assemblage was primarily or exclusively restricted to post-butchery disturbance, precludes consideration of the potential range of carnivore involvement with the assemblage as demonstrated by the research reported here. Additionally, the notion that initial access to carcasses by carnivores would relegate hominids to scrounging morsels of nutrients, ignores variability in the competitive context of carnivore feeding behavior, as well as differences in the abilities of flesh-specialist and bone-crunching carnivores to utilize carcasses. Clearly, information on the extent that Plio-Pleistocene hominids had early access to fleshy carcasses or late access to carcasses abandoned by carnivores is relevant for developing plausible models of early hominid behavior.

The predominance of either hunting or scavenging animals by early hominids implies distinctively different foraging strategies and social adaptations. The ability of Plio-Pleistocene hominids to hunt adult wildebeest-sized prey suggests that they had developed cooperative social skills that exceed those of extant non-human primates. Additionally, the systematic hunting of such prey by early hominids would indicate that the contribution of flesh to their diet was similar to that of modern hunter-gatherers. While this human-like view of early hominid behavior is appealing, it does not explain how early hominids acquired the skills necessary for hunting such prey.

In contrast, scavenging implies a less central role for animal products in the diet of Plio-Pleistocene hominids. If early hominids acquired most or all of their animal foods by scavenging abandoned carnivore kills, it is likely that their diet, foraging strategies and social adaptations were markedly different from that of later prehistoric or extant hunter-gatherers.

Opportunities to scavenge abandoned predator kills have been documented by several recent actualistic studies (Blumenschine, 1985; Cavallo & Blumenschine, 1989; Selvaggio, n.d.*b*). These studies indicate that abandoned carcasses are predictably located in specific environments and can persist for long periods of time prior to their discovery by bone-crunching carnivores. Methods of predator avoidance useful while foraging for plant foods could have served early hominids equally well for collecting abandoned carcass parts. Resolution of conflicting scenarios of carcass procurement by early hominids is fundamental for developing plausible models of early hominid behavior.

Naturalistic observations on carnivore behavior and data from experimental simulations of different scenarios of hominid and carnivore access to carcasses provide a controlled basis for interpreting the sequence of hominid and carnivore involvement with archaeological bones. Specifically, such data allow comparisons between the incidence and anatomical patterning of marks inflicted on bones during tissue removal by known agents under known conditions and similar marks on archaeological bones (cf. Binford, 1981). Experimental simulations of different scenarios of hominid and carnivore access to carcasses have already been conducted (Blumenschine, 1988; Selvaggio n.d.*a*). Data from these experiments will be applied to the FLK *Zinjanthropus* bone assemblage and reported elsewhere (Blumenschine, n.d.; Selvaggio & Blumenschine, n.d.).

Identification of the timing and sequence of hominid and carnivore access to bones in archaeological assemblages compels consideration of damage inflicted on bones during defleshing and disarticulation by various carnivores, as well as an evaluation of the patterning and relationship of carnivore damage to both cut marks and hammerstone percussion marks. Recognition of hammerstone fracture of long bones is essential in order to determine if marrow extraction was the primary focus of hominid carcass processing.

Present indications of the complexity of hominid and carnivore involvement with bones recovered from Plio-Pleistocene sites justifies the systematic consideration of multiple lines of evidence related to hominid carcass procurement and site formation processes. Controlled experimental simulations and naturalistic observations can enhance the development of ecologically realistic models of assemblage formation.

Summary and conclusions

Data presented compare naturalistic observations on carnivore feeding behavior and carcass abandonment with results of experimental butchery of long bones where carnivores had initial access to carcasses. Naturalistic observations indicate that disarticulation of limbs from carcasses and the availability of flesh scraps on long bones after carnivore abandonment of carcasses are related to competition among carnivores for carcass nutrients. More nutrients were available on carcasses abandoned by less than five carnivores where little or no competition was observed among the animals, compared to feeding episodes where nine or more carnivores competed for carcass nutrients. The competitive context of carnivore feeding behavior, hence the availability of long bone flesh versus marrow after carnivore abandonment of carcasses, can be inferred from marks inflicted by carnivores and butchery on the bones.

Data presented indicate that the incidence and relationship of carnivore tooth marks and stone tool butchery marks vary in a predictable way and reflect the contribution of carnivores and butchery in tissue removal. Among specimens bearing at least one mark, carnivore tooth-marked specimens are most frequently represented (65%) compared to percussion-marked specimens (46%) and cut-marked specimens (34%). These data reflect initial defleshing of bones by carnivores and the relative availability of marrow versus flesh after carnivore abandonment of carcasses. Variability within the sample is related to the condition of the bone upon carnivore abandonment. The highest ratio of cut marks to carnivore tooth marks ($\geqslant 7:1$) is found exclusively on bones abandoned by solitary animals where flesh scraps were commonly available. The highest ratio of percussion marks to cut marks ($\geqslant 5:1$) is exhibited on bones abandoned by nine or more animals where marrow was usually the only remaining nutrient. Where such data can be obtained from archaeological assemblages, the number of carnivores involved in defleshing carcasses can be estimated, hence the relative availability of flesh versus marrow to hominid scavengers can be inferred.

Naturalistic observations and experiments that model different scenarios of assemblage formation can decrease ambiguity about the timing and sequence of hominid and carnivore involvement within archaeological bone assemblages. Such studies, when applied to the archaeological record, can enhance sensitivity of the behavioral variability of agents that produced archaeological bone assemblages and thereby enrich inferences of early hominid behavior.

Acknowledgements

An early version of this paper was presented in a seminar at Rutgers University entitled Taphonomy and Archaeozoology, Spring 1992. I thank Rob Blumenschine and students in the seminar for their comments. Funding from the National Science Foundation, the Leakey Foundation and Rutgers University is gratefully acknowledged. I thank the Tanzanian Commission of Science and Technology for permission to conduct this research. I also thank Prof. Hirji, Director of the Serengeti Wildlife Institute and researchers at the Serengeti Wildlife Institute for advice and for sharing with me their expertise on carnivore behavior. I am especially grateful to Annie Vincent, Peter Jones, Diane Gifford-Gonzalez and Donald Moshe for assistance and logistical support during my stay in Africa. I thank Dr Peter Intarapanich of Southern Connecticut State University for statistical and editorial suggestions. I appreciate the useful comments of the anonymous reviewers and editorial suggestions by Nancy Sikes. This paper has benefited from discussions with Rob Blumenschine, Henry Bunn and Gary Haynes. Any errors or omissions are my own.

References

Behrensmeyer, A. K., Gordon, K. D. & Yanagi, G. T. (1986). Trampling as a cause of bone surface damage and pseudocutmarks. *Nature* **319,** 768–771.

Binford, L. R. (1981). *Bones: Ancient Men and Modern Myths.* New York: Academic Press.

Binford, L. R. (1984). *Faunal Remains from Klasies River Mouth.* New York: Academic Press.

Binford, L. R., Mills, M. G. L. & Stone, N. (1988). Hyena scavenging behavior and its implications for the interpretation of faunal assemblages from FLK 22 (the Zinj floor) at Olduvai Gorge. *J. Anthrop. Archaeol.* **7,** 99–135.

Blumenschine, R. J. (1985). *Early Hominid Scavenging Opportunities: Insights from the Ecology of Carcass Availability in Serengeti and Ngorongoro Crater, Tanzania.* Ph.D. dissertation, University of California, Berkeley.

Blumenschine, R. J. (1986). Carcass consumption sequence and the archaeological distinction of scavenging and hunting. *J. hum. Evol.* **15,** 639–659.

Blumenschine, R. J. (1988). An experimental model of the timing of hominid and carnivore influence on archaeological bone assemblages. *J. Archaeol. Sci.* **15,** 483–502.

Blumenschine, R. J. (n.d.). Percussion marks, tooth marks, and experimental determinations of the timing of hominid and carnivore access to long bones at FLK *Zinjanthropus,* Olduvai Gorge, Tanzania.

Blumenschine, R. J. & Selvaggio, M. M. (1988). Percussion marks on bone surfaces as a new diagnostic or hominid behavior. *Nature* **333,** 763–765.

Blumenschine, R. J. & Selvaggio, M. M. (1991). On the marks of marrow bones by hammerstones and hyenas: Their anatomical patterning and archaeological implications. In (J. D. Clark, Ed.) *Cultural Beginnings: Approaches to Understanding Early Hominid Lifeways in the African Savanna,* pp. 17–32. Bonn: Dr Rudolf Habelt, GMBH.

Bunn, H. T. (1983). Evidence on diet and subsistence patterns of Plio-Pleistocene hominids at Koobi Fora, Kenya and Olduvai Gorge, Tanzania. In (J. Clutton-Brock & C. Grigson, Eds) *Animals and Archaeology. Vol. 1 Hunters and their Prey*, pp. 107–141. Oxford: British Archaeological Reports International Series 163.

Bunn, H. T. (1988). Archaeological evidence for meat-eating by Plio-Pleistocene hominids from Koobi Fora and Olduvai Gorge. *Nature* **291,** 574–577.

Bunn, H. T., Harris, J. W. K., Isaac, G., Kaufulu, Z., Kroll, E., Schick, K., Toth, N. & Behrensmeyer, A. K. (1980). FxJj 50: An early Pleistocene site in northern Kenya. *World Archaeol.* **12,** 109–136.

Bunn, H. T. & Kroll, E. M. (1986). Systematic butchery by Plio/Pleistocene hominids at Olduvai Gorge, Tanzania. *Curr. Anthrop.* **5,** 431–452.

Bunn, H. T. & Blumenschine, R. J. (1987). On "Theoretical framework and test" of early hominid meat and marrow acquisition—A reply to Shipman. *Am. Anthrop.* **89,** 444–447.

Cavallo, J. A. & Blumenschine, R. J. (1989). Tree-stored leopard kills: Expanding the hominid scavenging niche. *J. hum. Evol.* **18,** 393–399.

Haynes, G. (1980). Evidence of carnivore gnawing on Pleistocene and recent mammalian bones. *Paleobiology* **6,** 341–351.

Klein, R. (1982). Age (mortality) profiles as a means of distinguishing hunted species from scavenged ones in Stone Age archaeological sites. *Paleobiology* **8,** 151–158.

Marean, C. W. & Spencer, L. M. (1991). Impact of carnivore ravaging on zooarchaeological measures of element abundance. *Am. Antiq.* **56,** 645–658.

Potts, R. (1983). Foraging for faunal resources by early hominids at Olduvai Gorge, Tanzania (J. Clutton-Brock & C. Grigson, Eds) *Animals and Archaeology. Vol. 1. Hunters and their Prey*, pp. 51–62. Oxford: British Archaeological Reports International Series 163.

Potts, R. (1988) *Early Hominid Activities at Olduvai*. New York: Aldine De Gruyter.

Potts, R. B. & Shipman, P. (1981). Cut marks made by stone tools on bones from Olduvai Gorge, Tanzania. *Nature* **291,** 577–580.

Selvaggio, M. M. (n.d.*a*) Identifying the archaeological traces of scavenging by early hominids from marks on bone surfaces. Ph.D. Dissertation Rutgers University.

Selvaggio, M. M. (n.d.*b*) Carnivore activity at FLK *Zinjanthropus*. Paper presented at the International Conference in Honor of Dr. Mary D. Leakey, August 1993, Arusha, Tanzania.

Selvaggio, M. M. & Blumenschine, R. J. (n.d.). Evidence from tooth marks and butchery marks for scavenging by early hominids at FLK *Zinjanthropus*, Olduvai Gorge, Tanzania.

Shipman, P. (1983). Early hominid lifestyles: Hunting and gathering or foraging and scavenging? (J. Clutton-Brock & C. Grigson, Eds) *Animals and Archaeology. Vol. 1. Hunters and their Prey*, pp. 31–49. Oxford: British Archaeological Reports International Series 163.

Shipman, P. (1986). Scavenging or hunting in early hominids: Theoretical framework and tests. *Am. Anthrop.* **8,** 27–43.

Sinclair, A. R. E. & Norton-Griffiths, M. (Eds) (1979). *Serengeti: Dynamics of an Ecosystem*. Chicago: Chicago University Press.

Vrba, E. S. (1980). The significance of bovid remains as indicators of environment and predation patterns. In (A. K. Behrensmeyer & A. P. Hill, Eds). *Fossils in the Making*, pp. 247–271. Chicago: University of Chicago Press.

Kathlyn M. Stewart
Canadian Museum of Nature,
P.O. Box 3443, Station D,
Ottawa, Ontario K1P 6P4, Canada

Received 24 September 1993
Revision received 4 March 1994
and accepted 5 March 1994

Keywords: early hominid
subsistence, fish, seasonality,
Olduvai Gorge.

Early hominid utilisation of fish resources and implications for seasonality and behaviour

While research into the diet and subsistence of early hominids has focussed primarily on medium to large size mammals, modern ethnographic and dietary evidence suggests that other food sources are of equal or greater importance in hunter-gatherer diets, particularly in seasonally stressful times of year. Fish is examined in this paper as an alternative food source for early hominids. Nutritional, ecological and ethnographic evidence indicates that fish would be a seasonally available, nutritious and easy to procure alternative food source for early hominids, particularly during periods when other food sources may be of poor quality. Carnivores and non-human primates rely on fish as a seasonal resource, and archaeological findings also document the importance of fish for Late Pleistocene hominid groups. Fish remains are associated with many early hominid sites, and five sites at Olduvai Gorge are examined here in detail. The patterns of fish exploitation seen in Late Pleistocene archaeological sites are manifested in three of the Olduvai Gorge sites, making a strong, although not absolute, case for early hominid fish procurement. The implications for early hominid behaviour of fish procurement are several, and include timing of the early hominid seasonal round to exploit spawning or stranded fish, and group size larger than the nuclear family unit, with greater social interaction. Further investigation must also be conducted on the possible differences in procurement strategy between the hominid species at FLKNN (*Homo habilis*) and BK (presumed *H. erectus*).

Journal of Human Evolution (1994) **27,** 229–245

Introduction

Research into the diet and subsistence of early hominids has focussed primarily on medium to large size mammals (e.g., Blumenschine, 1987; Blumenschine & Madrigal, 1993; Bunn *et al.*, 1988; Bunn & Ezzo, 1993; O'Connell *et al.*, 1988, 1992). Not only are these the most archaeologically visible of fossil food remains because of size, but direct evidence of early hominid modification, such as cutmarks, is usually only discernible on bones of those animals large enough to warrant butchery practices. Other potential food sources, including those of plants, invertebrates and smaller vertebrates have until very recently been only minimally studied by archaeologists (for a discussion see Sept, 1992).

Several lines of evidence suggest, however, that medium to large mammals formed a proportionally far smaller component of the early hominid diet than the amount of archaeological attention reflects. A study of the Hadza, a modern hunter-gatherer group in Tanzania, documented that hunters captured only one medium-large mammal on an average of every 6 days over a 274 day period (O'Connell *et al.*, 1988, 1992), while studies of the San, a modern hunter-gatherer group in the Kalahari, documented one medium-large animal caught every three days over a 180 day period (Tanaka, 1976). These numbers of kills are not sufficient to fully feed the approximately 50 people in each of the groups monitored. Further, studies of protein and fat deficiency in modern hunter-gatherer diets (e.g., Speth, 1987, 1989; Speth & Davis, 1976; Speth & Spielmann, 1983) have shown that at times of seasonal climatic stress, most mammalian sources of meat protein would be nutritionally deleterious when relied on as a major component of the diet. While very recent studies have suggested that the case for such nutritional stress has been overstated for early hominids (Blumenschine & Madrigal, 1993; Bunn & Ezzo, 1993), periods of drought or other environmental distress will even further

0047–2484/94/010229+17 $08.00/0

reduce the quality, quantity and/or availability of medium-large mammals (see Speth, 1987 for a discussion), forcing reliance on alternative food sources (see discussion below).

Alternative food sources documented in studies of the San, Hadza and other modern hunter-gatherer groups include the widespread use of plants and non-mammalian vertebrates including fish as a supplement or as a main component of the diet (e.g., Dyson & Fuchs, 1937; Lee, 1968; O'Connell *et al.*, 1988; Vincent, 1985). Among the San, plants and other non-mammalian food form between ca. 81% and 96% of the diet (Tanaka, 1976). While subsistence strategies in the modern Kalahari or elsewhere cannot be generalised wholesale to the environment or behaviour of early hominids (for example, see Foley, 1982), modern nutritional and ethnographic data points strongly to the importance of alternative food sources in the diets of hunter-gatherers.

One possible alternative food source available to early hominids is fish, an easily procured and nutritional food source. Fish procurement by modern traditional African fishers is well documented (see Stewart, 1989) as is the evidence for fish procurement in Late Pleistocene archaeological sites (discussion below). Fish is also relied on as a seasonally dependable source of food by a variety of vertebrates. Hyenas, leopards and canids among others have all been documented pulling fish from the water and eating them (e.g. Ewer, 1973; Kruuk, 1976; Turnbull-Kemp, 1967). There are also several anecdotal accounts of baboons taking fish from the water, or scavenging recently dead fish from lake or river shores (e.g., Goodall, 1971). These captures seem to occur either when the fish are spawning, in association with rains, usually March to May in eastern Africa, or when they are trapped in shallow receding pools during the dry season.

Fish remains are also associated with many early hominid sites (Figure 1), often in dense concentrations such as at sites in Olduvai Gorge. Unfortunately, fish bone assemblages associated with early hominid sites display little direct evidence of modifications such as cutmarks, and with few exceptions (see Clark, 1960) these have been given little attention in the archaeological literature, except to be regarded as washed in "background noise".

In this paper I discuss fish as a nutritional and seasonal alternative food source for early hominids, document its importance as a seasonal resource in the fossil and modern record, and examine the evidence for fish procurement by early hominids, with emphasis on fish remains from Beds I and II Olduvai Gorge sites. Behavioural implications of early hominid fish procurement are also discussed.

Fish as an alternative food source

Speth & Spielman (1983) have persuasively argued that seasonal deterioration in the quality of grazing and browsing matter renders the meat of game animals too lean to be of nutritional value to hominids. Dependence on such game can therefore have highly debilitating effects. During times of resource stress, which in eastern Africa are typically the mid-late dry season and beginning of the wet season, plant resources experience a loss of protein and many browsers/grazers become severely fat-depleted (Speth, 1989). Peoples dependent on these animals are therefore eating meat with reduced fat levels; to compensate, modern hunter-gatherers are reported to consume large quantities of lean meat. Such consumption of fat-depleted meat reportedly results in severe weight-loss and other nutrition-depletion problems (Wilmsen in Speth, 1989). In response to such nutrition depletion, modern hunter-gatherers are reported to turn both to other available fat-sufficient foods, and/or have developed a variety of techniques to derive the maximum possible fat and oil from available

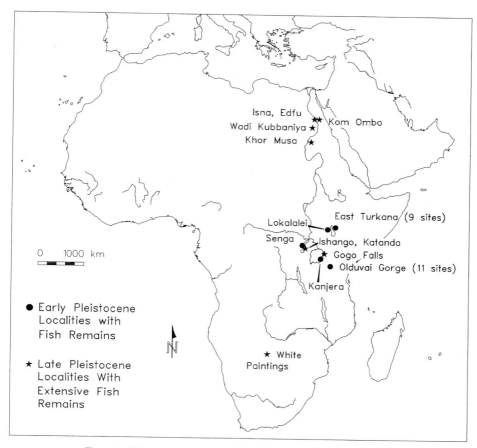

Figure 1. Pleistocene localities with fish remains mentioned in the text.

foods, including fish, in times of seasonal stress (e.g., Brelsford, 1946; Dyson & Fuchs, 1937; Lee, 1968; Tanaka, 1976; pers. obs.). Other more fat-rich parts of the mammalian body may also provide a limited source of fat (Bunn & Ezzo, 1993). While plant foods would normally provide an alternative food source, many too are depleted during the dry season, although tubers are a possible source, especially of carbohydrates (Vincent, 1985).

Fish as a group are a source of high-quality protein, comparable to meat protein, as well as a source of fat/oils, trace minerals, and vitamins (Pigott & Tucker, 1990). Of the most common freshwater fish, for example, one 40 cm long cichlid will provide almost 1 kg of meat, enough for a meal for a family. It is noteworthy that the average fat content of these fish equals or betters the 1–4% content of African ungulates reported by Speth (1989). However it should also be noted that once killed, fish meat spoils faster than most other meats.

As with mammals, the quality and content of these nutrients in fish varies seasonally with behaviour and food availability. Anecdotal and informal accounts by East African researchers state that at particular times of year certain fish groups are sought after by local fishers for their fat/oil reserves, with siluroids (catfish) and mormyroids well-known for these qualities (Brelsford, 1946; Jubb, 1967). Other researchers have observed that indigenous

East African fishers will throw back what are described as "thin" fish, citing lack of fat as the reason (Brelsford, 1946; also pers. obs.). Unfortunately, very few quantitative studies exist on seasonal changes in body composition of African freshwater fish. One method used in fisheries biology to measure the physical well-being of the body of a fish is condition (K), which is calculated from weight (w) and length (l) variables ($K = 100wl^{-3}$). Several studies have positively correlated the fish's condition or K with amount of fat present (e.g., Hyslop, 1986; Jocque, 1977). Changes in condition are often shown to be correlated with spawning, and studies on catfish (*Clarias*) and cichlids indicate a higher condition factor prior to spawning. A study on *Clarias anguillaris* in the Niger River showed an increase in condition through the dry season due to "diverting a greater proportion of energy to the laying down of food reserves towards the end of the (dry) season" (Hyslop, 1986). Cichlids studied at Lake Victoria showed better condition (K) during the dry season prior to spawning (Garrod, 1959). My own study on the cichlid *Oreochromis niloticus*, undertaken from 1985–1987 at Lake Turkana, Kenya (see Stewart, 1988) indicated a higher condition (K) factor in the weeks prior to spawning (January to March). It has further been suggested that the nuchal hump that male cichlids develop at spawning is actually a source of fat reserves (Lowe-McConnell, 1987).

Fish therefore provide a nutritious alternative to mammalian meat as a food source throughout the year. Further, certain groups of fish have increased fat deposits seasonally and may have served as a supplemental food source when mammal and other food sources are nutritionally depleted.

Procurement opportunities

Examination of both ethnographic and anecdotal accounts of traditional fishing in Africa indicates clearly that the intermeshing of season, aquatic habitat type and fish ecology are determinants of which groups of fish are easily procured without technology, and when. Of especial importance are the distinctions between high water and low water exploitation, and between lake and river exploitation. High waters occur with the onset of the rainy season, usually in March or April in eastern Africa, with a shorter, less reliable period in November. With the advent of first rains and flooding of lakes and rivers, the majority of riverine fish taxa migrate upriver to spawning grounds, as do most of the non-perciform taxa of the lakes. Actual spawning usually takes place in the shallow waters of floodplains.

Opportunities for procurement therefore exist both along rivers and especially at river mouths on the originating migration runs. Without equipment however, catching fish from the river itself is difficult, because river waters are deep and fast-flowing, and usually some implements such as weirs, fences, baskets and spears are needed. However, *Barbus*, a large minnow-like fish, can be clubbed or speared as they congregate in pools along their spawning migration. Ethnographic reports document numerous large fish taken this way (Boulenger, 1901).

In the floodplains, fish often spawn in shallow waters, and capture without additional equipment is easier. Siluriforms, particularly clariids, spawn in waters which are only a few cm deep. They are therefore very easy to procure, and many reports exist of fishers taking *Clarias* in large numbers often with only bare hands as they spawn at first rains (Greenwood, 1955; Jubb, 1967). Young fish are also a favourite prey of fishers, because they also move to floodplain zones at high waters and therefore are vulnerable to predation in the shallow waters.

Many lake taxa, primarily perciforms, spawn in the lake, and because they often migrate to deeper waters, their procurement without sophisticated technology is difficult. An exception are cichlids, a very numerous group, which often construct nests in the shallow newly-submerged lake/river floodplains at high waters. Because cichlids are very territorial, reports exist of fishers marking their nests and returning year after year to rob the nests (Brelsford, 1946).

As river and lake waters recede at the end of the wet season and through the dry season, procurement opportunities increase. In fact, the dry season is generally the most productive period for traditional fishers in Africa (e.g., Brelsford, 1946). Fish, particularly catfish, returning from upriver spawning migrations become stranded in shallow pools and channels, and can be trapped with minimal or no technology. Fish which inhabit the most inshore zones of seasonal or small perennial lakes are also susceptible to being stranded in pools when waters recede. Cichlids, for example, are particularly vulnerable when their shallow water nesting grounds are exposed by receding waters (e.g., Leakey, 1971). Therefore traditional fishers, who are familiar with locations and levels of seasonal lakes and streams, plan their movements to coincide with the desiccation and subsequent stranding of often hundreds of fish. These can be captured with spears or baskets, or even scooped up by hand. The Dinka, for example, spend the dry season in the Sudd swamp subsisting primarily on fish (in Beadle, 1981).

The optimal periods and locales for procuring fish with minimal or no technology are therefore river floodplains in the early long wet season and occasionally in the short wet season, particularly for riverine-spawning catfish; and seasonal lakes and rivers, and perennial inshore rivers or lakes in the dry season (usually November–February in eastern Africa), when waters recede and fish are trapped in isolated pools. Mainly cichlids and catfish, but also other inshore taxa and young fish are trapped at this time. The late dry/early wet season would presumably be of most importance to hunter-gatherer groups, as this is when fish are generally in peak condition; this also coincides with the period of poorest nutritional value of plants and terrestrial mammals.

Evidence from the Late Pleistocene

Numerous archaeological sites with fish remains (see Figure 1) indicate that Late Pleistocene hominids relied heavily on fish, suggesting that fish procurement has considerable time depth in Africa. Analyses of fish from these sites indicate patterns which could aid in identifying fish procurement in early hominid sites. Fish remains from the Nile River sites, in particular those at Wadi Kubbaniya, Isna, Edfu, Kom Ombo and Khor Musa (Figure 1), indicate that fishers procured fish selectively and seasonally. Dating from about 40 000 BP to the Holocene, over 40 Nile River sites have yielded hundreds of thousands of fish bones, often far outnumbering mammal bones (Churcher, 1972; Gautier & Van Neer, 1989; Greenwood, 1968; Greenwood & Todd, 1976; Van Neer, 1986; for a review and discussion of these sites see Gautier & Van Neer, 1989). The number of fish exploited at some sites was often phenomenal, as demonstrated by the depth of deposits; one site alone contained over 53 000 elements.

Over 90% of these fish remains derive from *Clarias* the catfish, and in over 50% of the sites *Clarias* remains comprise 99% to 100% of the totals. A total of eight other genera are represented at the sites, but only cichlids are represented in proportions greater than 5%. *Clarias* was probably obtained in two cycles, according to a discussion by Van Neer (Gautier & Van Neer, 1989). In the first cycle, *Clarias* were caught as they spawned on the floodplain of the Nile, shortly after the river flooded in the rainy season. They could be trapped by a variety of techniques, most of which required little or no technology. Most of these fish were between 30 and 100 cm in total length. In the second cycle, *Clarias*, and other fish, primarily *Tilapia*, were caught as the waters receded in the dry season, and fish were trapped in residual pools. These later sites are smaller than the earlier rainy season sites. It is suggested that sites were densely occupied during the early flood season to harvest the abundant spawning catfish, but as waters fell, people dispersed and exploited the fish in the residual pools on a more sporadic basis. The rainy season/dry season systematic exploitation of *Clarias* was therefore carried out annually over long periods of time, as demonstrated by the depth of deposits of many of these sites.

Similar patterns of repeated, intensive seasonal fish exploitation are seen in other Late Pleistocene African sites, including the Ishango and Katanda sites in Zaire (Greenwood, 1959; Stewart, 1989), the Gogo Falls site in Kenya (Marshall & Stewart, in press) and the White Paintings site in Botswana (Robbins *et al.*, in press) (Figure 1). The Ishango 11 and 14 sites are located along the Upper Semliki River, with Ishango 11 at the mouth of the Semliki River, where fishers could take advantage of fish making their seasonal migrations, while Ishango 14 is further upriver. In the main cultural layer of Ishango 11, dated at ca. 25 000 BP (Brooks *et al.*, 1990), fish elements were extremely densely concentrated (Stewart, 1989). Although there was a greater diversity of taxa represented at Ishango than at the Nile sites, *Barbus*, a large minnow-like fish, dominated, comprising 32·8% of the taxa from the main layer. The *Barbus* individuals averaged between 45 and 60 cm in length, with several very large individuals over 1 metre in length. Several anecdotal and ethnographic accounts exist of present day fishers spearing and/or clubbing large numbers of *Barbus* as they congregate at the river mouths on their seasonal migrations during the rainy season (e.g., Boulenger, 1901). The Ishango fossil remains are thought to represent a similar, repeated seasonal event, given the number of fish represented (Stewart, 1989). Hundreds of barbed bone points were recovered from the main layer at Ishango 11, and presumably these, possibly *in tandem* with weirs, were used to spear the fish. The size range of the *Barbus* individuals at Ishango represents a primarily mature population, probably on a spawning migration, and therefore these levels were interpreted as being a rainy season occupation (Stewart, 1989).

Ishango 14, located about 1 km downriver of Ishango 11, provides further evidence of seasonal fish exploitation. Although undated, it contains a similar cultural inventory to Ishango 11, and is considered to be contemporary with Ishango 11. The difference in location (riverine versus delta), is reflected in different proportions of taxa. While *Barbus* is still the dominant taxon, *Clarias* becomes the second most frequent. This may reflect the proximity of Ishango 14 to a *Clarias* spawning area on the flooded river banks where it could be captured while spawning (Stewart, 1989).

The Ishango sites represent clear exploitation of fish during the wet season while the fish were on spawning migrations or at breeding grounds. The repeated occupations at the Ishango sites indicate the predictability of these spawning runs, and the presumed reliance of the Ishango inhabitants on the spawning fish. As with the Nile sites, the Ishango sites also demonstrate long-term utilisation of a seasonally reliable and relatively easy to capture source of food, at a time when other food sources were nutritionally depleted.

The pattern of seasonal selective fishing continues to be evident in more recent sites. At Gogo Falls, near Lake Victoria, a variety of mammals and fish are exploited over several thousands of years (Marshall & Stewart, in press; Robertshaw, 1991). Of the over 10 000 fish remains, approximately 95% belong to the genus *Barbus*, with the remainder being primarily *Clarias* and some cichlids. Similar to Ishango, these remains appear to reflect repeated seasonal

Table 1 **The three most commonly-occurring fish taxa found in some Pleistocene sites, with a summary of proposed seasonality and least-effort method of procurement**

	High waters			Low waters		
	Bare hands	Clubs	Spears	Bare hands	Clubs	Spears
Barbus		+	+		+	+
Clarias	+			+		
Cichlids		+	+	+		

exploitation of the *Barbus* spawning migrations. Such exploitation is well-documented by fishers in this century in the affluent/effluent rivers around Lake Victoria (e.g. Dobbs, 1927).

A similar pattern exists at the White Paintings Rock Shelter, located in southern Africa. The site dates from >20 000 BP throughout the Holocene (Robbins *et al.*, in press) and contains a primarily mammalian fauna along with the remains of *Clarias* throughout the over 2 m of deposits. The nearest river source was the Okavango floodplains, located 17 km away; present day inhabitants, until recently, captured fish there in artificially constructed small earth dams (Robbins, pers. comm.). Presumably the White Paintings Rock Shelter inhabitants similarly exploited fish on a perennial basis when the river flooded its banks seasonally.

Analysis of fish remains from Late Pleistocene sites indicates selective, seasonal patterns of fish exploitation, often very intensive, which have re-occurred in areas of eastern and southern Africa over thousands of years.

Early hominid exploitation of fish resources

The seasonal availability, nutritional benefits and the demonstrated long-term seasonal exploitation of certain freshwater fish in Late Pleistocene sites suggests fish should be considered as a food source for early hominids. This suggestion is strengthened by the association of fish remains with many early hominid sites, which include Kanjera (Plummer, pers. comm.), Lokalalei (Kibunjia, pers. comm.), Senga (Harris *et al.*, 1987), nine sites at East Turkana (Harris, 1978) and 11 sites at Olduvai Gorge (Leakey, 1971) (Figure 1), although no definitive proof of hominid modification of the remains exists (with possible exceptions at BK, Olduvai Gorge). Faunal evidence just reported indicates that Late Pleistocene archaeological sites with fish remains are characterised by: (1) riverine or delta locations; (2) selective exploitation of seasonally spawning taxa; (3) taxa which can be easily procured with little or no technology; (4) low taxonomic diversity (also noted by Van Neer, 1986) because only seasonally available species requiring little technology to capture will be represented (see Table 1 for a summary), even given the low diversity resulting from ecological conditions; and (5) repeated occupation of sites, implying familiarity with spawning migrations. Given the unsophisticated nature of early hominid artifacts, these patterns seen in Late Pleistocene sites and their fish assemblages may also characterise early hominid sites and fish assemblages.

The best-preserved and most abundant fish remains associated with early hominid cultural and/or other faunal remains occur at Olduvai Gorge, where over 4000 well-preserved fish elements were recovered from 11 Bed I and II sites (Figure 2). Approximately half of these were identified from Bed I sites by Greenwood & Todd (1970), while the rest were analysed by the author, primarily from Bed II sites, using comparative collections at the National Museums of Kenya in Nairobi. Of the total number of fish elements recovered from the Olduvai Gorge

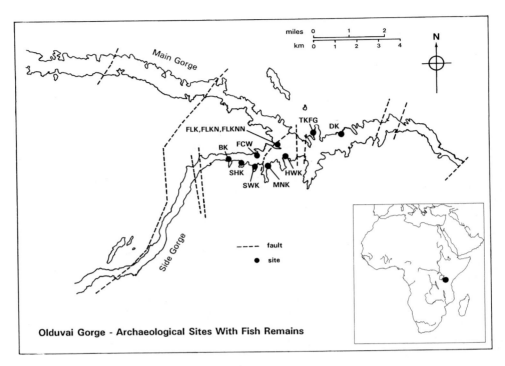

adapted from M.D. Leakey, 1979

Figure 2. Archaeological sites with fish remains at Olduvai Gorge, Beds I and II.

sites, 91·5% were concentrated at five sites: FLKNN Levels 2 and 3, and FLK-*Zinj* (Level 22) in Bed I, and MNK-Main, FLK and BK in Bed II. Because the FLK Bed II cultural and faunal remains have not been fully published, and the depositional context of the site is in doubt, I will not discuss the remains here. The fish remains at FLKNN-Level 3, FLK-*Zinj*, MNK and BK were associated with hominid and cultural remains. While this does not necessarily mean hominid exploitation, certain discrepancies exist in the composition of certain of the fish bone assemblages when compared with the composition of naturally deposited bone assemblages.

Only two taxa of fish occur throughout all Olduvai Gorge sites—*Clarias* sp., the catfish, and Cichlidae, also known colloquially as tilapia, a perch-like fish. It is not possible to identify these groups to a specific level based only on disarticulated elements. *Clarias* is a bottom-inhabiting inshore catfish which can grow to 2 m in length. It is generally tolerant of a wide variety of hydrological conditions, although less tolerant of saline and alkaline waters. Cichlids are shallow water dwellers, living within the 4 m depth contour. They are also highly tolerant of poor water conditions, but are less tolerant of de-oxygenated waters.

The low diversity of fish taxa in the Beds I and II sites is unusual compared with many modern East African lakes and rivers, and is in part likely due to the chemistry suggested for paleo-lake Olduvai. Hay's reconstruction of paleo-lake Olduvai as "saline, alkaline and rich in dissolved sodium carbonate-bicarbonate" (1976:53) suggests that it shared chemical features with similar present-day East African soda lakes, which have high salinity and alkalinity values, and impoverished aquatic faunas. Above a certain range of total salinity many common

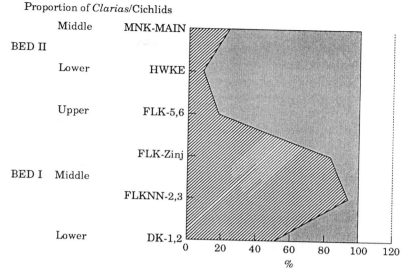

Figure 3. Changing proportions of *Clarias* (▨) and Cichlid (▦) elements through Beds I and II.

freshwater plants and animals are eliminated (Beadle, 1981). Modern soda lakes such as Lake Manyara with high salinity values [salinity=5·8% (g per liter) (Beadle, 1981)] contain only the most tolerant of fish, with only cichlids in the lake proper and clariids living on the fresher outer margins of the lake.

The composition of the fish assemblages is therefore in part affected by ecological factors, with proportions of taxa varying considerably between sites. All lake margin sites located in middle and lower Bed I contain both *Clarias* and cichlids, but are dominated by *Clarias* remains (Figure 3). In upper Bed I and in lower and middle Bed II, all but one (FCW) lake margin sites show a reversal in taxa, containing over 80% cichlids, with few *Clarias*. Because this pattern is found at sites associated both with and without artifacts, it must be interpreted as due to ecological circumstances. The presence of both *Clarias* and cichlids in lower and middle Bed I sites suggests the lake was somewhat fresher than in upper Bed I and lower/middle Bed II, thus supporting Hay's reconstruction of the palaeoecology. As the lake became smaller during Bed II deposition, its waters would have become more concentrated and even more saline and alkaline, probably forcing the less tolerant *Clarias* into associated streams. This process probably began in upper Bed I, in the interval above Tuff ID and below Tuff IF, because FLKN contains mainly cichlids. The lake was smallest in this interval in upper Bed I according to Hay (1976). At this time it would have been extremely saline, probably forcing *Clarias* out of the lake into the fresher stream channels. In upper Bed II, *Clarias* appears to have been present only in the fresher stream channels, such as represented at site BK.

A comparison of fish bone scatter frequencies (calculated as n bones/m^2) at the Olduvai Gorge sites with modern fish assemblages at Lake Turkana and Crater Lake (located on an island in Lake Turkana) indicates that most of the Olduvai Gorge bone frequencies are low (Figure 4) (for sampling of modern fish bone assemblages see Stewart, 1991). Studies of both naturally deposited and human made fish assemblages indicate that the latter are usually characterised by a greater frequency of bones (see Stewart, 1991). Most Olduvai Gorge assemblages therefore probably do represent natural deposition. However the bone scatter

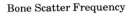

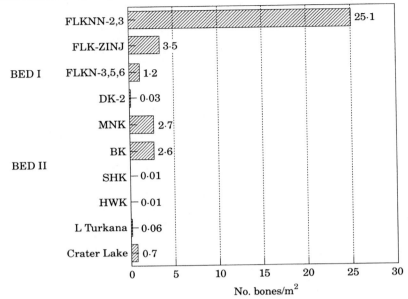

Figure 4. Fish bone scatter frequency of Olduvai Gorge sites compared with modern fish bone assemblages at Lake Turkana and Crater Lake, Kenya.

frequency from FLKNN Levels 2 and 3, FLK-*Zinj*, MNK and BK is considerably higher than both the other sites and the modern Lake Turkana and Crater Lake assemblages, suggesting their accumulations were not due strictly to natural depositional circumstances. I will discuss here the fish remains from Bed I sites FLKNN Levels 2 and 3, originally identified by Greenwood & Todd (1970), and from the Bed II sites of MNK and BK, identified by myself. I will also refer briefly to FLK-*Zinj*, identified both by myself and by Greenwood & Todd (1970), which has fewer fish elements than the other four sites.

A variety of data, including the presence of rootlet holes and reed casts in Tuff IB has led to the interpretation that FLKNN was situated in the lake margin zone ca. 1 km from the paleo-lake (Leakey, 1971). The lake, having no outlet, had greatly fluctuating levels (Hay, 1976). Level 3 is described as an occupation surface, with associated cultural and faunal remains, and remains of *Homo habilis* (Leakey, 1971). Level 2 contained only faunal remains with no associated cultural remains.

Both Levels 2 and 3 contain extremely high densities of fish bones, far greater than either fossil or modern naturally-deposited fish remains at Lake Turkana, or modern bones in an enclosed small lake (Figure 4). Over 85% of these individuals are *Clarias*, the catfish, with the remainder being cichlids (Table 2). The size and age profile constructed from the *Clarias* elements indicates that a range of sizes and ages of *Clarias* are represented in both Levels (Figure 5), not a specific segment of the population, such as a group of mature spawning fish. Cranial bones comprise the majority of elements in both levels, with vertebral elements poorly represented compared to an average skeleton, particularly in Level 3 (Table 3). While under-representation of catfish vertebrae in fossil sites is not infrequent (e.g., Van Neer, 1986), such low representation in sites with good preservation is unusual.

Table 2 Numbers of identified elements (NISP) and numbers (*n*) and percentages of Minimum Number of Individuals (MNI) of fish from Olduvai Gorge sites

	Clarias			Cichlids		
	NISP	MNI		NISP	MNI	
	n	*n*	%	*n*	*n*	%
FLKNN Level 2	203	59	95·2	41	3	04·8
FLKNN Level 3	605	87	86·0	135	14	14·0
FLKNN Levels 2/3*	498	137	93·9	112	9	06·1
FLK-*Zinj*	97	10	83·3	13	2	16·7
MNK	29	12	23·5	359	39	76·5
BK	170	39	88·6	7	5	11·4

*Treated as one unit.

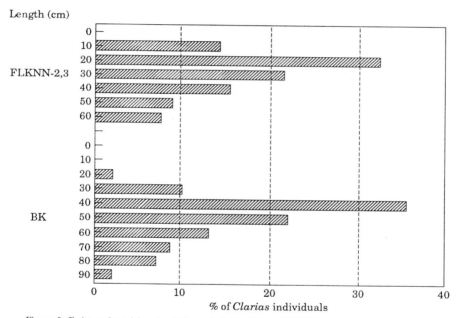

Figure 5. Estimated total length of *Clarias* individuals from FLKNN Levels 2 and 3, and BK.

There are several possible explanations for the anomalously large numbers of fish bones at FLKNN. Catastrophic events such as wind storms can cause mass fish die-offs and anomalous bone accumulations (e.g., Schafer, 1972), however such events are unlikely to occur in the same area through successive deposits. Water transport may also concentrate bones *post mortem* (e.g., Behrensmeyer, 1975), but the lack of size sorting and large number of bones makes this unlikely at FLKNN. The explanation which best fits the fossil and ecological data is that the bones represent large numbers of fish stranded by receding waters in the succeeding dry seasons, a scenario also suggested by Mary Leakey (1971). These fish were then preyed upon by carnivores and possibly hominids, although there is no direct evidence of this. This scenario is not infrequent in modern shallow lakes with seasonally fluctuating levels, such as is

Table 3 **Numbers (n) and percentages of Minimum Number of Elements (MNE) by skeletal category for Olduvai Gorge sites, as compared to average *Clarias* and cichlid skeletons**

	Clarias				Cichlids			
	Cranial		Axial		Cranial		Axial	
	n	%	n	%	n	%	n	%
FLKNN Level 2	61	74·4	21	25·6	1	04·8	20	95·2
FLKNN Level 3	304	89·9	34	10·1	2	04·8	20	95·2
FLKNN Levels 2/3	157	75·9	50	24·1	15	29·4	36	70·6
FLK-*Zinj*	75	88·2	10	11·8	0	0·0	10	100·0
MNK	23	92·0	2	08·0	143	54·2	121	45·8
BK	135	92·5	11	07·5	4	80·0	1	20·0
Average *Clarias*	56	48·3	60	51·7				
Average Cichlid	59	67·1	29	32·9				

reconstructed for paleo-lake Olduvai (Hay, 1976). The stranding hypothesis accounts for both the anomalously large and recurring concentration of bones in the FLKNN deposits, and for the range of sizes and ages represented, which would be expected in a stranded population.

Stranding of the fish at FLKNN would indicate a dry season accumulation. This statement is supported by the other aquatic-based fauna, in particular the avian taxa, where the most numerous groups in Levels 2 and 3 are winter or dry season residents (D. Matthiessen, pers. comm.). The presence in Level 3 of several turtle shells without associated skeletal elements has been interpreted by Louis Leakey and others to indicate hominid predation (Leakey, 1971). These turtles can be caught when lake waters are receding, before they aestivate, and their presence can be seen as further support for a dry season occupation. While the aquatic-based faunas at FLKNN were not necessarily accumulated at the same time as the mammalian faunas, the coincident occurrence of these large accumulations of bones implies real association.

Predation could account for the lack of fish vertebral elements in both Levels 2 and 3, in that bodies are either consumed on site or are detached from heads and removed, leaving the difficult-to-digest heads. Differences in taxonomic and skeletal element representation between Levels 2 and 3 may imply that different predators modified the assemblages. In Level 2 the fish are mainly *Clarias*, while in Level 3 cichlids make up 14% and *Clarias* the rest. Further, a greater absence of vertebrae in Level 3 may imply bodies detached from heavy, inedible heads, and removed off-site by hominids. While the Greenwood & Todd data do not explicitly quantify fish elements, calculation of a rough fragmentation index (ratio of whole to fragmented bones) indicates that the bones in Level 3 are more fragmented than those of Level 2, with numerous cranial fragments. Cranial fragmentation is characteristic of contemporary fish processing, with modern fishers severing and then smashing skulls with sticks or rocks to extract the brains. Thus the high degree of fragmentation may suggest hominid predation. The lesser fragmentation of cranial elements in Level 2 may represent an assemblage modified by carnivores, where only bodies were eaten. Experimental studies are needed to describe fish consumption by carnivores.

The site of FLK-*Zinj* is very similar to FLKNN Level 3 both in depositional context and in fish assemblage composition (Leakey, 1971). While total numbers of bones are comparatively low at FLK-*Zinj*, taxonomic proportions are similar to those of FLKNN-3, while the cranial to vertebral proportions are also similar (Tables 2, 3). This suggests similar depositional and predation events at both sites. Elements of *Australopithecus* and *Homo habilis* were associated with the FLK-*Zinj* assemblage.

The MNK site was located in the flood zone area of paleo-lake Olduvai (Hay, 1976). The sediments where cultural and faunal remains were recovered consist of a fine-grained reworked tuff. Hay (1976) has interpreted the depositional environment as an intermittently flooded lake zone. Cultural and faunal remains have been interpreted by Leakey (1971) as representing successive periods of occupation; she suggests the remains have not been secondarily transported.

The taxonomic representation at MNK is reversed from that of FLKNN and FLK-*Zinj*, with cichlids now dominant (Table 2). With the widely fluctuating levels postulated for paleo-lake Olduvai, cichlids with nests on the floodplain could have been stranded at the end of the dry season by receding waters. Cichlids are not airbreathers, and many accounts exist of large numbers of them isolated around lakeshores by receding waters (Coe, 1966; Leakey, 1971). The cichlid skeletal element representation at MNK is dissimilar to that of all other Olduvai Gorge sites, and more similar to that of an average skeleton (Table 3). There is no evidence of tooth- or cutmarks. I therefore suggest that, similar to the FLKNN Level 3 and FLK-*Zinj* assemblages, the MNK assemblage represents large numbers of fish which were stranded during receding waters, but unlike the other assemblages, were little disturbed *post mortem* by predators. The skeletal element representation of *Clarias* at MNK was very skewed, unlike an average skeleton, but the number of elements are too small to be significant.

BK, in upper Bed II, was interpreted as a stream channel site which was secondarily deposited, probably only a short distance from its original location, according to Mary Leakey (Leakey, 1971). Paleo-lake Olduvai at this time was reduced to small lakes and marsh, and BK was formed in eastern fluvial-lacustrine sediments. The site consists of a series of reworked tuffs, clays, silts and sand which have filled a river channel (Leakey, 1971). Leakey has suggested that the associated cultural and faunal remains, including the remains of at least 24 *Pelorovis* (long-horned buffalo) individuals, represent remains from a camp site which had been secondarily deposited, but only over a short distance. The fish remains were abundant and concentrated primarily in a single lithological unit several metres thick throughout the trenches. The redeposited nature of the site means that the artifacts and fauna were treated as one unit. The thickness of deposits and number of artifacts suggests a re-occupied site.

The fish fauna at BK is comprised of approximately 89% *Clarias* and 11% cichlids (Table 2). At a FLKNN Level 3 and at FLK-*Zinj*, there is a dominance of *Clarias* cranial elements, and few vertebral elements (Table 3), suggesting *post mortem* disturbance. The age/size profile of the *Clarias* individuals is strongly biased towards very large individuals, unlike at other sites including FLKNN (Figure 5), suggesting this represents a spawning population which was either stranded by receding stream waters, or preyed upon while spawning, or both. Two elements have toothmarks, probably of a carnivore, while two elements, a dermethmoid and a frontal fragment, have possible cutmarks (Figure 6). SEM work on modern cutmarks on fish bone is presently being conducted by Marie Selvaggio (Southern Connecticut State University), and it is hoped that this will clarify the nature of these marks. Carnivore and possibly hominid activity clearly therefore altered this assemblage; however, whether by actively procuring and then consuming the fish, or through scavenging the already-stranded fish, is unclear.

The fish elements at BK were almost certainly accumulated during the wet season during a spawning run. However, unlike at the FLKNN levels, where fish remains were in close

Figure 6. Photographs of (a) dermethmoid and (b) frontal elements with possible cutmarks from BK.

association with the mammal and reptile remains, at BK the secondary deposition may mean the fish were obtained separately from the *Pelorovis* remains, the most common mammal represented.

Implications for early hominid behaviour

The patterns of fish exploitation seen in Late Pleistocene sites are also evident in the early hominid sites of FLKNN Level 3, FLK-*Zinj* and BK at Olduvai Gorge, and combined with evidence of possible cutmarks on elements at BK, make a strong, although not absolute, case for early hominid fish procurement. The implications of early hominids procuring fish are several. First, because the availability of the fish peaks at the beginning of the rainy season, sometimes for only a few days, and at mid-dry season, hominids would need to time their seasonal round to coincide with these peaks. Because harvesting fish at this time would provide reasonably abundant food, hominid groups would probably be less mobile, focussing their activities around the drying ponds or streams. The potential abundance of fish could also result in larger groupings of people, resulting in greater social interaction. As well, because the catfish can be caught either with bare hands or sticks, they are accessible to all hominid group members, and not limited to only those with tools, or those not caring for altricial young.

Because these peak periods of fish availability coincide with times when most other food resources are scarce and/or nutritionally poor, it is clear that hominids would potentially be competing for these resources with the carnivores mentioned above (i.e., hyenas, canids and felids). However unlike hunting for mammals, when usually only a few individuals are captured at a time, there would be numerous fish during spawning or stranding, so actual on the spot competition with other predators would likely be lower. Potential predators would presumably time their access to the fish so as not to coincide with other competitors. These factors would also make fish a more attractive resource to hominids, when faced with competition with predators at scavenged carcasses.

Finally, there is a difference between the type of fish procurement undertaken at the lake margin of FLKNN Level 3 and FLK-*Zinj*, than at the stream site of BK, which may imply differences in strategy between hominid species. At FLKNN the fish were probably stranded by receding waters, so procurement would only involve scooping them up. However at BK the fishers may have timed their visit to the spawning migrations of the catfish, and actively captured the fish while spawning. The former event involves scooping up dead and dying fish, while the latter presupposes a certain level of knowledge of fish behaviour as well as ability to capture active spawning fish. Remains of *Homo habilis* were recovered from FLKNN, while remains of *Homo erectus* were recovered from upper Bed II sites, although not from BK. It is possible that these sites show a change in procurement strategy between the two hominid species, although other early hominid site fish remains must be analysed to further investigate this idea.

Summary

Nutritional, ecological and ethnographic evidence indicate that fish would be a seasonally available, nutritious and easy to procure food source for early hominids, particularly during periods when other food sources may be unavailable or of poor quality. Fish are relied on as a seasonal resource by carnivores and non-human primates, and archaeological reports also document the importance of fish as a seasonal resource for hominids during the Late Pleistocene and Holocene.

Fish remains from Late Pleistocene archaeological sites indicate that these sites were characterised by: (1) riverine or delta locations; (2) low taxonomic diversity; (3) selective exploitation of seasonally spawning taxa; (4) taxa which can be easily procured with little or no technology; (5) skewed skeletal element representation; (6) repeated occupation of sites; and (7) in some cases, bone modification. Fish remains are associated with several early hominid sites, with especially well-preserved and relatively large assemblages at Olduvai Gorge. While some of the seven characteristics listed are also common to naturally-deposited sites, the presence of five or six characteristics strongly suggests hominid modification. The first six characteristics are manifest in the Olduvai Gorge sites of FLKNN Level 3, FLK-*Zinj* and BK sites, with the exception of the depositional environment at FLKNN and FLK-*Zinj*, which is lake margin, rather than fluvial or deltaic. Further, some BK elements may have cutmarks, a certain sign of hominid modification. The MNK remains were thought to represent a non-modified naturally-deposited assemblage, due to its reasonably non-skewed skeletal element representation, lake margin setting and lack of evidence of bone modification. FLKNN Level 2 was thought to be modified only by carnivores.

Possible implications of early hominid fish procurement include timing of the seasonal round to exploit spawning or stranded fish, and larger group size with greater social interaction to exploit seasonally abundant fish. Further investigation must also be conducted on the possible differences in procurement strategy between the hominids associated with the FLKNN site (*Homo habilis*) and those associated with the BK site (presumed to be *H. erectus*).

Acknowledgements

I would like to thank the L. S. B. Leakey Foundation and the Social Sciences and Humanities Council of Canada for providing funding for this study. I would also like to thank the

Tanzanian Government for allowing me to analyse the Olduvai fish collections. Thanks are due to the National Museums of Kenya for providing me with facilities, and especially to the Palaeontology staff for their help. I am grateful to Dr Mary Leakey for discussions in Nairobi on the formation processes at Olduvai Gorge, and to Dr Richard Hay for further clarification of site locations. I also thank Jim Oliver and Nancy Sikes for helpful discussions, and to the three reviewers who made insightful suggestions on this paper.

References

Beadle, L. C. (1981). *The Inland Waters of Tropical Africa*. London: Longman.

Behrensmeyer, A. K. (1975). The taphonomy and paleoecology of the Plio-Pleistocene vertebrate assemblages east of Lake Rudolf, Kenya. *Bull. Mus. Comp. Zool. Harv.* **146**, 473–578.

Blumenschine, R. J. (1987). Characteristics of an early hominid scavenging niche. *Curr. Anthrop.* **28**, 383–407.

Blumenschine, R. J. & Madrigal, T. C. (1993). Variability in long bone marrow yields of east African ungulates and its zooarchaeological implications. *J. Archaeol. Sci.* **20**, 555–587.

Boulenger, G. A. (1901). *Les Poissons du Bassin du Congo*. Bruxelles: Publications de l'Etat Independent du Congo.

Brelsford, W. V. (1946). *Fishermen of the Bangweulu Swamps*. Livingstone: Rhodes-Livingstone Institute.

Brooks, A. S., Hare, P. E., Kokis, J. E., Miller, G. H., Ernst, R. D. & Wendorf, F. (1990). Dating Pleistocene archaeological sites by protein diagenesis in ostrich eggshell. *Science* **248**, 60–64.

Bunn, H. T., Bartram, L. E. & Kroll, E. M. (1988). Variability in bone assemblage formation from Hadza hunting, scavenging, and carcass processing. *J. Anthrop. Archaeol.* **7**, 412–457.

Bunn, H. T. & Ezzo, J. A. (1993). Hunting and scavenging by Plio-Pleistocene hominids: nutritional constraints, archaeological patterns, and behavioural implications. *J. Archeol. Sci.* **20**, 365–398.

Churcher, C. S. (1972). Late Pleistocene vertebrates from archaeological sites in the plain of Kom Ombo, Upper Egypt. *Life Sci. Contrib. R. Ontario Mus.* **82**. Toronto: Royal Ontario Museum.

Clark, J. D. (1960). Human ecology during Pleistocene and later times in Africa south of the Sahara. *Curr. Anthrop.* **1**, 307–324.

Coe, M. J. (1966). Biology of *Tilapia grahami* in Lake Magadi, Kenya. *Act. Trop.* **23**, 146–177.

Dobbs, C. M. (1927). Fishing in the Kavirondo Gulf, Lake Victoria. *E. Afr. J. Nat. Hist.*, 97–110.

Dyson, W. S. & Fuchs, V. E. (1937). The Elmolo. *J. Roy. Anthrop. Inst.* **67**, 327–336.

Ewer, R. F. (1973). *The Carnivores*. Ithaca: Cornell University Press.

Foley, R. (1982). A reconsideration of the role of predation on large mammals in tropical hunter-gatherer environments. *Man* **17**, 393–402.

Garrod, D. J. (1959). The growth of *Tilapia esculenta* Graham in Lake Victoria. *Hydrobiologia* **36**, 268–298.

Gautier, A. & Van Neer, W. (1989). Animal remains from the Late Paleolithic sequence at Wadi Kubbaniya. In (A. Close, Ed.) *The Prehistory of Wadi Kubbaniya Volume 2*, pp. 119–169. Dallas: Southern Methodist University Press.

Goodall, J. (1971). *In the Shadow of Man*. Boston: Houghton Mifflin.

Greenwood, P. H. (1955). Reproduction in the cat-fish *Clarias mossambicus* Peters. *Nature* **176**, 516–517.

Greenwood, P. H. (1959). Quaternary fish fossils. *Exploration du Parc National Albert, Mission J. de Heinzelin de Braucourt (1950)* Fasc. 4.

Greenwood, P. H. (1968). Fish remains. In (F. Wendorf, Ed.) *Prehistory of Nubia, Vol. I*, pp. 100–109. Dallas: SMU Press.

Greenwood, P. H. & Todd, E. J. (1976). Fish remains from Upper Paleolithic sites near Idfu and Isna. In (F. Wendorf & R. Schild, Eds) *Prehistory of the Nile Valley*, pp. 383–389. New York: Academic Press.

Greenwood, P. H. & Todd, E. J. (1970). Fish remains from Olduvai. In (L. S. B. Leakey & R. J. G. Savage, Eds) *Fossil Vertebrates of Africa Vol. 2*, pp. 225–241. London: Academic Press.

Harris, J. W. K. (1978). *The Karari Industry*. Unpublished Ph.D. Dissertation. Berkeley: University of California.

Harris, J. W. K., Williamson, P. G., Verniers, J., Tappen, M. J., Stewart, K. M., Helgren, D., de Heinzelin, J., Boaz, N. T. & Bellomo, R. V. (1987). Late Pliocene hominid occupation in Central Africa: the setting, context, and character of the Senga 5A site, Zaire. *J. hum. Evol.* **16**, 701–729.

Hay, R. L. (1976). *Geology of Olduvai Gorge*. Berkeley: University of California Press.

Hyslop, E. J. (1986). The growth and feeding habits of *Clarias anguillaris* during their first season in the floodplain pools of the Sokoto-Rima river basin, Nigeria. *J. Fish. Biol.* **30**, 183–193.

Jocque, R. G. (1977). Une etude sur *Clarias senegalensis* Val. dans la region du Lac Kossou (Cote d'Ivoire). *Hydrobiologia* **54**, 49–65.

Jubb, R. A. (1967). *Freshwater Fishes of Southern Africa*. Capetown: Balkema.

Kruuk, H. (1976). *The Spotted Hyena: A Study of Predation and Social Behaviour*. Chicago: University of Chicago Press.

Leakey, M. D. (1971). *Olduvai Gorge, Vol. 3*. Cambridge: Cambridge University Press.

Leakey, M. D. (1979). *Olduvai Gorge*. London: Collins.

Lee, R. B. (1968). What hunters do for a living, or how to make out on scarce resources. In (R. B. Lee and I. Devore, Eds) *Man the Hunter*, pp. 30–48. Chicago: Aldine.

Lowe-McConnell, R. H. (1987). *Ecological Studies in Tropical Fish Communities.* Cambridge: Cambridge University Press.

Marshall, F. B. & Stewart, K. M. (in press). Hunting, fishing and herding pastoralists of western Kenya: the fauna from Gogo Falls. *Archaeozoologia.*

O'Connell, J. F., Hawkes, K. & Blurton-Jones, N. G. (1988). Hadza scavenging: Implications for Plio-Pleistocene hominid subsistence. *Curr. Anthrop.* **29,** 356–363.

O'Connell, J. F., Hawkes, K. & Blurton-Jones, N. G. (1992). Patterns in the distribution, site structure and assemblage composition of Hadza kill-butchering sites. *J. Archaeol. Sci.* **19,** 319–347.

Pigott, G. M. & Tucker, B. W. (1990). *Seafood. Effects of Technology on Nutrition.* New York: Marcell Dekker, Inc.

Robbins, L. H., Murphy, M. L., Stewart, K. M., Campbell, A. C. & Brook, G. A. (in press). Barbed bone points, paleoenvironment, and the prehistory of fish exploitation in the western Kalahari Desert, Botswana. *J. Field Archaeol.*

Robertshaw, P. R. (1991). Gogo Falls: A complex site east of Lake Victoria. *Azania* **26,** 63–195.

Schafer, W. (1972). *Ecology and Palaeoecology of Marine Environments.* Edinburgh: Oliver and Boyd.

Sept, J. (1992). Archaeological evidence and ecological perspectives for reconstructing early hominid subsistence behaviour. In (M. B. Schiffer, Ed.) *Archaeological Method and Theory* **4,** 1–56.

Speth, J. D. (1987). Early hominid subsistence strategies in seasonal habitats. *J. Archaeol. Sci.* **14,** 13–29.

Speth, J. D. (1989). Early hominid hunting and scavenging: the role of meat as an energy source. *J. hum. Evol.* **18,** 329–343.

Speth, J. D. & Davis, D. D. (1976). Seasonal variability in early hominid predation. *Science* **192,** 441–445.

Speth, J. D. & Spielmann, K. A. (1983). Energy source, protein metabolism and hunter-gatherer subsistence strategies. *J. Anth. Archaeol.* **2,** 1–31.

Stewart, K. M. (1988). Changes in condition and maturation of the *Oreochromis niloticus* L. population of Ferguson's Gulf, Lake Turkana, Kenya. *J. Fish Biol.* **33,** 181–188.

Stewart, K. M. (1989). *Fishing Sites of North and East Africa in the Late Pleistocene and Holocene.* Cambridge Monographs in African Archaeology 34, BAR International Series 521.

Stewart, K. M. (1991). Modern fishbone assemblages at Lake Turkana, Kenya: a methodology to aid in recognition of hominid fish utilization. *J. Archaeol. Sci.* **18,** 579–603.

Tanaka, J. (1976). Subsistence ecology of Central Kalahari San. In (R. B. Lee & I. Devore, Eds) *Kalahari Hunter-Gatherers,* pp. 98–120. Cambridge: Harvard University Press.

Turnbull-Kemp, P. (1967). *The Leopard.* Capetown: Howard Timmins.

Van Neer, W. (1986). Some notes on the fish remains from Wadi Kubbaniya (Upper Egypt: Late Palaeolithic). In *Fish and Archaeology,* BAR International Series **294,** 103–113.

Vincent, A. S. (1985). Plant foods in savanna environments: a preliminary report of tubers eaten by the Hadza of northern Tanzania. *World Archaeol.* **17,** 131–147.

Henry T. Bunn
Department of Anthropology, University of Wisconsin, Madison, WI 53706, U.S.A.

Received 13 September 1993
Revision received 2 February 1994 and accepted 3 February 1994

Keywords: Plio-Pleistocene hominids, foraging strategies, butchery, cut marks, meat and marrow consumption, paleolandscape taphonomy and ecology.

Early Pleistocene hominid foraging strategies along the ancestral Omo River at Koobi Fora, Kenya

Reconstructions of hominid foraging activities at Koobi Fora must account for an uneven distribution of Plio-Pleistocene archaeological sites and hominid fossils. Near the 1·64 million year old stratigraphic level of the Okote Tuff complex, stone raw materials and tools are abundant in fluvial contexts at the Karari Ridge. In contrast, hominid fossils are abundant but stone raw materials are absent and tools are rare in fluvial and shallow, ephemeral lake margin contexts at Ileret and at the Koobi Fora Ridge. As the distance to the nearest source of stone raw materials increases, from local abundance at the Karari Ridge to 5 km at Ileret and to at least 15 km at the Koobi Fora Ridge, the archaeological visibility of hominid activities, as defined by sites with stone tools, decreases to essentially zero. Cut marks made with large stone tools during butchery occur on more than 60 fossil bones from Ileret and especially from the Koobi Fora Ridge. In the absence of associated stone tools, those define a new kind of stone age site. The fossils with cut marks represent diverse skeletal parts and taxa, especially hippopotamus, and occur over several square km as isolated specimens, as associated elements of one carcass, and as concentrations of conjoining pieces and multiple carcasses. Some of the defleshing cut marks occur on intact limb elements that were not utilized for marrow. Early *Homo erectus* probably carried large stone tools during their foraging visits to these sub-regions, using the tools in butchery but rarely discarding them. Alternative hominid foraging strategies along the ancestral Omo River and taphonomic processes of site formation are discussed.

Journal of Human Evolution (1994) **27,** 247–266

Introduction

During the past 25 years, the Koobi Fora region east of Lake Turkana in northern Kenya has provided a wealth of paleoanthropological evidence of Plio-Pleistocene human evolution, including hominid and other fossils, concentrations of stone artifacts and fossil bones, and geological data on the dating and paleoenvironmental context of the paleontological and archaeological evidence. Most archaeological research at Koobi Fora has logically emphasized the excavation of sites with large, dense concentrations of stone artifacts and bones, because of the initial need to describe large, presumably representative samples (Isaac *et al.*, 1971; Isaac & Harris, 1978). Isaac & Harris (1975, 1980) also recognized the importance of documenting lower density scatters of archaeological material from the landscape areas surrounding the sites, and they initiated a surface survey of that part of the record which they termed the scatter-between-the-patches. The defining characteristic of stone age sites and scatters until now has been the stone artifacts; associated bones have then been studied as possible evidence of hominid subsistence behaviour (Bunn, 1981, 1982; Bunn *et al.*, 1980; Isaac, 1971, 1978, 1984). The discovery at Koobi Fora of a new kind of stone age archaeological site, composed solely of bones bearing stone-tool cut marks and/or hammerstone percussion damage enables the documentation, even in the absence of any discarded stone artifacts, of sites on the ancient landscape where hominids foraged for carcasses (Bunn, 1981).

This paper introduces new bone evidence and preliminarily considers how it contributes to an understanding of hominid foraging strategies. The emphasis on butchered fossil bones and on regions of the paleolandscape where stone tools are rare provides new data from areas with abundant hominid fossils but little or no previous archaeological evidence of hominid

0047–2484/94/010247+20 $08.00/0

behaviour. Surface survey and excavation within a restricted stratigraphic interval centered at the 1·64 million year old base of the Okote Member provide landscape archaeological evidence for evaluating alternative models of hominid foraging strategies from three broadly penecontemporaneous sub-regions at Koobi Fora: the Koobi Fora Ridge (Areas 101, 103), the Karari Ridge (Areas 130, 131), and Ileret (Areas 1A, 5, 8A).

Previous research at Koobi Fora

Plio-Pleistocene sedimentary deposits of the Koobi Fora Formation are extensively exposed within an approximately 2500 km^2 area to the east of Lake Turkana (Figure 1). Consisting of 560 meters of fluvial, lacustrine, and volcaniclastic sediments, the Koobi Fora Formation along with correlative deposits to the west and north of Lake Turkana provides a spectacular paleoenvironmental record of a semi-arid East African Rift Valley lake basin from approximately 4–0·6 m.y. (Figure 2). Major discoveries and changing interpretations of the paleontological, archaeological, and geological evidence are well documented in a voluminous literature (e.g., Brown & Feibel, 1986, 1988; Bunn, 1982; Coppens et al., 1976; Feibel, 1988; Harris, 1983, 1991; Harris, 1978; Harris & Isaac, 1976; Isaac, 1971, 1978, 1984, n.d.; Isaac et al., 1971; Isaac & Harris, 1975, 1978, 1980; Kaufulu, 1983; Kroll, n.d.; Leakey & Leakey, 1978; Schick, 1984; Stern, 1991; Toth, 1982; Walker & Leakey, 1978; Wood, 1992). Of particular interest here are: (1) the geological work in the 1980s by Brown and Feibel that has significantly changed the reconstructions of the paleoenvironmental context in which the archaeological evidence occurs, and (2) the history and results of previous paleoecological and archaeological attempts to define the habitat preferences and foraging patterns of Plio-Pleistocene hominids.

Recent geological studies at Koobi Fora

Geological studies at Koobi Fora by Brown and Feibel (1986, 1988; Feibel, 1988) led to a redefinition of the Koobi Fora Formation and of the predominant paleoenvironmental features of the Turkana Basin. Instead of a basin dominated by a large permanent lake, as it is today, Brown and Feibel described it as dominated by the perennial ancestral Omo River flowing north to south through the Koobi Fora region with a series of mostly smaller, short-lived lakes farther to the west (Figure 2). Earlier geological research at Koobi Fora in the 1970s (e.g., Coppens et al., 1976; Findlater, 1978) yielded reconstructions of a fluctuating but permanent, large lake dominating the basin from Plio-Pleistocene to modern times. A perennial river with intermittent tributaries flowing from the northeast transported stone cobbles, suitable for stone tools, only part of the way to the lake into which it emptied in the vicinity of the Koobi Fora Ridge in Area 103. Both archaeological and paleoecological perspectives on hominid adaptations have relied on the paleoenvironmental framework developed during the 1970s (e.g., Coppens et al., 1976; Leakey & Leakey, 1978). The abundant stone-artifact bearing sites along intermittent paleostream channels at the Karari Ridge were described in relation to a 15–20 km distance to the paleolake shore, and the distribution of such "inland" sites and of hominid fossils in lake margins at the Koobi Fora Ridge and at Ileret was contrasted (Harris, 1978; Isaac & Harris, 1978). Similarly, paleoecological researchers described patterns in the landscape distribution of different hominid taxa in relation to the stream channel and the lake margin habitats (Behrensmeyer, 1975; see also Walker & Leakey, 1978).

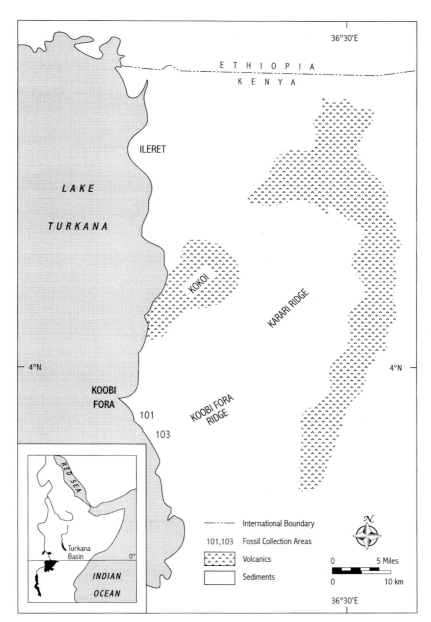

Figure 1. The Koobi Fora landscape east of Lake Turkana showing three sub-regions: Koobi Fora Ridge (Area 101, Area 103), Karari Ridge, and Ileret.

The distribution of the archaeological evidence needs to be re-evaluated, by accounting for the proximity of a large perennial river to most of the areas having stone artifacts and/or fossil bones with cut marks or percussion notches. The previous lake-dominated model of the early Pleistocene landscape at Koobi Fora invited the working hypothesis that water-dependent

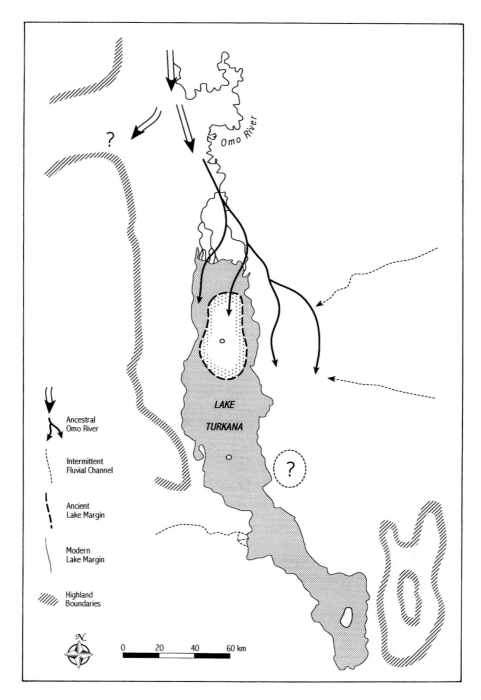

Figure 2. The route of the ancestral Omo River through Koobi Fora at 1·65 million years ago, as reconstructed by Feibel. Adapted from Feibel (1988).

hominids and other taxa were seasonally constrained to the use of inland, intermittent stream areas at the Karari Ridge in the wet season, and perhaps the reliance on lake margin habitats only in the dry season (e.g., Bunn, 1987). The new, river-dominated model of the Koobi Fora landscape, on the other hand, largely releases hominids and other water-dependent taxa from problems of water availability. Hominids could have foraged year-round along a wide band on either side of the Omo along the entire length of its course through the Koobi Fora region. Perhaps the marginal areas along the eastern rim of the basin would have been out of reach except during wet seasons, but the sites along the Karari Ridge would have been accessible to foraging hominids year-round. For example, Feibel (personal communication, 1990) has suggested that the main channel of the ancestral Omo River may have been only a kilometer or less from the FxJj 50 site (Bunn *et al.*, 1980). Although water availability along the Omo implies an element of similarity among the different sub-regions of the study area, the penecontemporaneous microhabitats, the resources available to hominids, and the known archaeological record from the Koobi Fora Ridge, the Karari Ridge, and Ileret, are different (Figures 1 and 2). Alternative hominid foraging strategies specific to the river-dominated Koobi Fora landscape are discussed below. Landscape archaeological research at Koobi Fora is also useful for evaluating published models of Plio-Pleistocene hominid foraging strategies.

Previous archaeological and paleoecological studies at Koobi Fora

Archaeological investigations at Koobi Fora under the joint leadership of Richard Leakey and Glynn Isaac began with excavations of sites from two stratigraphic intervals: an older set, the so-called Lower Member sites from the KBS Tuff, which was then dated to approximately 2·61 m.y., and a younger set, mainly from the Karari Ridge and associated with the Okote Tuff Complex, which was dated to approximately 1·5–1·6 m.y. (Isaac, 1976; Isaac *et al.*, 1971, 1976). The sites from the KBS Tuff, consisting of FxJj 1, 3, and 10, provided Isaac with evidence for his home base and food-sharing model and with enough diversity in assemblage composition for a basic classification of the sites as home base (FxJj 1), butchery (FxJj 3), and artifact-only (FxJj 10) sites (Isaac, 1971, 1978; Isaac & Harris, 1978). Large-scale excavation of the Karari sites was undertaken as a dissertation project by J. W. K. Harris (1978; Harris & Herbich, 1978; Harris & Isaac, 1976; Isaac & Harris, 1978) and provided (1) the basic definition of the Karari Industry as a regional variant of the Developed Oldowan and (2) the large, well-preserved assemblages of artifacts and fossil bones for follow-up, problem-oriented studies of lithic technology (Toth, 1982), subsistence behaviour and taphonomy (Bunn, 1982), archaeological site structure and spatial analysis (Kroll, 1994, n.d.; Kroll & Isaac, 1984), geoarchaeological site formation (Kaufulu, 1983), experimental site formation (Schick, 1984), and plant food availability in analog habitats (Sept, 1984, 1986). The excavation of FxJj 64, an apparent mini-site (Isaac *et al.*, 1981), and of FxJj 50, one of the largest and best preserved sites at Koobi Fora (Bunn *et al.*, 1980), completed the excavation work of the 1970s.

In addition to excavating large-scale sites, there was a complementary program of surface survey. Recognizing that much of the stone artifact record at Koobi Fora consists of surface scatters of low density that have already eroded from Plio-Pleistocene deposits, Isaac & Harris (1975) initiated a study of the scatter-between-the-patches to document the composition and distribution of surface artifact occurrences through outcrop transects at selected stratigraphic intervals (for further discussion of landscape or off-site approaches, see Dunnell & Dancey, 1983; Foley, 1981). In the early 1970s, Behrensmeyer (1975) conducted surface transects for studying fossil bones for paleoecological reconstructions of hominid habitat preferences; in the late 1970s, she also recovered excavated samples of fossil bones, hominid and other animal footprints, and some low densities of stone artifacts at the Koobi Fora Ridge, at the Karari Ridge (FxJj 64 started out as Behrensmeyer's paleontological excavation no. 18), and at Ileret (Behrensmeyer, 1985; Behrensmeyer & Laporte, 1981). The surface transect study of the scatter-between-the-patches was pursued as the dissertation work of Stern (1991, 1993), who focused on a transect along several kilometers of the lower Okote Member in Area 131 at the Karari Ridge. All of this work established the potential of the Koobi Fora region for site-oriented studies of hominid activities and for investigations of hominid foraging strategies in diverse habitats across a horizontal area of paleolandscape many times larger than the exposed paleolandscapes at other well known field areas, including Olduvai Gorge (Hay, 1976; Leakey, 1971) and Olorgesailie (Isaac, 1977; Potts, 1989).

There are two fundamental questions about the archaeological evidence and the research design for investigating hominid foraging strategies. First, how accurately does the material excavated from dense site accumulations reflect hominid foraging patterns? Second, how precise a time resolution is achievable from surface transect data, and are there alternative approaches that would yield better results? Stern (1991, 1993) concluded that the archaeological record consists of palimpsests of variable density that accumulated over vast time intervals, that, except in density, the large site assemblages are equivalent in composition to the surface scatters, that the surface scatters provide the most representative archaeological materials, and that the most precise time resolution achievable for her time-averaged, surface scatters data from the several meter stratigraphic interval of the lower Okote Member is approximately 70,000 years.

The fundamental problem with the surface transect approach is that all of the stone artifacts and fossils are recovered out of context, lack precise stratigraphic control, and are of unknown derivation. Each isolated artifact recovered on the modern ground surface could represent (1) the undisturbed residue of an isolated hominid act at that location, as advocates of surface transect studies tacitly assume, (2) the disturbed residue from an originally high density site that was dispersed and transported to that location by fluvial processes, or (3) the eroding edge of a large buried site (Bunn & Kroll, 1993). Even without knowing which of these or other alternatives is most representative, the surface finds do, however, have obvious value. Surface artifacts and fossils can often be attributed with reasonable confidence to an approximate stratigraphic provenance and microhabitat, as has been done with the hominid fossils at Koobi Fora (Leakey & Leakey, 1978), and the presence of cut-marked bones in surface contexts establishes the reality of hominid foraging in paleolandscape areas that have not yielded stone artifacts (Bunn, 1981; see below). However, defining the context, derivation, and full behavioural meaning of the surface archaeological remains, and assessing the significance of dense site accumulations will require a sustained program of landscape excavation. Until that is implemented, the promising potential for landscape archaeology at Koobi Fora will remain largely untapped.

Alternative models and new archaeological data

Several alternative foraging models have been proposed using archaeological and modern analog evidence. These have been usefully reviewed by Potts (1991), and all of the supporting evidence and arguments will not be reiterated here. Isaac (1971, 1978, 1983, 1984) developed the home base/central place model by working from the premise that the Plio-Pleistocene archaeological sites at Koobi Fora and Olduvai are high-density anomalies formed through the transport of artifacts and food by hominids who practiced a gender-based division of labour and cooperatively shared the food carried back to home bases. Isaac (1981) discussed the archaeological patterns in relation to mapped foraging patterns of modern hunter-gatherers and modern apes. Binford (1984) adopted the ape foraging pattern, labeled it routed foraging, and concluded that it epitomized hominid capabilities in the Plio-Pleistocene. Potts (1984, 1988, 1991) developed the stone cache model, arguing that hominids at Olduvai stockpiled stone raw material at strategic locations in anticipation of future needs and used the caches during brief butchery episodes. More recently, Blumenschine & Masao (1991) have challenged the basic premise that the home base/central place sites are high density anomalies, arguing from preliminary landscape excavations at Olduvai (that yielded densities of material similar to one of the excavations by Mary Leakey that she interpreted as a living site) that "the available results are adequate to invite consideration of an alternative working hypothesis to the living site interpretation of HWKE level 1 and, by extension, to other type C Oldowan sites from similar depositional environments . . ." and that "home bases, or repeatedly visited focal locations for *multiple* hominid activities, have not been shown to exist during basal Bed II times."

New data on butchered fossil bones from Koobi Fora contributes to evaluations of these models. Archaeological samples are derived from both surface and excavated contexts covering broad areas of the paleolandscape from three sub-regions of the Koobi Fora study area: (1) the Koobi Fora Ridge in Area 101 and Area 103; (2) the Karari Ridge in Area 130 and Area 131; (3) Ileret in Areas 1A, 5, and 8A (Figures 1 and 3). This provides broadly penecontemporaneous data centered at the base of the 1·64 m.y. Okote Member, from widely spaced areas of the paleolandscape along the approximately 80 km, north-south course of the ancestral Omo River. The principal emphasis of the research and of the following discussion is the recovery of fossil bones with cut marks and/or percussion damage [i.e., notches (Bunn *et al.*, 1980; Bunn, 1981) or striated pits (Blumenschine & Selvaggio, 1988)] from areas of the paleolandscape where exquisite preservation of morphological details on the surfaces of fossil bones makes this feasible. As discussed below, previous research established the existence of unique archaeological evidence composed of isolated and concentrated fossil bones with cut marks and/or percussion notches at localities at the Koobi Fora Ridge and at Ileret that have not yielded stone artifacts. In both sub-regions, which have received relatively little archaeo-logical attention (understandably, given the minimal densities of stone artifacts) compared to the Karari Ridge, systematic survey for surface material is being conducted in collaboration with the Koobi Fora Field School; a complementary excavation strategy would document the microhabitats and associations of the surface finds. With abundant artifacts but relatively poor bone preservation at the Karari Ridge, most available data are from excavations in the 1970s that are interpreted as large home base/central place sites. Landscape excavations away from those large, high density sites would provide artifact and fossil data for measuring the congruence between surface transect data and samples excavated from the same deposits. They would also provide a faunal database for paleoecological and taphonomic comparisons with the Koobi Fora Ridge and Ileret data and with the bone assemblages from previous excavations of large home base/central place sites at the Karari Ridge. Stern and M. Rogers are independently pursuing such projects at certain localities at the Karari Ridge. The relatively poor bone preservation at the Karari, however, reduces the probability of finding landscape examples of butchered fossil bones there, in contrast to the Koobi Fora Ridge and Ileret sub-regions where meaningful samples are already known from previous surface surveys.

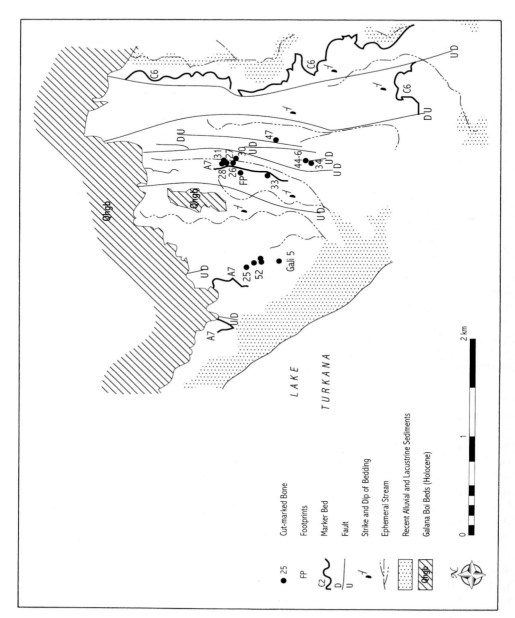

Figure 3. Fossil collection Area 103 at the Koobi Fora Ridge showing the distribution of fossil bones with cut marks in relation to geological features. Most of the fossils with cut marks occur at or just below the base of the Okote Tuff Complex and marker bed A7. Catalogue numbers are beside fossil specimens; for identification and description, see Table 1. Adapted from unpublished geological map by Feibel.

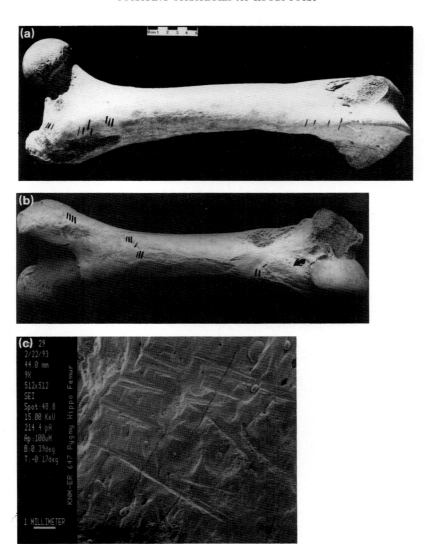

Figure 4. Defleshing cut marks on femur of pygmy hippopotamus (*Hippopotamus aethiopicus*; KNM-ER 647). Anterior (a) and posterior (b) views with cut marks highlighted. Additional cut marks occur on medial surface of distal epicondyle and between head and greater trochanter. Arrow on posterior of distal shaft indicates cut marks shown in SEM micrograph (c).

The Koobi Fora Ridge sub-region

In 1979, my research at the Koobi Fora Ridge began with the discovery of multiple cut marks on an essentially complete fossil pygmy hippo femur (*Hippopotamus aethiopicus*; museum catalogue no. KNM-ER 647) in the paleontology collections at the National Museums of Kenya (Figure 4). The museum catalogue indicated that the specimen had been collected in 1970 during geological and paleontological work by Behrensmeyer in Area 103 near the location of *Homo* sp. mandible KNM-ER 730. Later in 1979, a group of us, including Behrensmeyer and Isaac, returned to the KNM-ER 730 locality, and Behrensmeyer indicated

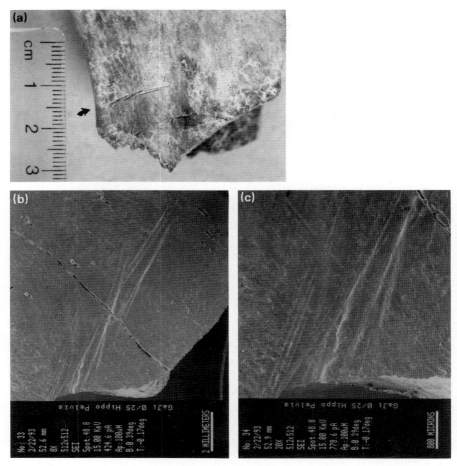

Figure 5. Defleshing cut marks on hippopotamus pelvis (GaJi 0/25). Arrow (a) indicates cut marks shown in SEM micrographs (b & c).

that the pygmy hippo femur had been found approximately 30 meters further along the outcrop. Approaching that spot, a conspicuous concentration of fossil bones eroding onto the outcrop caught everyone's attention, and several more cut-marked specimens, including a bovid pelvis and several pieces of a giraffe metapodial, were quickly found. A subsequent 4 m^2 test excavation established that the concentration continued as a discrete horizon into an undisturbed sand layer, and approximately 35 smaller specimens of the same taxa were recovered. Although none of the small sample of excavated bones was cut-marked, the site, GaJi 5 (Bunn, 1981), warrants more excavation. No stone artifacts were observed on the outcrop surface or in the excavation. A return visit to the GaJi 5 vicinity of several days duration in 1984 with Isaac and other Koobi Fora researchers yielded additional exquisitely preserved specimens with cut marks, including isolated finds of a pelvis (Figure 5), atlas, and tibia of hippopotamus and a tibia of a suid, at penecontemporaneous localities spaced kilometers apart within the Area 103 outcrops. As an instructor at the Koobi Fora Field School (co-sponsored by the National Museums of Kenya and the Harvard Summer School), which is directed by Harry Merrick and Assistant Director Craig Feibel, I have supervised student participation in surveys for cut-marked and percussion-notched bones in 1989, 1990, 1992, and 1993. This has consistently produced new specimens at the Koobi Fora Ridge in Areas 101 and 103 and at Ileret, yielding the current inventory of approximately 60 cut-marked and percussion-notched fossil bones (Table 1).

These reconnaissance surveys were guided by topographic considerations rather than by strict adherence to a precise stratigraphic level. For example, a morning was spent prospecting through stratigraphic layers exposed along several kilometers of one side of a particular drainage, while the afternoon was dedicated to the opposite side, and so on. Most of the ridgetops that define the sides of the roughly parallel, north-south drainages are capped by resistant layers of the Koobi Fora Tuff (Figure 3). Thus, the several square kilometers of outcrop that have been preliminarily surveyed (half of the available exposures in Area 103) sample intermittent, terrestrial environments in the lowest several meters of the Okote Member and the highest several meters of the underlying KBS Member. Older, lacustrine deposits of the KBS Member further to the east are fossiliferous, but unproductive archaeologically. Fossil specimens with hominid modifications were marked with cairns, pin-pointed on aerial photos, and then collected following protocols of the National Museums of Kenya.

Surface occurrences of cut-marked bones range from isolated specimens on fossiliferous outcrops, multiple, conjoinable specimens of the same hammerstone-fractured element, multiple elements of a single carcass, and multiple, associated specimens from several individual animals (Table 1). Interestingly, a non-random pattern emerged in the approximate stratigraphic provenance of the several dozen fossils with cut marks or percussion notches. Most (>90%) are derived from a restricted level several meters below the base of the Koobi Fora Tuff, even though the enclosing layers are highly fossiliferous and were searched with equal intensity (Figure 3). Based on surface indications, the specimens with cut marks are probably derived from transgressive sands that represent small isolated channels of this downstream section of the ancestral Omo River (Feibel, 1988). The main channels were tree-lined and flooded seasonally to form shallow water, swampy or ponded areas across large areas of the surrounding grassy floodplains (Feibel, 1988). The majority of the butchered bones in Area 103 probably samples a time interval ranging up to several thousand years. As discussed below, the evidence from basal Okote Member deposits at the Karari Ridge and at Ileret, in contrast, is probably a few tens of thousands of years younger than most of the Area 103 evidence. No pretense of precise contemporaneity or reconstruction of the movement and foraging activities of any hominid group is claimed. Rather, the aim is to obtain a representative archaeological sample from the three sub-regions during this time interval.

The value of this unique collection stems from the truly exceptional conditions for fossil preservation across a landscape in which Early Pleistocene hominids butchered carcasses using stone tools but apparently curated the tools and discarded them rarely. The cut-marked specimens reveal a significant number of butchered hippopotamus bones and the dismemberment and defleshing of particular skeletal elements (e.g., atlas and axis vertebrae, pelvis, femur) of hippopotamus of which there are no examples in the Plio-Pleistocene record elsewhere. The cut marks are as well (or better) preserved as the best examples from the Bed I Olduvai site of FLK *Zinjanthropus* (Bunn & Kroll, 1986; Leakey, 1971; Potts, 1988), and they meet all available criteria for stone tool cut marks: they are *deeply* incised, V-shaped in cross-section, straight-walled, and longitudinally microstriated grooves; they occur singly and in multiple, parallel sets in recessed anatomical locations that are logical by-products of butchery but improbable

Table 1 Butchered fossil specimens from Koobi Fora Ridge and Ileret

Site	Cat. no.	Taxon	Skeletal part	Description
GaJi 5	KNM-ER 647	*Hippopotamus aethiopicus*	Femur, complete	Multi-cuts, posterior to neck between head and greater trochanter, along anterior and posterior of shaft, posterior of shaft near distal epiphysis; Figure 4
GaJi 5	1	Bovid size 3A	Ilium	Multiple cuts, oblique, lateral surface
GaJi 5	70	Bovid size 3	Rib, proximal	Oblique, possible cut
GaJi 5	79	Mammal size 3	Limb shaft	Multiple cuts, transverse
GaJi 5	101	*Hippopotamus aethiopicus*	Ulna shaft	Multiple cuts, transverse
GaJi 5	103	*Giraffa* size 5	Metatarsal shaft	Multiple cuts, oblique, near proximal epiphysis
GaJi 5	104	*Giraffa* size 5	Metatarsal shaft	Multiple cuts, oblique, near proximal epiphysis
GaJi 5	121	Bovid size 3A	Ischium	Multiple cuts, transverse, lateral surface
GaJi 5	153	Mammal size 4–5	Rib shaft	Multiple cuts, oblique
GaJi 5	184	Suid size 3A	Tibia, distal epiphysis + shaft	2 Cuts, transverse, anterior surface
GaJi 5	230	Bovid size 3A	Mandible, articular condyle	Multiple cuts, oblique, lateral surface, below condyle
GaJi 0/	25	Hippo	Pelvis 1/2, frag.	Multiple cuts
GaJi 0/	26	Hippo	Atlas vertebra, complete	Multiple cuts
GaJi 0/	27	Hippo	Tibia, complete	Multiple cuts
GaJi 0/	28	Suid size 3A	Tibia, distal epiphysis + shaft	Multiple cuts
GaJi 0/	29	Mammal size 5	Rib shaft	Multiple cuts
GaJi 0/	30	Mammal size 3	Femur shaft	Multiple cuts
GaJi 0/	31	*Hippopotamus aethiopicus*	Humerus, distal epiphysis + shaft	Multiple cuts
GaJi 0/	32	*Hippopotamus aethiopicus*	Radio-ulna, complete	Multiple cuts
GaJi 0/	33	Bovid size 3B	Mandible, ramus frag.	Multiple cuts
GaJi 0/	34	*Hippopotamus aethiopicus*	Radio-ulna, proximal epip. + shaft	Multiple cuts
GaJi 0/	37	Mammal size 3–4	Humerus shaft	Multiple cuts
GaJi 0/	38	*Giraffa* size 5	Metapodial shaft	Multiple cuts
GaJi 0/	39	Hippo	Rib shaft	Multiple cuts
GaJi 0/	40	Hippo	Rib shaft	Multiple cuts
GaJi 0/	41	Suid size 3A	2nd Phalanx	Multiple cuts

Site	No.	Taxon	Element	Marks
GaJi 0/	42	Bovid size 2	Femur shaft	Percussion notch
GaJi 0/	43	Bovid size 3A	Humerus, proximal epip. + shaft	Cut
GaJi 0/	44	Hippo	Rib shaft, proximal	Multiple cuts
GaJi 0/	45	Hippo	Rib shaft, proximal	Multiple cuts
GaJi 0/	46	Hippo	Rib shaft	Cut
GaJi 0/	47	Giraffa size 5	Metatarsal shaft, proximal 1/2	Multiple cuts, percussion; 67 associated pieces includes 14 with cuts and 5 conjoining
GaJi 0/	48	Bovid size 3A	Tibia shaft	Multiple cuts
GaJi 0/	49	Bovid size 3A	Radius, complete	Cut
GaJi 0/	50	Bovid size 2	Tibia shaft	Multiple cuts
GaJi 0/	52	Mammal size 5	Scapula blade	Cut
GaJi 0/	56	Hippo	Phalanx	Multiple cuts
GaJi 0/	57	Equid	Calcaneum	Multiple cuts
GaJi 0/	58	Bovid size 3A	Metapodial, distal epip. + shaft frag.	Cut
GaJi 0/	59	Hippo	Axis vertebra, complete	Multiple cuts
GaJi 0/	60	Hippo	Scapula	Multiple cuts
GaJi 0/	61	Hippo	Scapula	Multiple cuts
GaJi 0/	62	Hippo	Tibia, distal epip. + shaft	Cut
FwJj 0/	Ileret8A	Hippo	Humerus, distal epip. + shaft	Multiple cuts
FwJj 0/	Ileret5	Hippo	Rib shaft	Multiple cuts
FwJj 0/	Ileret5	Antilopini size 1	Metacarpal, proximal epip. + shaft	Multiple cuts
FwJj 0/	Ileret8A	Antilopini size 1	Metacarpal, proximal epip. + shaft	Multiple cuts
FwJj 0/	Ileret8A	Giraffa size 5	Metapodial shaft	Multiple cuts
FwJj 0/	Ileret1A	Hippo	Rib shaft	Multiple cuts
FwJj 0/	Ileret1A	Antilopini size 1	Tibia	Multiple cuts

1. Site designations follow protocols used by the National Museums of Kenya. GaJi and FwJj denote 15 × 15 mile areas of the landscape at the Koobi Fora Ridge and Ileret, respectively. GaJi 0/25 is specimen no. 25 from the National Museum catalogue of surface finds from the GaJi area.

2. Sizes 1 to 5 refer to animal size groups based on live weight (e.g., Bunn, 1986).

by-products of alternative processes, such as trampling. The absolute width and less acute cross-sectional V-shape of most of the cut marks, compared to experimental cut marks produced with small unmodified flakes and to characteristic cut marks at home base/central place sites at Koobi Fora and Olduvai, indicate that hominids on the Koobi Fora Ridge were carrying and using flaked cores or very large flakes for butchering (Figures 4, 5; Bunn, 1981). Macroscopic, light microscopic, and scanning electron microscopic (SEM) techniques, and anatomical considerations based on modern butchery observations guide analysis of the cut-mark evidence.

The Karari Ridge and Ileret sub-regions

The Karari Ridge sub-region is situated near the confluence of the axial drainage system of the perennial ancestral Omo River and the higher gradient, marginal streams that intermittently carried stone cobbles into the area from the eastern basin margins. It is, thus, a landscape in which natural occurrences of channel gravels facilitated hominid access to stone raw material (Harris, 1978). Bone preservation in the excavated Karari Ridge sites is adequate for the identification of surface modifications and fracture patterns, but the bone is leached and chalky and does not stand up well to prolonged exposure on the modern outcrop surface (Bunn, 1982; Bunn et al., 1980). This re-emphasizes the limitations of a surface landscape approach and the greater promise of an excavated landscape approach at localities at the Karari Ridge having adequate bone preservation. All of the reported cut-marked and percussion-notched specimens from the Karari Ridge are from the excavated home base/central place sites (Bunn, 1981, 1982, n.d.; Bunn et al., 1980).

The Ileret sub-region is situated to the north and upstream along the ancestral Omo River in basal Okote times, with the nearest source of stone raw materials (the Suregei Plateau) approximately 5 km to the east (Figures 1 and 2). One significant artifact and bone bearing site, FwJj 1, has been excavated, and it is broadly similar in composition to the central place sites of the Karari Ridge. The surrounding landscape areas contain a low density scatter of stone artifacts and generally well-preserved fossil bone, thus offering a contrast to both the Koobi Fora and Karari Ridges. At its best, bone preservation is equal to Area 103, although some of the low, gentle outcrops at Ileret retain fossil bone that has been exposed on the outcrops for decades or more of weathering and is too badly damaged for identification of bone surface modifications. The discovery of cut marks in Area 103 prompted similar prospecting at Ileret, and a hippo distal humerus with cut marks was found by Barbara Isaac in 1979 (Table 1). In collaboration with the Koobi Fora Field School in 1989–93, we have now recovered seven fossil bones with cut marks from Ileret, although no concentrations of cut-marked bones have been located. Several of the specimens are from the outcrops surrounding FwJj 1. The remainder is from Areas 1A and 5.

Discussion

The butchered fossil bones from the Koobi Fora Ridge and Ileret provide some preliminary answers to general questions provoked by the uneven distribution of Early Pleistocene hominid fossils and archaeological sites across the Koobi Fora landscape. What were the activities of hominids in landscape areas where hominid fossils are abundant but stone raw materials and tools are rare? The cut marks on diverse taxa and skeletal elements indicate that hominids foraged for carcasses and used stone tools to butcher them. To achieve that at the Koobi Fora Ridge, hominids carried stone tools with them for at least 15 km, the approximate distance to the nearest source of stone raw material in channel gravels to the east or northeast. From the near absence of stone tools at the Koobi Fora Ridge in Areas 101 and 103 (in 25 years of searching the outcrop surfaces, only two stone flakes of probable Plio-Pleistocene age are known to the author), it appears that hominids conserved the stone that they had transported and only rarely discarded it. The relatively wide and less acute cross-sectional shape of most of the cut marks at the Koobi Fora Ridge, compared to the narrower width and very acute shape of cut marks at home base/central place sites at the Karari Ridge where hominids typically used small unmodified flakes for defleshing bones, indicates that hominids at the Koobi Fora Ridge and at Ileret usually used large cutting tools, either flaked cobbles or large flakes, for butchery.

The butchered bones raise new questions about hominid mobility, foraging strategies, and site formation. First, the stratigraphic distribution of cut-marked bones at the Koobi Fora Ridge can be viewed in relation to the changing paleogeography reconstructed by Brown and Feibel (1988; Feibel, 1988). Why is there a prevalence of cut marks at and just below the base of the Okote Member at approximately 1·65 m.y.? The paleogeographic reconstructions partially explain the pattern. At 2·0 m.y., the Koobi Fora Ridge area was covered by a deep lake; thereafter, ephemeral lakes intermittently covered the area. A series of fluvial cycles, with abundant terrestrial fauna, footprints, and cut-marked bones, characterizes the period leading up to the base of the Okote Member. By 1·6 m.y., the area was again covered by lake water (Brown & Feibel, 1988; Feibel, 1988). However, lacustrine conditions do not fully explain the distribution of cut marks. There are fluvial deposits with abundant terrestrial fauna both below and above the basal Okote level that do not appear to contain evidence of hominid foraging for carcasses. Moreover, the deposits at Ileret with abundant terrestrial fauna contain much less evidence of carcass utilization by hominids than the Koobi Fora Ridge. Perhaps some of the channels of the ancestral Omo River posed an impassable barrier to east-west movement by hominids, at times preventing them from reaching some of the landscape areas that have yielded a rich fossil record but little in the way of archaeological evidence. Certainly the seasonal flooding of braided Omo channels at the Koobi Fora Ridge, which produced shallow, ephemeral lakes (Feibel, 1988), would have rendered large areas unsuitable for hominid foraging on a seasonal basis.

The available sample of butchered bones suggests that hominid foraging strategies may have differed in several ways between the Koobi Fora Ridge and the Karari Ridge. First, there is a taxonomic difference in the samples, with a higher proportion of butchered hippopotamus bones relative to bovid bones at the Koobi Fora Ridge. Forty-three percent (18 of 42 skeletal elements, not counting associated fragments of the same original bone; Table 1) of the butchered skeletal elements from the Koobi Fora Ridge in Areas 101 and 103 are identifiable as hippopotamus, and 26% (11 of 42 elements) are identifiable as bovid. In contrast, the collection of all cut-marked specimens ($n=19$) from the excavated home base/central place sites at the Karari Ridge, includes 5% (1 of 19) hippopotamus and 74% (14 of 19) identifiable as bovid. If the obvious problem of small sample size and the potential for taphonomic loss of less durable, non-hippopotamus bones can be ruled out as major causes of the observed pattern, then ecological and behavioural explanations can be considered. For example, the lush paleohabitat of the Koobi Fora Ridge may have supported a larger hippopotamus population than the Karari, and opportunistic hominid foragers may have utilized carcasses in proportion to their availability in the different areas.

Alternatively, the pattern at the Koobi Fora Ridge may result from sampling a different component of a foraging strategy that was taxonomically similar to the Karari Ridge. In other words, hominids may have utilized similar taxa in similar proportions in the two sub-regions, but the available archaeological record may document different stages in the utilization process. As discussed below, the pattern at the Koobi Fora Ridge may document some single-event locations of the butchery and consumption of meat and/or marrow that did not involve significant transport of carcass parts. In contrast, the pattern at the Karari Ridge, based as it is on home base/central place sites, may document a later stage in the process of carcass utilization in which hominids transported more bovid bones and fewer hippopotamus bones to the known sites. As at the Koobi Fora Ridge, hominids at the Karari Ridge often might have defleshed and abandoned hippopotamus bones at the locations of acquisition, rather than habitually transporting them to home base/central place sites. Such evidence may not have been preserved at the Karari Ridge, or archaeologists may not have detected it.

The butchered bones reveal differing strategies of carcass utilization regarding the types of tissues that hominids actually extracted for consumption. Defleshing cut marks characterize both the Koobi Fora Ridge/Ileret sample (Table 1) and the Karari Ridge sample (Bunn, 1981, 1982, n.d.; Bunn et al., 1980), and they document meat removal and consumption in all sub-regions. Virtually all limb bones at the Karari Ridge sites were broken for marrow. In contrast, four of the seven hippopotamus limb bones in the Koobi Fora Ridge/Ileret sample were not broken enough to penetrate the marrow cavity and were not utilized for marrow; yet the same hippopotamus bones exhibit clear well-preserved defleshing cut marks (e.g., Figure 4). Even though the marrow fat would have yielded more calories per gram than lean meat, the large quantities of meat available from a hippopotamus limb might have encouraged hominids to practice a quantity-over-quality strategy in butchering some of the hippopotamus carcasses.

All of these varied archaeological patterns and the questions they raise about hominid foraging strategies in the Koobi Fora landscape invite reconsideration of a familiar and fundamental question. What did the large home base/central place sites mean to the hominids who contributed to their formation? As Kroll (1994) points out, recent hypotheses address this question by emphasizing different components of hominid behaviour, including social relationships, resource distribution and use, subsistence activities, and site location under secure shade trees. Landscape approaches have even questioned the idea of home base/central place sites as distinct focal points for repeated hominid activities. The butchered bones from the Koobi Fora Ridge and Ileret contribute to these discussions by documenting a novel hominid foraging strategy that involved the butchery of large animal carcasses but only rare discard of stone artifacts.

Hominid daily mobility at the Koobi Fora Ridge and at Ileret only rarely involved the transport of carcass parts to central places or the recurrent use of such locations. If the only difference in foraging activities between the Koobi Fora Ridge and the Karari Ridge was the need to conserve rather than discard stone raw materials and tools, then more sites like GaJi 5 would be expected in areas where hominids repeatedly butchered carcasses. Instead, and in contrast to the Karari Ridge, hominids apparently practiced a more ape-like, feed-as-you-go strategy during brief foraging visits for carcasses within the marshy or shallow floodplain lake conditions of the Koobi Fora Ridge. As indicated in Figure 4 and Table 1, feeding behaviour included the defleshing of large limb bones without breakage for marrow processing. Such a strategy is partially consistent with the working hypotheses by Blumenschine and Masao (1991) and by Potts (1989) for lake margin foraging at Olduvai and Olorgesailie, respectively. Preliminary landscape studies at Olduvai and Olorgesailie indicate a variable but continuous distribution of stone artifacts and a prevalence of flakes (detached pieces) rather than core tools (flaked pieces) in lake margin habitats located several (2–5) km from raw material sources, and the authors identify carcass processing and predator avoidance as causative factors for the archaeological patterns. Unlike the relatively short distances between lake margins and stone raw materials at Olduvai and Olorgesailie, the distances between the exposures at the Koobi Fora Ridge (Area 103) and the nearest channel gravels are several times greater, probably more than 15 km. These sources of stone raw materials were within the foraging range of particular hominids who butchered carcasses in Area 103, but the transport distances involved might have constrained stone artifact discard and led hominids to use predominantly core tools for most of the cutting activities. For hominids whose behavioural repertoire included caching stone for anticipated future use, Area 103 would have been an ideal area. The apparent absence of caches or isolated artifacts is not supportive of the stone cache hypothesis. At Ileret and at the Karari Ridge, more proximal sources of stone raw materials coincide with higher frequencies of stone artifact discard across the landscape.

To understand foraging strategies on the Koobi Fora Ridge and Ileret, the derivation of the few concentrations and the isolated scatters of cut-marked bones requires documentation. Several alternatives exist: (1) The occurrences document the undisturbed locations of carcass-processing activities by hominids, involving both (a) transported parts of multiple carcasses at recurrently visited central place sites and (b) untransported carcass parts at single-event feeding locations; (2) the occurrences document fluvial transport, involving (a) reconcentration of initially separate and unrelated fossils, (b) dispersal of fossils that were initially concentrated by hominids at central place sites, and/or (c) movement of isolated bones away from single-event feeding locations; (3) the occurrences document concentration and dispersal of bones by carnivore scavengers.

Because the known concentrations of cut-marked specimens involve associations of bones with different fluvial transport potential at GaJi 5 and conjoining, hammerstone-broken pieces (primary pieces with matrix cemented over fracture surfaces and over some of the cut marks) at the giraffe metatarsal locality (cat. no. 47), it is likely that at least some of the time hominids recurrently transported carcass parts to particular locations (GaJi 5) and that fluvial processes did not disperse bones from locations of hominid butchery (GaJi 5 and giraffe metatarsal locality). To establish how many of the isolated occurrences document single-event butchery or feeding locations, additional excavations and microstratigraphic work will be required. That would reveal whether or not fluvial processes would have been sufficient to transport the isolated cut-marked specimens, possibly from fluvially-dispersed fossil concentrations. In other words, if fluvial transport were ruled out, the work would help determine how many of the isolated cut-marked specimens are locations of brief hominid activity across the landscape.

Summary and conclusions

The discovery of more than 60 fossil bones with butchery marks from sub-regions of the Early Pleistocene landscape at Koobi Fora that have yielded only rare stone artifacts defines a new kind of stone age archaeological site. Nearly all of the butchered specimens exhibit cut marks made with stone knives during defleshing activities. Two specimens exhibit percussion notches from marrow extraction. Several hippopotamus limb bones were defleshed but were not broken for marrow extraction. Hominids evidently carried stone tools with them for butchering carcasses in areas of the landscape where stone raw materials were otherwise unavailable, and the stone tools were then conserved for reuse and rarely discarded. One concentration of butchered bones, site GaJi 5, consists of 11 bones with cut marks from a minimum of five different carcasses. The remainder of the butchered bones occurs as isolated specimens across several square kilometers of the paleolandscape.

At least some of the isolated specimens demonstrate the archaeological visibility of single-event locations of butchery and/or consumption of carcass portions. That suggests that the conventional division of Plio-Pleistocene sites into only several types, such as home bases, butchery sites, and stone knapping sites, oversimplifies what hominids did and what archaeologists can hope to detect from well-preserved paleolandscapes. It will be necessary to obtain new data through the completion of surface surveys for butchered bones and through an excavation strategy that samples these same landscape areas. That would enable the formulation and testing of more comprehensive models of Early Pleistocene hominid foraging strategies.

Acknowledgements

An earlier version of this paper was presented at the 62nd Annual Meeting of the American Association of Physical Anthropologists at a symposium titled "Early Hominid Behavioral Ecology: New Looks at Old Questions", and I thank the co-organizers, J. Oliver, N. Sikes, and K. Stewart, for inviting me to participate. I also thank E. Kroll, N. Sikes, and three anonymous reviewers for providing useful comments that have improved the final version of the paper. I am most grateful to the Koobi Fora Field School, especially Dr Harry Merrick, Director, and Dr Craig Feibel, Assistant Director, for advice, encouragement, and logistical support. Feibel also generously gave me permission to use in Figure 2 the paleogeographic reconstruction from Feibel (1988) and in Figure 3 the geological map of the Koobi Fora Ridge from his unpublished Master's thesis. I thank the National Museums of Kenya for permission to participate in the Koobi Fora Field School and for providing access to the fossil collections. Initial funding for work reported herein was provided by a grant from the National Science Foundation to Glynn Isaac. Since 1984, my participation has been generously supported by faculty research grants from the University of Wisconsin Graduate School and by a University of Wisconsin Vilas Associateship grant.

References

Behrensmeyer, A. K. (1975). The taphonomy and paleoecology of Plio-Pleistocene vertebrate assemblages east of Lake Rudolf, Kenya. *Bull. Mus. Comp. Zool.* **146,** 473–578.

Behrensmeyer, A. K. (1985). Taphonomy and the paleoecologic reconstruction of hominid habitats in the Koobi Fora Formation. In (Y. Coppens, Ed.) *L'Environment des Hominides au Plio-Pleistocene,* pp. 309–324. Paris: Masson.

Behrensmeyer, A. K. & Laporte, L. F. (1981). Footprints of a Pleistocene hominid in northern Kenya. *Nature* **289,** 167–169.

Binford, L. R. (1984). *Faunal Remains from Klasies River Mouth.* London: Academic Press.

Blumenschine, R. J. & Selvaggio, M. M. (1988). Percussion marks on bone surfaces as a new diagnostic of hominid behaviour. *Nature* **333,** 763–765.

Blumenschine, R. J. & Masao, F. T. (1991). Living sites at Olduvai Gorge, Tanzania? Preliminary landscape archaeology results in the basal Bed II lake margin zone. *J. hum. Evol.* **21,** 451–462.

Brown, F. H. & Feibel, C. S. (1986). Revision of lithostratigraphic nomenclature in the Koobi Fora region, Kenya. *J. Geol. Soc., Lond.* **143,** 297–310.

Brown, F. H. & Feibel, C. S. (1988). "Robust" hominids and Plio-Pleistocene paleogeography of the Turkana Basin, Kenya and Ethiopia. In (F. E. Grine, Ed.) *Evolutionary History of the "Robust" Australopithecines,* pp. 325–341. New York: Aldine de Gruyter.

Bunn, H. T. (1981). Archaeological evidence for meat-eating by Plio-Pleistocene hominids from Koobi Fora and Olduvai Gorge. *Nature* **291,** 574–577.

Bunn, H. T. (1982). Meat-Eating and Human Evolution: Studies on the Diet and Subsistence Patterns of Plio-Pleistocene Hominids in East Africa. Ph.D. Dissertation, University of California, Berkeley.

Bunn, H. T. (1986). Patterns of skeletal representation and hominid subsistence activities at Olduvai Gorge, Tanzania, and Koobi Fora, Kenya. *J. hum. Evol.* **15,** 673–690.

Bunn, H. T. (1987). Comment on "Characteristics of an early hominid scavenging niche" by R. J. Blumenschine. *Curr. Anthrop.* **28,** 394–396.

Bunn, H. T. (n.d.). The bone assemblages. In (G. Ll. Isaac, Ed.) *Koobi Fora Research Project, Volume 5. Archaeology.* Oxford: Clarendon Press.

Bunn, H. T. & Kroll, E. M. (1986). Systematic butchery by Plio-Pleistocene hominids at Olduvai Gorge, Tanzania. *Curr. Anthrop.* **27,** 431–452.

Bunn, H. T. & Kroll, E. M. (1993). Comment on "The structure of the Lower Pleistocene archaeological record: a case study from the Koobi Fora Formation" by N. Stern. *Curr. Anthrop.* **34,** 216–217.

Bunn, H., Harris, J. W. K., Isaac, G., Kaufulu, Z., Kroll, E., Schick, K., Toth, N. & Behrensmeyer, A. K. (1980). FxJj 50: an Early Pleistocene site in northern Kenya. *World Archaeol.* **12,** 109–136.

Coppens, Y., Howell, F. C., Isaac, G. Ll. & Leakey, R. E. F., Eds (1976). *Earliest Man and Environments in the Lake Rudolf Basin.* Chicago: University of Chicago Press.

Dunnell, R. C. & Dancey, W. S. (1983). The siteless survey: a regional scale data collection strategy. *Adv. Archaeol. Meth. Theory* **6,** 267–287.

Feibel, C. S. (1988). Paleoenvironments of the Koobi Fora Formation, Turkana Basin, northern Kenya. Ph.D. Dissertation, University of Utah.

Findlater, I. C. (1978). Stratigraphy. In (M. G. and R. E. F. Leakey, Eds) *Koobi Fora Research Project, Volume 1. The Fossil Hominids and an Introduction to their Context, 1968–74*, pp. 14–31. Oxford: Clarendon Press.

Foley, R. (1981). Off-site archaeology: an alternative approach for the short-sited. In (I. Hodder, G. Isaac & N. Hammond, Eds) *Patterns of the Past*, pp. 157–183. Cambridge: Cambridge University Press.

Harris, J. M., (Ed.) (1983). *Koobi Fora Research Project, Volume 2. The Fossil Ungulates: Proboscidea, Perissodactyla and Suidae.* Oxford: Clarendon Press.

Harris, J. M., (Ed.) (1991). *Koobi Fora Research Project, Volume 3. Stratigraphy, Artiodactyls and Paleoenvironments.* Oxford: Clarendon Press.

Harris, J. W. K. (1978). The Karari Industry: its place in East African Prehistory. Ph.D. Dissertation, University of California, Berkeley.

Harris, J. W. K. & Herbich, I. (1978). Aspects of early Pleistocene hominid behaviour east of Lake Turkana, Kenya. In (W. W. Bishop, Ed) *Geological Background to Fossil Man*, pp. 529–547. Edinburgh: Scottish Academic Press.

Harris, J. W. K. & Isaac, G. Ll. (1976). The Karari Industry: Early Pleistocene archaeological evidence from the terrain east of Lake Turkana, Kenya. *Nature* **262,** 102–106.

Hay, R. L. (1976). *Geology of the Olduvai Gorge: A Study of Sedimentation in a Semi-Arid Basin.* Berkeley: University of California Press.

Isaac, G. Ll. (1971). The diet of early man: aspects of archaeological evidence from Lower and Middle Pleistocene sites in Africa. *World Archaeol.* **2,** 278–299.

Isaac, G. Ll. (1976). Plio-Pleistocene artifact assemblages from East Rudolf, Kenya. In (Y. Coppens, F. C. Howell, G. Ll. Isaac & R. E. F. Leakey, Eds) *Earliest Man and Environments in the Lake Rudolf Basin*, pp. 552–564. Chicago: University of Chicago Press.

Isaac, G. Ll. (1977). *Olorgesailie: Archeological Studies of a Middle Pleistocene Lake Basin in Kenya.* Chicago: University of Chicago Press.

Isaac, G. Ll. (1978). The food-sharing behavior of proto-human hominids. *Sci. Am.* **238,** 90–108.

Isaac, G. Ll. (1981). Stone Age visiting cards: approaches to the study of early land-use patterns. In (I. Hodder, G. Ll. Isaac & N. Hammond, Eds) *Patterns of the Past*, pp. 131–155. Cambridge: Cambridge University Press.

Isaac, G. Ll. (1983). Bones in contention: competing explanations for the juxtaposition of early Pleistocene artifacts and faunal remains. In (J. Clutton-Brock & C. Grigson, Eds) *Animals and Archaeology, Volume 1: Hunters and their Prey*, pp. 1–20. Oxford: British Archaeological Reports, International Series 163.

Isaac, G. Ll. (1984). The archaeology of human origins: studies of the Lower Pleistocene in East Africa 1971–1981. *Adv. World Archaeol.* **3,** 1–87.

Isaac, G. Ll., (Ed.) (n.d.) *Koobi Fora Research Project, Volume 5. Archaeology.* Oxford: Clarendon Press.

Isaac, G. Ll. & Harris, J. W. K. (1975). The scatter between the patches. Paper presented at the Kroeber Anthropological Society Meetings, Berkeley, California.

Isaac, G. Ll. & Harris, J. W. K. (1978). Archaeology. In (M. G. Leakey & R. E. F. Leakey, Eds) *Koobi Fora Research Project, Volume 1. The Fossil Hominids and an Introduction to their Context, 1968–1974*, pp. 64–85. Oxford: Clarendon Press.

Isaac, G. Ll. & Harris, J. W. K. (1980). A method for determining the characteristics of artefacts between sites in the Upper Member of the Koobi Fora Formation, East Lake Turkana. In (R. E. Leakey & B. A. Ogot, Eds) *Proceedings of the 8th Panafrican Congress of Prehistory and Quaternary Studies*, Nairobi, pp. 19–22. Nairobi: TILLMIAP.

Isaac, G. Ll., Leakey, R. E. F. & Behrensmeyer, A. K. (1971). Archaeological traces of early hominid activities, east of Lake Rudolf, Kenya. *Science* **173,** 1129–1134.

Isaac, G. Ll., Harris, J. W. K., Crader, D. (1976). Archaeological evidence from the Koobi Fora Formation, In (Y. Coppens, F. C. Howell, G. Ll. Isaac & R. E. F. Leakey, Eds) *Earliest Man and Environments in the Lake Rudolf Basin*, pp. 533–551. Chicago: University of Chicago Press.

Isaac, G. Ll., Harris, J. W. K. & Marshall, F. (1981). Small is informative: the application of the study of mini-sites and least effort criteria in the interpretation of the Early Pleistocene archaeological record at Koobi Fora, Kenya. In (J. D. Clark & G. Ll. Isaac, Eds) *Las Industrias mas Antiguas*, pp. 101–19. *X Congresso, Union Internacional de Ciencias Prehistoricas y Protohistoricas*, Mexico City.

Kaufulu, Z. M. (1983). The Geological Context of Some Early Archaeological Sites in Kenya, Malawi, and Tanzania: Microstratigraphy, Site Formation and Interpretation. Ph.D. Dissertation, University of California, Berkeley.

Kroll, E. M. (n.d.). The Anthropological Meaning of Spatial Configurations at Plio-Pleistocene Archaeological Sites in East Africa. Ph.D. Dissertation to be filed. University of California, Berkeley.

Kroll, E. M. (1994). Behavioral implications of Plio-Pleistocene archaeological site structure. *J. hum. Evol.* **27,** 107–138.

Kroll, E. M. & Isaac, G. Ll. (1984). Configurations of artifacts and bones at Early Pleistocene sites in East Africa. In (H. J. Hietala, Ed) *Intrasite Spatial Analysis in Archaeology*, pp. 4–31. Cambridge: Cambridge University Press.

Leakey, M. D. (1971). *Olduvai Gorge, Volume 3. Excavations in Beds I and II*. Cambridge: Cambridge University Press.

Leakey, M. G. & Leakey, R. E. F., Eds (1978). *Koobi Fora Research Project, Volume 1. The Fossil Hominids and an Introduction to their Context, 1968–1974*. Oxford: Clarendon Press.

Potts, R. B. (1984). Home bases and early hominids. *Am. Sci.* **72,** 338–347.

Potts, R. B. (1988). *Early Hominid Activities at Olduvai*. New York: Aldine de Gruyter.

Potts, R. B. (1989). Olorgesailie: new excavations and findings in Early and Middle Pleistocene contexts, southern Kenya Rift Valley. *J. hum. Evol.* **18,** 477–484.

Potts, R. B. (1991). Why the Oldowan? Plio-Pleistocene toolmaking and the transport of resources. *J. Anthrop. Res.* **47,** 153–176.

Schick, K. D. (1984). Processes of Paleolithic Site Formation: An Experimental Study. Ph.D. Dissertation, University of California, Berkeley.

Sept, J. M. (1984). Plants and Early Hominids in East Africa: A Study of Vegetation in Situations Comparable to Early Archaeological Site Locations. Ph.D. Dissertation, University of California, Berkeley.

Sept, J. M. (1986). Plant foods and early hominids at site FxJj 50, Koobi Fora, Kenya. *J. hum. Evol.* **15,** 751–770.

Stern, N. (1991). The Scatters-Between-the-Patches: A Study of Early Hominid Land Use Patterns in the Turkana Basin, Kenya. Ph.D. Dissertation, Harvard University.

Stern, N. (1993). The structure of the Lower Pleistocene archaeological record: a case study from the Koobi Fora Formation. *Curr. Anthrop.* **34,** 201–225.

Toth, N. P. (1982). The Stone Technologies of Early Hominids at Koobi Fora, Kenya: An Experimental Approach. Ph.D. Dissertation, University of California, Berkeley.

Walker, A. & Leakey, R. E. F. (1978). The hominids of East Turkana. *Sci. Am.* **239,** 54–66.

Wood, B. (1992). Origin and evolution of the genus *Homo*. *Nature* **355,** 783–790.

J. S. Oliver
Anthropology Section, Illinois State Museum, Research and Collections Center, 1011 East Ash St., Springfield, Illinois 62703, U.S.A.

Received 22 November 1993
Revision received 15 March 1994
and accepted 15 March 1994

Keywords: Hominid- and carnivore-induced bone damage frequencies, food transport and processing, altricial young, predator risk-reduction strategies.

Estimates of hominid and carnivore involvement in the FLK *Zinjanthropus* fossil assemblage: some socioecological implications

Studies of Plio-Pleistocene zooarchaeological assemblages indicate that early *Homo* transported and processed carcass-parts with tools for meat and marrow. Because hominid and carnivore damage frequencies have not been estimated, however, debate continues on the *extent* and *nature* of hominid and carnivore involvement in many assemblages, thereby hindering reconstructions of early *Homo* socioecology. Analysis of hammerstone- and carnivore-induced damage frequencies on all large mammal limb pieces in the FLK *Zinjanthropus*, Olduvai Gorge, Tanzania, fossil assemblage provides *direct estimates* of hominid and carnivore involvement. Three patterns are clear: (1) hominid-induced damage on 75% of the limb MNE demonstrates that early *Homo habitually* processed carcass-parts; (2) carnivore damages and fractures occur on 54% and 8·5% of the limb MNE, respectively, indicating a significant carnivore role in assemblage modification, but not breakage; (3) the lack of carnivore damage on 60% of hominid-damaged specimens implies that early *Homo* had regular access to meat-rich carcasses.

Because social carnivores also habitually transport food, the biological and ecological forcing mechanisms behind these behaviors are examined for implications for early *Homo* socioecology. Anti-predator defense is a primary factor in the evolution of sociality in the Carnivora (Gittleman, 1989). Thus, food transport from a carcass by a foraging group is an anti-predator strategy that minimizes interspecific competition at carcasses. Similarly, maintenance of a refuge area to which food is transported to mother and altricial young reflects a risk-reduction strategy to protect the young. Early *Homo* shared these ecological and biological conditions, occupying with its altricial young (Stanley, 1992) semi-open habitats with high predator densities. Thus, the habitual food transport and processing documented at FLK *Zinjanthropus* suggest that early *Homo* was subject to these same socioecological mechanisms that require similar group division into two functional units (foraging and care-giving) when infants are present. Processed animal remains at places like FLK *Zinjanthropus* may not represent provisioning or food sharing, per se, but merely the intersection of two anti-predator strategies. Nevertheless, the ecological and biological conditions, and prerequisite subsistence behaviors were apparently in place for the development of social behaviors originally envisioned by Isaac (1978*a*).

Journal of Human Evolution (1994) **27**, 267–294

Introduction

Because of their implications for other aspects of hominid biological and social evolution, the subsistence strategy and diet of Plio-Pleistocene hominids, particularly early *Homo*, are major concerns in paleoanthropology. Specifically, current models variously characterize early *Homo**

Revision of paper presented at the 62nd meeting of the American Association of Physical Anthropologists, Toronto in the Symposium, "Early hominid behavioral ecology: New looks at old questions".

*Though other hominids (e.g., *Australopithecus robustus*) may have had the anatomical ability to manufacture and use stone tools (Susman, 1988), I assume that sites containing large numbers of tools, flaking debris and bone such as FLK *Zinjanthropus*, were created by early *Homo*. The basis for this assumption is that while at least two hominid species (*Homo erectus*/*ergaster* and *Australopithecus robustus*) survived the extinction of all other hominids in the early Pleistocene, only one distinct stone-tool industry, the Acheulean, is recognized for the 500,000 year period that these two hominids were contemporaries. This suggests that habitual stone-tool manufacture was a species-specific behavior. The creator of smaller sites and sites from earlier times with less well-defined artifactual material may be open to question, however.

as meager scavengers of carnivore-ravaged carcasses (e.g., Binford, 1981; Potts & Shipman, 1981; Shipman, 1986*a,b*), as seasonal (e.g., Blumenschine, 1986, 1987) or year-round scavengers (e.g., Cavallo & Blumenschine, 1989; Marean, 1989) of skeletally intact carcasses, and as successful procurers of meaty carcasses through either hunting or confrontational scavenging (e.g., Bunn, 1981, 1982, 1986; Bunn & Kroll, 1986; see also, Potts, 1982, 1988). The cut mark and bone frequency patterns on which some of these models are based and tested, however, do not fully resolve questions about the taphonomic importance of carnivores and hominids in Plio-Pleistocene zooarchaeological assemblages, and the hominid mode of meat acquisition. This situation hinders attempts to define archaeological behaviors, demonstrate specific site functions, and define socioecological and evolutionary mechanisms that lay behind these behaviors.

· The inability of present taphonomic data from Plio-Pleistocene sites to resolve these problems is due to the lack of quantified estimates of hominid- and carnivore-induced damages, and a reliance on bone frequency data which at best only provide indirect information on the relative importance of hominids and carnivores. For example, though cut marks demonstrate hominid involvement in a number of East African Plio-Pleistocene faunal assemblages, the cut mark frequencies are quite low (Potts & Shipman, 1981; Potts, 1982, 1988; Bunn *et al.*, 1980; Bunn, 1981; Bunn & Kroll, 1986), making it difficult to extrapolate hominid involvement to the rest of the assemblage for which whose damages and causal agents are unknown. A case in point is the fossil assemblage from FLK *Zinjanthropus*, Olduvai Gorge, Tanzania. This tool- and fossil-rich site, located between Tuff IB and IC in the middle of Bed I in level 22 at the FLK locality (Leakey, 1971), dates between 1·86 and 1·76 Ma (Walker *et al.*, 1991). Published tables indicate only a small number of fossils display cut marks (<5% of the assemblage; $n=172$; Bunn & Kroll, 1986), yet almost all studies of this assemblage conclude that hominids were the major agent of bone accumulation and modification (Bunn, 1982, 1983; Bunn & Kroll, 1986; Potts & Shipman, 1981; Potts, 1982, 1988; Shipman, 1986*a*). This conclusion rests heavily on (1) the association of stone tools and bones with cut-marks, (2) the presence of other documented, but unquantified hominid-induced damages, i.e., hammerstone-induced damage, and (3) a variety of bone frequency data that may allow qualitative assessments of taphonomic agents, but which suffer from problems of equifinality.

Binford (1981), for example, demonstrated this problem with equifinality in his analysis of bone assemblages created by carnivores and Eskimos. He noted that carnivores and humans often create sites with similar bone frequency patterns and suggested that "the analysis of the condition of the bones in an assemblage may permit the identification of the agents responsible for it . . ." (Binford, 1981:252). Thus, analyses of skeletal frequency patterns may yield equivocal results and cannot *directly* identify the agent responsible for the observed frequency patterns. The condition of bone, including bone damages, can directly identify the agents involved in assemblage accumulation. Consequently, it is necessary to document the extent of hominid involvement in assemblage accumulation and modification on a bone-by-bone basis before competing models of meat acquisition and carcass-part transport by early *Homo* can be assessed. Bunn (1989:301; see also Bunn, 1991) also recognizes the importance of such assessments of zooarchaeological faunas: ". . . the study of bone fracture patterns offers a likely but complex means for diagnosing the predominant agents of bone accumulation or modification of entire assemblages." Thus, proponents of different models of Plio-Pleistocene hominid subsistence agree that detailed analysis of bone damages, particularly bone fracture patterns, is a necessary first step in the identification of the taphonomic agents active in the accumulation and condition of the FLK Zinj assemblage. Nevertheless, direct estimates of the level of carnivore and hominid involvement are not available.

Accurate evaluation of specific site function models, i.e., the "home-base" inferred by Isaac (1978*a,b*) and the "stone cache" model proposed by Potts (1982, 1988), or the "central place" revision of the home-base model by Isaac (1984) and Bunn (1982, 1986; Bunn & Kroll, 1986), is also partly dependent on documenting the extent and nature of hominid and carnivore involvement in fossil assemblage accumulation and modification. All proposed site functions require substantial hominid involvement with carcass-parts, but as with evaluation of the amount of meat procured, any discussion of early *Homo* site function requires some understanding of the *nature* of hominid and carnivore involvement. Home base, central place, and stone cache site functions may take on different meanings depending on the amount of animal tissue at the site. For example, a central feature of Potts' (1982, 1988) stone cache model is that a large quantity of animal tissue would be attractive to carnivores, making the site hazardous for hominids to occupy for very long. Lesser amounts of attractive refuse would be created, however, if little meat was transported to the site or if transported carcass-parts were heavily processed. Similarly, extensive carnivore damage on bone without hominid-induced modification might reduce the plausibility of the home-base model. Thus, it is difficult to argue convincingly about the frequency of carcass-part transport, meat-eating, food-sharing, stone-caching, or interactions with carnivores without direct estimates of hominid and carnivore involvement in the Plio-Pleistocene zooarchaeological assemblages.

In this paper the initial results of a quantitative analysis of limb bone fractures and damages in the FLK Zinj assemblage (limb NISP=1914†) provide an estimate of the relative contributions of hominids and carnivores to the fossil assemblage. On-going work (Oliver, n.d.), including quantitative analysis of fracture features, damage overprinting, and comparisons of damage patterns in the Amboseli hyaena den with those of FLK Zinj, are not discussed in detail here. This assessment of carnivore- and hominid-induced damages on limb pieces is assumed to be representative of hominid and carnivore involvement in the entire assemblage because between 52% (limb NISP=1457, Bunn & Kroll, 1986) and 68% (limb NISP=1914, this study†) of the identifiable bones in the FLK Zinj assemblage are limb pieces (exclusive to teeth). Thus, quantification of hominid and carnivore damage patterns will permit a characterization of hominid and carnivore interactions and activities at this one point on the Plio-Pleistocene landscape.

As indicated above, these data have implications for current models of Plio-Pleistocene hominid behaviour and site function. Additionally, they are relevant to identifying specific socioecological variables and forcing mechanisms that influenced the function and creation of Plio-Pleistocene archaeological sites. Therefore, after quantifying the extent of hominid and carnivore involvement in the processing of carcass-parts represented at FLK Zinj and outlining the character of these activities, the possible socioecological implications of the bone accumulation are examined in light of the behavior of various carnivores who also transport and process food. Though this approach is not new (see for example Schaller & Lowther, 1969; Potts, 1982), a re-examination of the social and ecological contexts of carnivore food transport and processing behavior is warranted in view of data provided here as well as that by Stanley (1992) on the terrestrial lifestyle of early *Homo* and the altricial condition of their infants (see also Shipman & Walker, 1989 for a discussion of early hominid post-natal growth

---

†Bunn (1982) identified 1457 limb bone pieces. I identified an additional 457 limb pieces in the large bone fragment bags. Most were identified on the basis of cortical bone thickness and diagnostic dynamic loading and impact damage including impact marks, bulbs of percussion, and auxiliary flake scars.

rates and the behavioral ramifications of altricial infants). Specifically, the nature and extent of hominid involvement in the accumulation and modification of the FLK Zinj fossil assemblage suggest that the behavioral response to early *Homo* to the high ecological costs of having altricial young in semi-open habitats with high predator densities was driven by considerations of anti-predator defense (rather than social bonds or attractiveness of a home-base, or energetic concerns associated with tool and carcass-part transport).

Methods

Limb bone pieces ($n = 1914$) from FLK *Zinjanthropus* were examined by the author for three classes of bone damage: (1) hammerstone-induced impact and fracture features; (2) carnivore-induced fracture features and tooth marks; and (3) cut marks. Damage features accepted as diagnostic to agent and described in the literature (e.g., Bonnichsen, 1979; Binford, 1981; Blumenschine & Selvaggio, 1988; Bunn, 1989; Haynes, 1980, 1981, 1982, 1983; Hill, 1989; Johnson, 1985; Shipman, 1981; Shipman & Rose, 1983) as well as a limited number of hammerstone fracture and carnivore feeding experiments (Oliver, n.d.) are used to identify the three classes of bone damage. Damages not diagnostic to agent were recorded as indeterminate or questionable, but are not discussed here.

With few exceptions, the identification of agents of bone modification is not simply a matter of comparing specific damages on fossils with specific damages produced in actualistic studies. There are two reasons why this is so. First, if the mechanics of damage creation are similar, outwardly similar modifications may result. Thus, part of the method for identifying taphonomic agents is careful consideration of site context, and the potential geological and biological processes active at the site (Oliver, 1989). For example, hammerstone-induced damage requires a stone to impact bone and a high energy source to move the bone and stone together. Since fine-grained, ash fall deposits buried the FLK Zinj bone and lithic assemblages (Hay, 1976*a,b*), geological processes can be eliminated as possible agents of hammerstone-like damages, as well as pseudo-cut marks. Furthermore, distribution and frequency analyses of the lithic assemblage indicates that fluvial transport was not a significant agent of material accumulation or dispersal (Kroll & Isaac, 1984). However, lower energy processes like trampling may create pits and lineations, which may resemble cut marks and/or tooth marks (Oliver, 1989), a situation that often makes the identification of individual damages problematic. Another mimic of carnivore tooth marks not given adequate consideration is early *Homo* feeding behavior. Given the demonstrable hominid involvement with the fossil assemblage (Bunn, 1981, 1982; Bunn & Kroll, 1986; Potts & Shipman, 1981; Potts, 1982, 1988) it is likely that hominid teeth created at least some tooth marks.

Furthermore, taphonomic agents rarely leave diagnostic traces on all bone specimens with which they interacted. Thus, another critical part of the methodology for assigning agency to specific damages is assessment of damage context (Oliver, n.d.). In this analysis it is not individual damages that are important, but the relationships of damages to one another and the damage's spatial and frequency relationships to the bone specimen. For example, because both carnivores and hammerstones create loading notches and negative flake scars, identification of fracture agent requires examination of other features including character of negative flake scar, presence or absence and appearance of opposing loading points, presence or absence of tooth pits or scores near either loading point, etc. To attain this level of detail, the unit of analysis for this study is the fracture line. Focus on individual fracture lines forced examination of damages on, and associated with, those lines, as well as the relationship between different damages.

Element identification and MNE and MNI determinations

Bone identifications (taxa, element, portion, segment, side), and animal size class‡ determinations for 1299 FLK *Zinjanthropus* limb specimens were provided by Bunn (1982, pers. comm., 1993). Thus, "specimen identification" for this study consisted of matching catalogue numbers recorded for this study with those recorded by Bunn (Bunn, 1982; pers. comm., 1993). Almost 93% of the specimens examined by Bunn (1982) were matched with specimens examined in this study. MNI and MNE estimates also followed those calculated by Bunn (1982:476–492). The match frequency for limb specimens used by Bunn (1982) in his MNE and MNI calculations is also high for all limb elements (MNE mean match frequency=96%; MNI mean match frequency=97·4%). Specimens for which a matching specimen was not located in this study were subtracted from the MNI or MNE estimates and when calculating MNI and MNE damage percentages.

Method of inspection

All specimen surfaces were examined under a high intensity (12,500 ft candles @ 6 in) microscope illuminator using a low magnification (2–6 ×) binocular visor or 8 × hand lens. Problematic damages such as very fine linear striae or small pitting near a fracture loading point were examined with a binocular microscope (8–20 ×) using the same high intensity illuminator. As argued elsewhere (Bunn, 1983; Oliver, 1986, 1989; Blumenschine & Marean, 1993), inspection of bone under low magnification is sufficient for identifying the damage agent for most features. High intensity light, which highlights minute textural differences, is critical to assessing bone modifications where the presence and context of the finest damage feature may reveal the agent.

Tooth-marks and carnivore-induced fractures

Identification of carnivore tooth marks is more problematic than the identification of cut marks and hammerstone impact marks. First, in spite of a few studies that identify geological processes of bone modification (e.g., Lyman, 1989; Oliver, 1986, 1989; Thorson & Guthrie, 1984), the variability of damages created by many geological processes (e.g., fluvial transport, sediment compaction, debris flows, etc.) remains poorly documented. Additionally, a number of biological processes, notably trampling (e.g., Behrensmeyer *et al.*, 1986, 1989; Fiorillo, 1989; Oliver, 1989; Olsen & Shipman, 1988), may produce small pits and linear marks similar to those produced by mammalian teeth (which not only come in a variety of shapes and sizes, but which change size and shape with age). It should also be noted that less obvious damages will always be confidently identified in experimental studies where all potential taphonomic agents can be controlled for, or observed directly. As a consequence, faint or isolated lineations or pits were not counted as tooth marks (but were recorded as indeterminate) unless other damages directly implicated carnivores.

Carnivore damage is well described in the taphonomic literature (e.g., Binford, 1981; Bonnichsen, 1979; Haynes, 1980, 1981, 1982, 1983; Shipman & Rose, 1983) and evidence of carnivore activity used in this study includes the following damages based on this literature and on personal observations. Tooth marks, including scores and pits, are minor damages restricted to the bone surface, while punctures, furrows and fracture may create obvious damages. Carnivore damage features outlined below are not individually diagnostic, but rather define a suite of characteristics that are examined and compared to hominid-induced damages.

‡Animal size classes by weight (lb): 1=5–50; 2=50–250; 3=250–750; 4=750–2000; 5=2000–6000.

Tooth pits are circular to oblong or slightly irregular depressions in which at least part of the bone appears compressed. At low magnification, minute cortical pieces at the junction with the depression walls may appear pushed down, oriented toward the base of the depression. A tooth pit may have a linear aspect as it becomes a tooth score. Pitting may be concentrated on epiphyseal ends and on broken bones near the exposed medullary cavity due to the carnivore's attempt to extract marrow or lever pieces back to expose more of the marrow cavity (pers. observ.). Tooth pits are distinguished from hammerstone impact marks by their regular shape and a lack of micro-striae on the pit walls (Blumenschine & Selvaggio, 1988). A tooth score is a broad, u-shaped groove in which the groove walls lack micro-striae and appear compressed or torn (Haynes, 1981; Shipman, 1981). Cut marks are differentiated from tooth scores by their more v-shaped appearance, the presence of micro-striae on groove walls, and a lack of torn or compressed bone at the juncture of the groove and the bone surface. Finally, licking of marrow cavities may produce polish that is often associated with heavy tooth pitting and scoring (Haynes, 1981).

Because a number of (undocumented) processes may produce similar marks, isolated lineations and pits are difficult to define as tooth marks (see above). Accordingly, pits and scores define carnivore activity only if (1) the two co-occur on the same area of bone, (2) one or the other occur in large numbers so that there is little doubt that potential mimics could not have produced the marks, and/or (3) they are associated with other more diagnostic carnivore-induced damages such as furrows, punctures and lever-up breaks.

Damages in which the bone is broken and often removed (as opposed to compressed and torn) defines the second class of carnivore-induced damages. This class of more severe damage includes furrows, punctures and bone fractures. Furrowing describes an area of cancellous bone with numerous large grooves (the furrow) that indicate the removal of large portions of bone. Tooth pits and scores are frequently found on adjacent cortical bone. Punctures are circular to oblong areas of bone failure in which the puncture walls are comprised of small pieces of fractured bone oriented downward to the puncture base. Carnivore-induced fractures display variable fracture features and may or may not display a well-defined loading point.

Gnawing on limb diaphyses may create a loading point comprised of a semi-circular tooth-notch above a flake scar (Bonnichsen, 1979; Binford, 1981; Haynes, 1981; Potts, 1982, 1988). Carnivore-induced notches typically have a semi-circular, non-jagged outline. The flake scar below a tooth notch consists of a single scar that lacks numerous hinges or evidence of serious bone comminution. Opposing loading points with a tooth-notch and flake scar may or may not be present, but in the Amboseli hyaena den assemblage they appear to be absent or poorly defined on limb bones from larger (size 3+) bovids (Oliver, n.d.). Carnivore-induced flakes often display one gently sloping fracture surface (the one adjacent to the flake scar created on the diaphysis) and a more perpendicular face that apparently defines initial failure of the diaphysis. Tooth pits or scores may or may not be found near the loading point on the cortical surface. Carnivores also tend to produce distinctive fractured pieces. Specifically, lever-up pieces are often dagger-like shaft fragments that may lack loading points, but which show tooth pits and/or scores at one end of the piece and hinge fracture at the opposite end that lacks tooth pitting and scoring (Figure 1).

Caution must be exercised in interpreting the presence and frequency of tooth marks because not every tooth pit or score on fossil bone necessarily identifies the agent as a carnivore. Although clearly some damages on large animal limbs (animal size class 2–5) such as extensive furrowing, tooth notches on long bones and levered limb shaft pieces are diagnostic of large carnivores (dog size and larger), the gnawing of smaller animals, notably

Figure 1. Lever-up pieces from (a) the Amboseli hyaena den, and (b) the FLK *Zinjanthropus* fossil assemblage. On both specimens note the long "dagger-like" shape, the lack of loading points on either of the two main fracture lines, the hinge fracture at one end that lacks tooth marks and the presence of tooth marks at the opposite end.

hominids, may create damages such as tooth pits and scores. Chimpanzees (*Pan troglodytes*), for example, strip meat from bone with their teeth (Teleki, 1973) and may pit and score bone in the process. Similarly, gnawing on bones is a common practice among western and non-western peoples alike. The Hadza, for example, gnaw on bones of small animals, as well as those of size 3 animals while attempting to remove adhering pieces of flesh (Oliver, 1993). Until actualistic data become available documenting differences between primate and carnivore tooth-inflicted damage, I assume observed tooth pits and scores identify carnivores, but recognize that early *Homo* created at least some tooth marks.

Cut mark identification

Cut marks are narrow, linear, v-shaped grooves with sub-parallel to parallel micro-striae present on the groove walls (Potts & Shipman, 1981; Bunn, 1981; Shipman, 1981). Typically, the broad, more flat appearance, general lack of micro-striae, torn groove edges at the cortical surface, and frequent compressed appearance, particularly at the proximal section (where damage creation was initiated) of the score distinguishes carnivore tooth marks from cut marks. Pseudo-cut marks resulting from trampling, sediment compaction or movement were distinguished from tool-induced marks using established criteria (Behrensmeyer *et al.*, 1986, 1989; Fiorillo, 1989; Oliver, 1986, 1989; Olsen & Shipman, 1988). Specifically, damages particularly diagnostic of trampling include (1) patches of irregularly oriented striae around a larger v-shaped groove, (2) shallow grooves with micro-striae interrupted by torn or compressed areas of other patches of striae, (3) groove-tails that show a marked change in orientation, or a compressed or torn appearance. Further evidence of trampling is the abundance of randomly oriented lineations.

Hammerstone-induced fractures and impact marks

Hammerstone-induced fractures are defined by either impact marks, loading features on the fracture surface, and/or fracture surface features (Binford, 1981; Bonnichsen, 1979; Blumenschine & Selvaggio, 1988; Bunn, 1989; Johnson, 1985). Features noted in a limited set of hammerstone experiments (Oliver, n.d.) and personal observations are also utilized.

Hammerstone impact features have been illustrated (e.g., Bonnichsen, 1979) as well as described in detail by Blumenschine & Selvaggio (1988). High energy geological processes also create similar impact damages (Oliver, 1989). Dynamic loading with a hammerstone typically creates either irregularly shaped depressions with torn edges of bone, or dense patches of micro-striae that define a slightly depressed area of cortical bone. Experimental observations (Blumenschine & Selvaggio, 1988; Bonnichsen, pers. comm., 1981; pers. observ.) show that the degree of mark irregularity is dependent on the surface texture of the hammerstone. Fine-grained, rounded stream cobbles will produce more uniform depressions than coarse-grained, battered blocks of quartz. With battered or angular-edged hammerstones, not only is the depression irregularly shape, the depression floors and walls often contain numerous micro-flakes and small, v-shaped valleys. Irregular depressions such as these thus define hammerstone loading points. Impact damage to the cortical surface is often visible, however, away from the loading point proper and represent glancing blows. Impact damages are almost invariably adjacent to a fracture surface that displays one or more other loading point features (see below). Small patches of micro-striae on the cortical surface usually define impact on small impact flakes, cones and cone fragments.

Other loading point features on fracture surfaces include flake scars, cone fragments, and stress features such as hackle marks, or fringe and are to differing degrees also indicative of hammerstone-fracture (Bonnichsen, 1979; Johnson, 1985; Bunn, 1989). In particular, impact cones and cone fragments have not been observed in experimental studies of hyaena-induced bone fractures (pers. observ.), and only rarely in the Amboseli hyaena den material (Oliver, n.d.). Thus, cones and cone fragments are taken to be indicative of hammerstone fracture in this study. Cones are usually small features measuring only a few millimeters in total length, though occasionally the major portion of a fracture line defines a partial cone and measures several centimeters in overall length. Usually cones are concave down from the loading point, but occasionally a small cone (and/or stress ridges) emanate from near the medullary wall directly below and are convex to the loading point. These inverse cones or stress features demonstrate bone failure from the interior medullary surface due to high impact velocity and have not been observed on carnivore-fractured bone. Complete cones and cone fragments are small impact fragments that, like impact flakes, often retain small patches of cortical bone that may exhibit hammerstone impact marks. A suite of other features often helps distinguish impact cones from those produced by carnivores. Typical carnivore-induced flakes, such as those illustrated by Potts (1988:97) and Hill (1989:175), appear thicker and longer than impact flakes and display a larger platform with a semi-circular shape in which a rounded bulb area on the gently sloping dorsal facet opposes a flat ventral surface that does not display any suggestion of a flake scar. Hammerstone-induced flakes may outwardly resemble carnivore flakes, but impact marks are often present on the platform. Additionally, due to the broader impact area and dynamic nature of the impact, hammerstone impact flakes can be sliver-like, may not retain the cortical surface (platform), and usually display flake scars or bulbs on both ventral and dorsal surfaces.

Though flake scars below loading points are easily, and often regularly, produced by some carnivores (e.g., Binford, 1981; Haynes, 1983; Hill, 1989; Potts, 1982, 1988), hammerstone-induced flake scars are typically more heavily comminuted and display numerous distal and lateral hinges due to the creation of numerous flakes and flake fragments (Bonnichsen, pers. comm., 1981; pers. observ.; see also Binford, 1981; Bunn, 1989). In addition to impact marks, the cortical surfaces above impact flake scars often display a concentric ring of flakes or flake scars, concentric fracture lines around the loading point or radiating fracture lines. None of these features have been observed in bones modified by hyaenas (pers. observ.). Other features associated with hammerstone fracture, such as cones, stress ridges, irregularly shaped notches and impact marks are also often present on the fracture line immediately adjacent to or below the loading point.

Results

Tabulation of hominid- and carnivore-induced damages observed on limb bone pieces in the FLK Zinj fossil assemblage (Tables 1 and 2) reveals high frequencies of all damage types. In all limb bone samples, hammerstone damage is more common than either cut marks or carnivore damage.

Carnivore damage: tooth marks and fractures
There are three significant aspects to the carnivore damage frequency data, and one notable pattern in element damage distributions. First, carnivore damage frequencies increase with greater levels of skeletal part identification (Tables 1 and 2) and indicate that recognition of carnivore damage is dependent on specimen size. This pattern underscores the difficulty in identifying taphonomic agents based on isolated damages that, because of small damage and

Table 1 Carnivore, cut mark and hammerstone damage frequencies on FLK *Zinjanthropus* limb bone specimens identified by **Bunn (1982)** to element, size class, and as indeterminate limb bone shaft fragment

Size class and element	NISP	Carnivore damage n	Carnivore damage %	Carnivore break n	Carnivore break %	Hammerstone damage n	Hammerstone damage %	Cut mark n	Cut mark %	Total hominid damage n	Total hominid damage %	Hominid and carnivore damage n	Hominid and carnivore damage %
Size 1+2													
Humerus	22	15	68·2	4	18·2	9	40·9	10	45·5	12	54·5	7	31·8
Radius-ulna	18	9	50·0	0	0·0	9	50·0	7	38·9	11	61·1	6	33·3
Metacarpal	21	6	28·6	0	0·0	12	57·1	4	19·0	13	61·9	6	28·6
Femur	19	5	26·3	0	0·0	10	52·6	5	26·3	11	57·9	1	5·3
Tibia	43	16	37·2	7	16·3	15	34·9	12	27·9	21	48·8	9	20·9
Metatarsal	27	14	51·9	0	0·0	19	70·4	5	18·5	19	70·4	10	37·0
Sub-Total	150	65	43·3	11	7·3	74	49·3	43	28·7	87	58·0	39	26·0
Size Class ID only	69	13	18·8	1	1·4	17	24·6	7	10·1	24	34·8	3	4·3
Total Size 1+2	219	78	35·6	12	5·5	91	41·6	50	22·8	111	50·7	42	19·2
Size 3+4+5													
Humerus	68	33	48·5	4	5·9	35	51·5	25	36·8	45	66·2	23	33·8
Radius-ulna	84	31	36·9	3	3·6	43	51·2	29	34·5	50	59·5	16	19·0
Metacarpal	29	9	31·0	0	0·0	18	62·1	5	17·2	18	62·1	7	24·1
Femur	41	17	41·5	4	9·8	22	53·7	10	24·4	24	58·5	8	19·5
Tibia	95	27	28·4	5	5·3	41	43·2	28	29·5	51	53·7	18	18·9
Metatarsal	22	7	31·8	3	13·6	11	50·0	6	27·3	13	59·1	3	13·6
Sub-Total	339	124	36·6	19	5·6	170	50·1	103	30·4	201	59·3	75	22·1
Size Class ID only	440	55	12·5	11	2·5	136	30·9	46	10·5	156	35·5	23	5·2
Total Size 3+4+5	779	179	23·0	30	3·9	306	39·3	149	19·1	357	45·8	98	12·6
Total													
Humerus	90	48	53·3	8	8·9	44	48·9	35	38·9	57	63·3	30	33·3
Radius-ulna	102	40	39·2	3	2·9	52	51·0	36	35·3	61	59·8	22	21·6
Metacarpal	50	15	30·0	0	0·0	30	60·0	9	18·0	31	62·0	13	26·0
Femur	60	22	36·7	4	6·7	32	53·3	15	25·0	35	58·3	9	15·0
Tibia	138	43	31·2	12	8·7	56	40·6	40	29·0	72	52·2	27	19·6
Metatarsal	49	21	42·9	3	6·1	30	61·2	11	22·4	32	65·3	13	26·5
Total Element Id.	489	189	38·7	30	6·1	244	49·9	146	29·9	288	58·9	114	23·3
Total Size Class ID	509	68	13·4	12	2·4	153	30·1	53	10·4	180	35·4	26	5·1
Indet. LBN FRAG	916	64	7·0	6	0·7	420	45·9	34	3·7	439	47·9	15	1·6
TOTAL LIMB	1914	321	16·8	48	2·5	817	42·7	233	12·2	907	47·4	155	8·1

Table 2 Damage frequencies for specimens used by Bunn (1982) to estimate MNE and MNI values for limbs in the FLK *Zinjanthropus* fossil assemblage

Size class and element	MNE	Carnivore damage		Carnivore break		Hammerstone damage		Cut mark		Total hominid damage		Hominid and carnivore damage	
		n	%	n	%	n	%	n	%	n	%	n	%
Size 1+2													
Humerus	6	4	66·7	1	16·7	5	83·3	5	83·3	5	83·3	3	50·0
Radius-ulna	4	3	75·0	0	0·0	3	75·0	3	75·0	4	100·0	3	75·0
Metacarpal	6	3	50·0	0	0·0	5	83·3	2	33·3	5	83·3	3	50·0
Femur	9	2	22·2	0	0·0	3	33·3	3	33·3	5	55·6	1	11·1
Tibia	11	6	54·5	2	18·2	3	27·3	4	36·4	6	54·5	1	9·1
Metatarsal	10	4	40·0	0	0·0	6	60·0	2	20·0	6	60·0	2	20·0
Total	46	22	61·1	3	8·3	25	69·4	19	52·8	31	86·1	13	36·1
Size 3													
Humerus	14	6	42·9	2	14·3	10	71·4	6	42·9	10	71·4	4	28·6
Radius-ulna	16	6	37·5	0	0·0	7	43·8	9	56·3	12	75·0	2	12·5
Metacarpal	10	5	50·0	0	0·0	7	70·0	3	30·0	8	80·0	8	80·0
Femur	13	7	53·8	0	0·0	5	38·5	2	15·4	6	46·2	1	7·7
Tibia	17	8	47·1	2	11·8	9	52·9	11	64·7	13	76·5	4	23·5
Metatarsal	6	3	50·0	2	33·3	0	0·0	1	16·7	1	16·7	1	16·7
Total	76	35	50·0	6	8·6	38	54·3	32	45·7	50	71·4	20	28·6
Size 1+2+3													
Humerus	20	10	50·0	3	15·0	15	75·0	11	55·0	15	75·0	7	35·0
Radius-ulna	20	9	45·0	0	0·0	10	50·0	12	60·0	16	80·0	5	25·0
Metacarpal	16	8	50·0	0	0·0	12	75·0	5	31·3	13	81·3	11	68·8
Femur	22	9	40·9	0	0·0	8	36·4	5	22·7	11	50·0	2	9·1
Tibia	28	14	50·0	4	14·3	12	42·9	15	53·6	19	67·9	5	17·9
Metatarsal	16	7	43·8	2	12·5	6	37·5	3	18·8	7	43·8	3	18·8
TOTAL	122	57	53·8	9	8·5	63	59·4	51	48·1	81	76·4	33	31·1

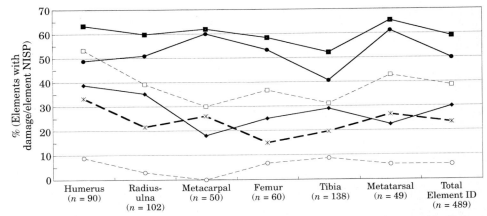

Figure 2. Frequency of hominid and carnivore damage on FLK *Zinjanthropus* limb pieces identified to element. (□) carnivore damage (38·7%); (○) carnivore break (6·1%); (●) hammerstone damage (49·4%); (◆) cut mark (29·9%); (■) total hominid damage (58·9%); (×) hominid+carnivore damage (23·3%).

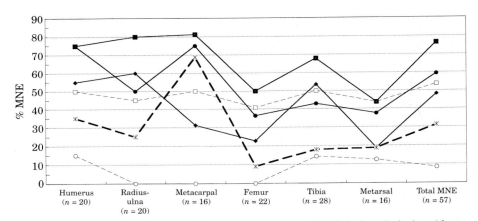

Figure 3. Frequency of MNE with hominid and carnivore damage on FLK *Zinjanthropus* limb pieces (elements used in MNE estimates after Bunn, 1992). (□) carnivore damage (53·8%); (○) carnivore break (8·5%); (●) hammerstone damage (59·4%); (◆) cut mark (48·1%); (■) total hominid damage (76·4); (×) hominid+carnivore damage (31·1%).

specimen size, lack an anatomical context or definable relationship to other damages that might aid identification.

Carnivore damage frequencies, which range between 25·8% for all limb specimens (Table 1) and 53·8% for specimens used in MNI and MNE estimates (Table 2, Figure 3), make it clear that carnivores played a significant role in modifying limb bones in the FLK Zinj assemblage. The overall carnivore damage frequency does not describe the character of carnivore activity, but some clues are revealed in a comparison of various damages. Although overall carnivore damage frequencies are high, in none of the limb bone samples is the frequency of carnivore-broken elements (Figure 1b) greater than 15% (Tables 1 and 2), and it is usually much less (Figures 2 and 3). This pattern demonstrates that although carnivores did damage limb bones, they were not extensively involved in the fragmentation of those bones.

Figure 4. Cut marks on a *Kobus sigmoidalis* femur shaft fragment from the FLK *Zinjanthropus* fossil assemblage. A hammerstone impact mark (chipped area) overlays one mark.

Finally, with the exception of size 1+2 femora, carnivore damage frequencies are fairly uniform for all limb elements (Table 2, Figure 3). This uniformity suggests that carnivores gnawed on all elements with equal intensity and showed no preference for particularly meaty or marrow-rich bones. This is counterintuitive given most carnivore's predilection for meaty, marrow-rich parts.

Cut marks

More cut marks were observed in this study (Table 1, Figure 4) than in previous studies of the FLK *Zinjanthropus* limb assemblage (Potts & Shipman, 1981; Potts, 1982, 1988; Bunn, 1981, 1982; Bunn & Kroll, 1986). The observed differences in cut mark frequencies, however, relate to methodological differences in which bone fragments are examined and whether or not the marks are examined by SEM (Bunn & Kroll, 1986). Potts & Shipman (1981), for example, restricted their analysis to elements identifiable to taxa, while this observer and Bunn (1981, 1982) examined all limb pieces. Cut mark frequencies reported here ($n=235$; Table 1) for identified elements are 38·7% greater than the microscopically (SEM) examined sample reported by Bunn & Kroll (1986; $n=91$), but the macroscopically identified cut marks reported by Bunn & Kroll ($n=219$) are only 6·2% less than that reported here. This close agreement in cut marks identified macroscopically in two independent studies demonstrates the distinctiveness of the cut marks in the assemblage.

Initial examination reveals at least two significant patterns in these data. First, the difference in cut mark frequencies between the upper, meaty limb elements and the most distal, least meaty elements shows that early *Homo* focused tissue cutting on more meaty carcass-parts, a pattern that suggests the presence of meat on bone and which was previously noted by Bunn

(1982; Bunn & Kroll, 1986). Utilizing a smaller sample of taxonomically identifiable bone, others (e.g., Potts & Shipman, 1981; Potts, 1982, 1988; Shipman, 1986a,b) have argued that the distribution of most marks on shafts indicates removal of meat scraps, tendons or other tissue. Similarly, Blumenschine (1986, 1987) indicates that any cut mark only indicates tissue removal, not tissue quantity. Nevertheless, had hominids scavenged carnivore-ravaged kills for meat scraps, tendon and hide, then one would expect (1) a more uniform distribution of cut marks than observed (Bunn, 1981, 1982; Bunn & Kroll, 1986; this study) and (2) a concentration of tooth marks on meaty elements rather than the observed uniform distribution across all elements (see above).

Preferential placement of cut marks on meaty limbs is particularly true of the size 1+2 limb material. Note, for example, that size 1+2 metacarpals have less than half the percentage of cut marks than that of the humeri (Table 1, 19·0% vs. 45·5%). For the size 3+4+5 material, however, the apparent focus on meatier limbs holds true only for the forelimb, though the loss of femoral ends (Bunn, 1982; Bunn & Kroll, 1986) may explain this apparent pattern.

Hammerstone and total hominid-induced damage

As previously noted (Bunn, 1982; Bunn & Kroll, 1986; Potts, 1982, 1988) hammerstone-induced damages are abundant in the FLK Zinj faunal assemblage. Damages such as impact marks and loading point scars (Figure 5a, 5b) are identifiable, while others are quite faint and require careful examination of each fracture line and associated features (Figure 5c). Due in part to the number of fragments created at the loading point by hammerstone impact and the unique character of hammerstone loading, even very small specimens frequently retain hammerstone impact features. Small impact flakes that retain the cortical surface (the platform) often display hammerstone impact marks, typically in the form of patches of micro-striae. Additionally, it is the size and shape of many flakes (thin flakes with flake scars on dorsal and ventral surfaces, and small platform), cones and cone fragments that define hammerstone impact (Figure 6).

There are two patterns in hammerstone-induced damage frequencies that are worth noting. First, 50% of all limb pieces identified to element (n=244, Table 1, Figure 2) and 63 of 106 (59·4%) of the MNE in the assemblage (Table 2, Figure 3) display hammerstone damage. Thus, hominids not only played a *major* role in the accumulation and modification of the FLK Zinj faunal assemblage, they played the role *habitually*. Although some variation is apparent, hammerstone damage frequencies tend to be greater than 50% for all limb pieces identified to element (Table 1, Figure 2). Note, however, that hindlimb MNE exhibits less hammerstone damage than the forelimbs (Table 2, Figure 3). In particular, the femora used in MNE calculations consistently have low hammerstone damage frequencies, a pattern that may be a result of the apparent post-fracture loss of epiphyses noted by Bunn (1982; Bunn & Kroll, 1986). Combining hammerstone and cut mark damages creates more uniformity in hominid-induced damage frequencies on forelimb and hindlimb bones (Table 2, Figure 3; see below). This is taken to indicate roughly uniform processing of all limb elements transported to FLK Zinj.

The combined hammerstone and cut mark frequency data also support previous conclusions (based on cut mark and bone frequency analyses) that hominids were the primary taphonomic agents at FLK Zinj (Bunn, 1982; Bunn & Kroll, 1986; Potts, 1982, 1988). In this study, for example, 47% of the 998 limb pieces identified at least to size class display evidence of hominid involvement (Table 1). Almost 60% of the limb pieces identified to element bear some evidence of hominid involvement (n=288, Table 1, Figure 2), while 76·4% of all pieces

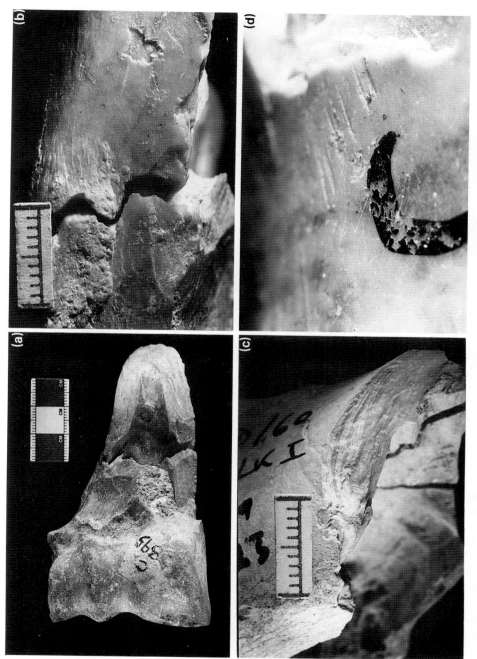

Figure 5. Examples of hammerstone-induced damages. Opposing hammerstone loading points and spiral fracture on a *Connochaetes* sp. humerus (a). The lateral loading point area (b) displays a radiating crack, numerous comminuted impact flake scars, one impact mark that is a patch of striae and another that displays striae at the edges of a chipped area which is crossed by a carnivore tooth score (right side of figure). Other hammerstone loading points such as this on a *Kobus sigmoidalis* distal humerus (c) are defined by cones or partial cones which are often associated with impact marks. On other specimens the only indication of fracture agent are small patches of hammerstone impact striae (d) or loading points that display cones, comminuted flake scars.

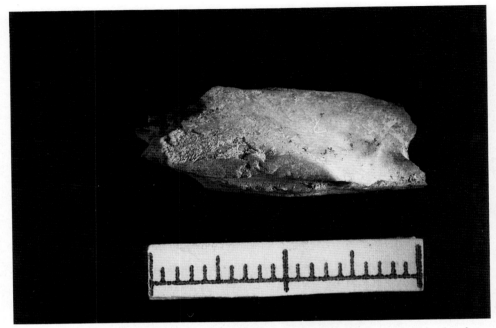

Figure 6. An example of a partial impact cone displaying hertzian cone features on the ventral surface, auxiliary flake scars on the dorsal surface, and hammerstone impact marks on the platform.

used in the MNE and MNI estimates exhibit cut marks and/or hammerstone damage ($n=81$, Table 2, Figure 3).

Co-occurrence of hominid- and carnivore-induced damage

Previous studies of the FLK *Zinjanthropus* faunal assemblage noted that hominid and carnivore damages co-occur on many bones (Bunn, 1982; Bunn & Kroll, 1986; Potts & Shipman, 1981; Potts, 1982, 1988). The frequency of co-occurrence was not quantified, however. Data presented in Tables 1 and 2 (Figures 2 and 3) show that the frequency of hominid and carnivore damage co-occurrence is only 8% for all 1904 limb pieces, 14% for specimens identified at least to size class, 23% for specimens identified to element, and 31% for specimens used in the MNE and MNI estimates. In other words, about 40% of the specimens identified to element and showing hominid-induced damage also show carnivore damage (Figure 7). Thus, 60% ($n=48$) of the FLK Zinj fossil assemblage limb MNE that display hominid-induced damage ($n=81$) *do not* exhibit carnivore damage (Table 2, Figure 3).

Finally, note that carnivore- and hammerstone-induced fractures comprise only 56% of the specimens identified to element (Table 1; 274 out of 489) and 68% of the MNE (Table 2; 72 out of 106). This leaves the fracture agent for a significant number of specimens unknown at this time.

Discussion

The above damage frequency data help set boundary conditions for an accurate assessment of hominid and carnivore behaviors that created the FLK *Zinjanthropus* fossil assemblage. These

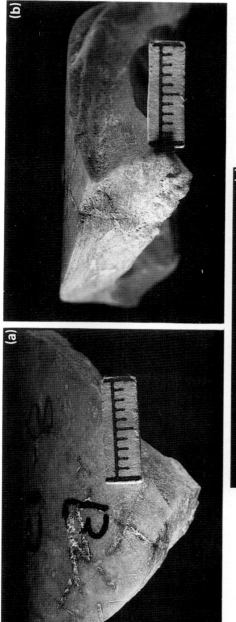

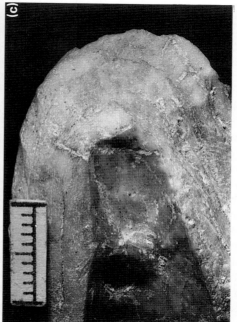

Figure 7. The cortical surfaces of many specimens display non-overlapping cut and tooth marks (a), but the same specimen (b) may also exhibit tooth scores on fracture surfaces. Regardless of the presence or absence of hammerstone-induced damages, this co-occurrence of carnivore damages on fracture surfaces and cut marks indicates the fracture was hominid-induced. (c) Co-occurrence of hominid-damage and carnivore-induced damages on specimens where the overprinting of a tooth-mark on hammerstone-induced fracture surface clearly demonstrates carnivore scavenging of hominid food refuse.

data provide quantified estimates of the degree of hominid and carnivore involvement in assemblage creation, as well as the nature of hominid-carnivore competitive interactions, including the sequence of access to carcass-parts. Moreover, data presented here provide a firm archaeological basis for exploring the socioecological implications of habitual animal food transport by early *Homo*.

The character and extent of carnivore and hominid involvement in assemblage creation

The carnivore damage frequencies suggest a number of conclusions concerning the extent of carnivore and hominid involvement in assemblage creation and support many previous interpretations of the assemblage based on cut mark and bone frequency analyses. Specifically, hominid-induced damage on 75% of limb MNE (Table 2, Figure 3) demonstrates that early *Homo* habitually transported carcass-parts to FLK Zinj where the limbs were intensively processed for meat and marrow. Carnivore damage on 38·7% of the limbs identified to element (Table 1, Figure 2) shows that carnivores played a significant role in modifying the bones. Significantly, however, the lack of carnivore-fractured bone suggests that hominids were not only responsible for the majority of bone accumulation, but also did not abandon many fresh, complete bones. Finally, the comparative lack of carnivore-fractured bone indicates that most carnivore activity took place either before carcass-part transport or following bone breakage by hominids.

Quantification of the hominid *vs.* carnivore access sequence is the subject of on-going work (Oliver, n.d.). Preliminary observations are instructive and suggest that early *Homo* had some access to meat-rich carcasses. For example, if carnivores had the sole first access to meaty carcasses (and thus complete bones) which were later scavenged by hominids (e.g., Blumenschine, 1986, 1987; Shipman, 1986*b*), then one might expect elements with comparable amount of meat and marrow to display similar carnivore damage frequencies. For example, the femora and humeri, which are marrow-rich and often preferred by large carnivores (Kruuk, 1972; Schaller, 1972; van Lawick & van Lawick, 1971; Mech, 1970), should show a greater incidence of tooth marking than the metapodials. This predicted pattern may be amplified by other feeding habits and by carcass anatomy. Scavenged carcasses at Amboseli National Park (Kenya) retained skin from the distal radius-ulna and distal tibiae, through the metapodials, to the phalanges (pers. observ., 1990; see also Behrensmeyer, 1993), thereby insulating those elements from damage at the death site. Carnivore damage frequencies do not show a concentration of damage on upper elements, nor are the frequencies depressed for metapodials. Rather, tooth-mark frequencies are uniform for all elements (Tables 1 and 2, Figures 2 and 3), indicating that similar amounts of meat were present on all bones when carnivores were active. To achieve this condition, much of the meat from the upper limbs must have been removed before carnivore tooth-marking. Also, as argued by Bunn (1981, 1982; Bunn & Kroll, 1986), the distribution of cut marks on meaty parts also suggests the presence of meat to cut and implies early access by hominids.

The hominid and carnivore co-occurrence frequencies are also instructive in determining the access sequence. If early *Homo* was a habitual scavenger of meat-poor, skeletally-intact, carnivore-ravaged carcasses, then one might expect a high frequency of specimens to exhibit both carnivore and hominid damage. Though a large number of specimens identified to element show both damage types (indicating that scavenging of meat-poor carcasses may have been an important, but not dominant, mode of carcass acquisition), this co-occurrence frequency is only 39·9% (114 of 288) that of bones that display evidence of hominid involvement (Table 1, Figure 2). Thus, the lack of carnivore damage on 60% of the specimens that display hominid involvement suggests that these specimens had significant amounts of meat when hominid processing of carcass-parts began (and that they were not scavenged after hominid processing).

Finally, preliminary damage context analysis reveals that a large number of the hammerstone-fractured bones also display carnivore tooth marks on hammerstone-induced fracture surfaces or on the medullary walls of hammerstone-fractured bone, thus demonstrating carnivore scavenging of early *Homo* food refuse (Figure 7b–c; Oliver, n.d.). Of the 114 specimens with evidence of hominid and carnivore involvement, at least 63 (55·3%) show evidence of carnivore scavenging after hammerstone breakage while only 2 (2%) may show evidence of hominid utilization after carnivore activity (cut marks overprinted on carnivore tooth marks; see also Potts & Shipman, 1988). The access sequence has not been determined for 49 specimens. The frequency of carnivore scavenging of hammerstone broken bones suggests that carnivore scavenging of bone previously processed by hominids was also responsible for some of the remaining carnivore damages for which the access sequence could not be determined.

Comparisons of types of carnivore damage, as well as the extent of hominid-induced damage, reveal other aspects of the character of hominid-carnivore competitive interactions. Most carnivore damage is minor, consisting of tooth pits and scores; more severe damages such as flaking, chipping, polish and furrowing are rare. This is consistent with fracture data presented above which indicate that carnivores were active in assemblage creation, but that they were not the major bone fracture agent. Although the extent of carnivore damage is significant, the damage severity is slight. Thus, because the frequency of carnivore-induced fracture is low, we may again infer that tooth-mark creation occurred mainly during meat stripping activities before carcass-part transport by early *Homo*, or during refuse scavenging after hominid utilization. As argued above, cut and tooth mark distributions, as well as hominid and carnivore damage co-occurrence frequencies all suggest the latter sequence of events is more likely. Carnivores did not gnaw bones at FLK Zinj with any great intensity, because few complete, juicy bones were available to carnivores after processing by early *Homo* (as indicated by hammerstone-induced damage on 75% of the MNE). This ability to completely process and consume meaty carcass-parts implies that early *Homo* was "secure" in this location during its occupation.

The carnivore damage data also suggest the type or size of carnivore involved. First, the comparative lack of carnivore-induced fractures and heavy pitting or scoring of epiphyseal ends suggest that the carnivore responsible for much of the damage was not a large, habitual bone-crushing animal like the hyaena, but rather a smaller carnivore that could make use of small meat scraps adhering to small bone fragments. The presence of micro-mammals at FLK Zinj that Andrews (1983) has interpreted as the remains of genet scat also imply scavenging by other small carnivores. Thus, it seems reasonable to conclude that the abandoned food refuse not only acted as an attractant for medium to large carnivores, but created a new microhabitat that attracted rodents, who in turn attracted other, smaller carnivores. This certainly seems to have been the case for many later archaeological sites (e.g., the Cherokee Sewer Site, Iowa; Semken, 1980, pers. comm., 1993; R. W. Graham, pers. comm., 1993).

These carnivore damage frequencies also corroborate the inference (Binford, 1981, 1988; Blumenschine & Marean, 1993; Potts, 1982, 1988) that carnivores were an important taphonomic agent at FLK Zinj. However, the nature of this involvement appears quite different from that previously envisioned. For example, the high frequency of carnivore damage juxtaposed with the low incidence of carnivore-induced fracture and the high incidence of hammerstone damage demonstrates that FLK Zinj was not an accumulation of temporally separated and spatially restricted carnivore kills later scavenged by hominids as argued by Binford (1981). Carnivore damage frequencies instead support Bunn's (1982; Bunn & Kroll, 1986), Potts' (1982, 1988) and Blumenschine & Marean's (1993) conclusions that hominids accumulated the bulk of the FLK Zinj assemblage which was later scavenged by carnivores.

Blumenschine & Marean (1993) base their conclusions on a study of epiphyseal loss and tooth-mark frequencies in simulated assemblage modified only by hyaenas, as well as epiphyseal loss and tooth-mark frequencies in hammerstone-fractured assemblages subsequently modified by hyaenas. In their experiments low epiphysis:shaft fragment ratios occur under conditions of low competition where on-site gnawing destroys epiphyses, thereby creating a high incidence of tooth-marked limb specimens. That is, in the absence of competitors (including conspecifics who may harass and steal carcass-parts), hyaenas tend to feed on carcass remains at the site. Blumenschine & Marean (1993, Figure 16.6) therefore conclude that the high frequency of carnivore tooth marks, in conjunction with the low limb epiphysis:shaft ratios at FLK Zinj, suggest that competition for carcass-parts was minimal between carnivores, and presumably between hominids and carnivores. The low frequency of carnivore-fractured bone makes it unlikely that on-site gnawing by large carnivores (e.g., hyaena) is responsible for the lack of epiphyseal ends documented by Bunn (1982; Bunn & Kroll, 1986). Rather, the lack of these meaty epiphyseal ends and low frequency of carnivore-induced fractures (6·1% of the specimens identified to element, Table 1, Figure 2) suggests some competition between larger scavengers or scavengers and early *Homo* during, or shortly after site occupation. (Alternatively, hominids may have transported nutrient-rich epiphyses for processing elsewhere.) Minimal competition among large carnivores on-site should have resulted in a greater frequency of carnivore-induced fractures. Thus, the high frequency of carnivore damage (38·7% of specimens identified to element, Table 1, Figure 2), may suggest that smaller carnivores unable to easily fracture bone created the bulk of the tooth marks under conditions of minimal competition after hominid abandonment. This is consistent with the hominid followed by carnivore access sequence noted above.

All measures of damage to limb pieces demonstrate that hominid-induced damages are more common than carnivore-induced damages (Tables 1 and 2, Figures 2 and 3). Because limb fragmentation from hammerstone blows may vary for a number of reasons (e.g., hammerstone weight and shape, bone density, individual hominid strength), and because fragmentation influences frequency patterns, tabulation of damage frequencies on the entire limb NISP may inflate estimates of hominid involvement. Therefore, damage frequency tabulations are based on bones used in MNE and MNI estimates (see Bunn, 1982 for a listing of these specimens). Of the 106 limb elements, at least 76·4% (81) bear evidence of hominid activity; 48·1% (*n*=51) of the MNE estimate display cut marks and 59·4% (*n*=63) exhibit hammerstone damage (Table 2). Thus, in conjunction with previous studies of cut marks and bone frequency patterns (Bunn, 1981; Bunn, 1982; Bunn & Kroll, 1986; Potts & Shipman, 1981; Potts, 1982, 1988), the hammerstone fracture frequency data demonstrate that FLK Zinj was indeed a site of *habitual* carcass processing by early *Homo*, the refuse from which was later scavenged by a variety of carnivore species.

Overall, limb bone fragmentation in the FLK Zinj assemblage appears similar to that documented for modern hunter-gatherers, notably the Hadza (Bunn, 1989; Oliver, 1993). That is, bones were not merely hit with hammerstone at the midshaft and the halves separated, as illustrated in earlier experimental studies of bone fracture (e.g., Bonnichsen, 1979; Johnson,

1985). Rather, many bones exhibit numerous impact marks along the shaft, others show fractures through the epiphyseal ends and others display numerous impact marks at or near the epiphyseal end. During bone processing the Hadza frequently pound the epiphyseal ends to gain access to bloody, fat-rich cancellous tissue. This activity largely reduces some epiphyseal ends to unidentifiable fragments of cancellous bone (Oliver, 1993). Thus, the presence of split epiphyseal ends, epiphyseal fragments, numerous impact marks, and impact marks at or near epiphyseal ends in the FLK Zinj assemblage suggests that the character of consumption behavior undertaken by early *Homo* may have been similar to that of the modern Hadza. That is, the severity of the damages is more than required to remove marrow; the severity of hammerstone-induced damage demonstrates concern with, and consumption of, the blood-rich cancellous ends. This intense processing may explain part of the loss of epiphyseal ends noted previously by Bunn (1982; Bunn & Kroll, 1986).

Socioecological implications of early Homo

The above quantitative estimates of hominid and carnivore involvement make it clear that early *Homo* was the primary agent of bone accumulation and modification at FLK Zinj. What then were the socioecological mechanisms that lay behind the evolution of these novel activities (the habitual transport and processing of carcass-parts for meat and marrow). Given that many carnivores transport food and that many of these live in environments similar to those inhabited, or at least frequented by, Plio-Pleistocene hominids, e.g., riparian woodlands, grassy woodlands, and open grasslands (see Kappleman, 1984; Shipman & Harris, 1988; Sikes, 1994; Potts, 1988; Plummer & Bishop, 1994), it is reasonable to examine carnivore behavior for insights into the biological and ecological mechanisms that determine food transport behavior.

Carnivores are one of the few animals besides modern humans and birds that habitually transport food to a discrete location for consumption (Eisenberg, 1981; see also Clutton-Brock, 1991). Data on carnivore behavior and biology, particularly food transport and intra- and interspecific competition (e.g., Ewer, 1973; Kruuk, 1972; Mech, 1970; Rasa, 1984; Rood, 1986; Schaller, 1972) suggest that there are apparently two major factors that work together to promote food transport to established dens or other protected locations: (1) the presence of altricial young, and (2) intensely competitive and dangerous interactions with other predators at food resources (i.e., carcasses). Moreover, the anti-predator behavioral responses that result from these two conditions, including the establishment of dens, food transport and various forms of group defense, are themselves a result of two environmental factors: (1) living in open to semi-open habitats with (2) high predator densities. Martin (1989), for example, suggests that sociality in many carnivores may have evolved as a response to the establishment of expansive grasslands in the late Pliocene. In general, studies of carnivore sociality frequently cite the need for anti-predator defensive strategies as a likely explanation for the evolution of carnivore sociality (Gittleman, 1989 and references therein). This is particularly true for smaller species that cannot defend themselves individually from larger competitors (Ewer, 1973; Kruuk, 1972) and for species living in open habitats where vulnerability to predation increases due to a lack of cover (Lamprecht, 1981). (The other explanation favoured for the evolution of sociality, particularly for larger carnivores, is the increased hunting efficiency afforded by cooperative foraging behavior (Gittleman, 1989; Kruuk, 1972; Lamprecht, 1981; Schaller, 1972).

As noted by many others (e.g., Eisenberg, 1981; Schaller & Lowther, 1969; Lovejoy, 1981), the common social context for food transport and processing among many carnivores is the presence of young in the group, particularly altricial young. Because of their biological characteristics (limited strength, endurance and mobility, poorly developed senses and rapid post-natal growth rate), however, altricial young carry a high ecological cost to the infant, mother, and other group members. Specifically, altricial infants are highly vulnerable to predation and have high nutritional and energy demands to meet rapid post-natal growth rates that, in turn, act to create foraging and energetic problems for the mother and other group members. Specific solutions to these ecological and energetic problems vary, but three broad options can be outlined: (1) infants may be carried on foraging trips; (2) infants may be sequestered in a secure location while other group members, including the mother, forage; and (3) infants may be cared for by the mother and/or other care-givers at a secure location.

An extreme option, the one preferred by other mammals that do *not* bear altricial young, is the movement of young with the adults on daily foraging rounds. No carnivore transports their altricial young on foraging trips or hunts. Even the lion (*Panthera leo*) leaves its young in a protected location while it hunts (Schaller, 1972). With adequate group size, however, lions can carry infants to secured carcasses. For all smaller carnivores, foraging with altricial young would subject infants to increased risk of predation and either force the rest of the group to reduce the frequency, length or speed of its foraging trips, or force the mother to keep up thereby further increasing her nutritional and energetic demands. In this regard it is notable that among yellow baboons (*Papio cynocephalus*) living in open to semi-open habitats, mothers with clinging (precocial) infants often appear stressed during foraging trips with the troop and often lag behind, conditions that Altmann (1980) believes may increase both mother's and infant's vulnerabilities to predation. Mothers that bear altricial young have to carry their infants, who, because of lack of strength, endurance or fur, cannot cling. This further increases their biological stress and vulnerability to predation.

A second option is sequestering the infant alone in a protected location while other group members, including the mother, forage. For example, though spotted hyaena (*Crocuta crocuta*) mothers initially remain with their infants for several weeks, they gradually begin to leave them alone at the den for longer and longer periods of time, often for several days. This option extracts a cost, however; Kruuk (1972) notes that this may be the time of greatest infant mortality, due largely to predation by other carnivores. Gittleman (1989) also draws attention to the fact that carnivores frequently prey on each other.

Finally, the mother and perhaps other care-givers may remain with the infant while others forage. Most carnivores living in open to semi-open areas display variations of this latter behavior (Eisenberg, 1981; Ewer, 1973; Gittleman, 1989; Kingdon, 1977; Kruuk, 1972; Moehlman, 1989; Rasa, 1984; Schaller, 1972). Several aspects of food transport among carnivores are worth noting. First, the sociobiological context of food transport among carnivores is to provision altricial young, and less often the mother and other helpers (typically the infant's elder siblings). The canids, in particular the wolf (*Canis lupus*) and wild dog (*Lycaon pictus*), exemplify provisioning behavior in carnivores (Eisenberg, 1981; Ewer, 1973; Moehlman, 1989; Gittleman, 1989) in the frequency of food brought not only to the infant, but also to the mother and other care-givers. Other carnivores, including the lynx (*Lynx rufus*), dwarf mongoose (*Helogale parvula*) and brown hyaena (*Hyaena brunnea*) also provision their infants. Among the dwarf mongoose, for example, the alpha male sometimes brings insects and rodents to the termite mound-den and presents them to weaning infants (Rasa, 1984). Similar behaviors may also be responsible for bone accumulations at spotted hyaena dens (Hill, 1975; Kingdon, 1977). As indicated above, lions seem to be the exception to this rule amongst the large carnivores by almost without exception moving their young to the kill. As the top predator, their size may insulate lions from risks associated with transporting altricial young.

Second, this food transport behavior among most carnivores apparently requires a daily division of the group into at least two functional units: an infant care-giving unit and a foraging unit. Altricial young are not taken foraging until they are at least able to keep up with adults, particularly among species in open to semi-open habitats with high predator densities (e.g., *L. pictus, C. aureus, C. mesomelus, H. brunnea, H. parvula*). Instead, the mother, and perhaps her helpers, remain with the altricial young at the den. Again, the canids are notable for having helpers that remain with the mother at the den (Ewer, 1973; Kingdon, 1977; Moelhman, 1989), though non-mother care-givers also remain with dwarf mongoose infants (Rasa, 1984; Rood, 1986). Though it is rarely habitual, delayed consumption also occurs in several species, notably wolves and wild dogs, when excess food not eaten by the infants or mother is eaten by the members of the transport group. Although not provisioning behavior, per se, several felids, including cheetahs (*Acinonyx jubatus*), bring live prey to their young for them to kill. In addition to "teaching" the young about hunting, this behavior keeps the young in a controlled, protected setting, and presumably reduces the risk of predation. The presence of altricial young and their provisioning often impacts the foraging behavior of others besides the mother. For example, when young are present among dwarf mongoose, the foraging unit tends to return to the den earlier in the afternoon (Rasa, 1984). Similarly, reduction in foraging distance occurs among wild dogs (Frame *et al.*, 1979).

Finally, studies by Kruuk (1972), Schaller (1972), van Lawick & van Lawick (1971) and others further emphasize the great risks associated with interspecific competition for food at animal carcasses. Most carnivores, particularly smaller carnivores, bolt down meat and/or retreat with carcass-parts to avoid this intense competition at animal carcasses and reduce their risk of injury and death. The ability to ingest so much food so quickly and the associated physiological changes in the stomachs of some carnivores, particularly the canids (e.g., Mech, 1970), may be an evolutionary anti-predator response to this competition. Thus, dangerous competition at kill sites acts as a biological mechanism that promotes transporting food away from the carcass.

In summary, the conditions that seem to promote food transport among carnivores are (1) altricial young, (2) living in open habitats to semi-open habitats characterized by (3) high predator densities. That is, the response to these biological and ecological boundary conditions is a suite of anti-predator behaviors. Altricial young, because of their vulnerability to predation, their high nutritional and energetic requirements and the consequent demands they make on their mother (and other care-givers), act as the biological force pulling food resources and other group members to the den or central place, while competition at food resources provides an impetus to retreat to more secure areas. The evolution of dens and food transport, therefore, appears to be a common and effective response to these conditions.

Early *Homo* apparently shared many of these same ecological and biological conditions. Using a limited set of biometric measurements and projected post-natal growth curves Stanley (1992) argues convincingly that early *Homo* infants were as altricial as modern humans. Even if they were not as altricial as modern human infants, post-natal growth curves (see Stanley, 1992; Figures 5 and 7) indicate they were closer to the modern human condition than that of precocial non-human primates. Second, the ancient lake shore habitats at Olduvai supported a diverse predator guild. Potts (1988), for example, notes that in the Bed I faunal assemblages the percentage of carnivore individuals (MNI) represented ranges from 1–21%. The bovids represented at FLK Zinj indicate that open-woodlands and grasslands dominated the area (Gentry & Gentry, 1978a,b; Kappleman, 1984; Potts, 1988), although recent biometric (Plummer & Bishop, 1994) and isotopic analyses (Sikes, 1994) indicate a more closed habitat at FLK Zinj. Though more wooded than previously thought, the presence of numerous open woodland to grassland taxa, e.g., alcelaphines and antilopines, certainly suggest considerable nearby open ground and habitat diversity.

Since early *Homo* may have shared the same fundamental biological and ecological boundary conditions that seem to regulate carnivore's denning and food transport behaviors, these mechanisms may have been at least as significant for early *Homo*. Thus, there may have been similar socioecological "rules" for rearing altricial early *Homo* young in open, predator-rich habitats. These "rules" may include the following: (1) Groups will divide into two foraging sub-groups in order to (a) reduce predation risks to the infant and mother, and (b) reduce the mother's energy load. (2) Mother, infant and other care-givers will tend to reside in a core area where protected locations are present and/or risks of predation may be lower. (3) Non-caregivers will tend to forage in areas away from the caregivers so as (a) not to compete for food resources in refuge areas, and (b) gain access to resources not available in the core area (e.g., animal carcasses). (4) Acquired carcass-parts will tend to be transported (a) away from the kill to avoid intense competition to (b) more wooded habitats or other refuge areas where predator densities are lower and physical means of protection are available (e.g., trees).

While preliminary in nature, the above socioecological explanations for the transported and processed bone and stone present at FLK Zinj and other Plio-Pleistocene sites may be tested with the archaeological data. For example, if early *Homo* had already expanded its habitat range dramatically over that of other hominids as argued by Foley (1987, 1989) and Sikes & Ambrose (1993), then the dual-unit foraging model outlined above suggests that there should be at least two different types of archaeological sites in the Plio-Pleistocene record both in terms of environmental setting and archaeological content. Specifically, Isaac's (1984) type C sites with large numbers of artefacts and bone indicative of intensive and/or repetitive occupation would have been located in more protected, wooded areas. In contrast, type B and particularly type A sites, with little fragmented bone or stone tool refuse suggestive of limited activities (Isaac, 1984), might be located in more open areas, removed from core, protected locations. Although limited, paleoenvironmental data support this interpretation. For example, Sikes' (1994) isotopic analysis of the paleosol suggests a riparian woodland or grassy woodland at FLK Zinj. Pollen (Bonnefille & Riollet, 1980) material also suggest a woodland or gallery forest near the site. The fauna exploited by hominids at FLK Zinj fauna gives rather mixed signals (Gentry & Gentry, 1978a,b; Potts, 1988), but indicates exploitation of a variety of habitats. The presence of numerous alcelaphines and suids suggest a particularly strong exploitation of more open habitats. Nevertheless, Plummer & Bishop's (1994) taxon-free analysis, while based on all levels at the FLK locality, suggests more closed conditions at FLK. Taken as a whole the faunal, isotopic and pollen records suggest that early *Homo* occupied a grassy woodland as FLK Zinj while utilizing faunal resources from a variety of habitats.

Implicit in this model is that early *Homo* was exploiting resources across ecotonal boundaries where limited arboreal habits (e.g., sleeping?), lower predator densities, and refuge areas provided by semi-open woodlands to woodlands defined a core habitat occupied by all group members, but favoured at certain times of the year by those at particular risk to predation, e.g., altricial young and their care-takers. More open habitats apparently provided other resources, that because of increased risks and limited anti-predator defensive mechanisms, were transported to refuge areas.

Summary

Data presented here demonstrate that early *Homo* was responsible for the accumulation and modification of at least 75% of the FLK *Zinjanthropus* fossil limb assemblage (based on MNE estimates). This suggests early *Homo* habitually transported carcass-parts to places like FLK Zinj for processing and consumption. Carnivore damage on approximately 54% of the MNE substantiates the conclusion of Binford (1981), Potts (1982, 1988) and Blumenschine & Marean (1993) that carnivores played a substantial role in the modification of the assemblage. However, the low frequency of specimens exhibiting evidence of carnivore-induced fracture suggests that small carnivores may have been an important scavenging agent.

The nature of carcass-part acquisition, sequence of hominid *vs.* carnivore access to carcasses, and the character of the demonstrable overlap of carnivore and hominid activity at FLK Zinj can also now be outlined. Specifically, the lack of carnivore damage on 60% of identified elements with hominid-induced damages, as well as the even distribution of carnivore damage on bovid limb elements, indicates that hominids had access to meaty carcasses. Hammerstone damage frequencies demonstrate that early *Homo* intensively processed these parts for marrow and cancellous tissue. An abundant micro-fauna and the lack of significant evidence of carnivore-induced fracture contrasted with the lack of meaty epiphyseal ends suggests that both large and small carnivores then scavenged the hominid food refuse.

Definition of the ecological and biological forcing mechanisms that apparently regulate food transport and processing in social carnivores permits a preliminary assessment of the socioecological implications of these carnivore behaviors for early *Homo*. Anti-predator defense is regarded as a primary factor in the evolution of sociality in the open-habitat carnivora (Gittleman, 1989). Thus, the transport of carcass-parts away from the kill to refuge areas, e.g., the den, represents a series of inter-related, anti-predator strategies. Intense interspecific competition at carcasses causes most carnivores to bolt and/or remove food to a refuge area, while anti-predator defense is a major function of dens for vulnerable altricial young who raise the mother's energy and nutritional requirements thereby potentially elevating her vulnerability to predation.

Early *Homo* occupied grassy woodlands (e.g., Potts, 1988; Sikes, 1994) near open to semi-open habitats (e.g., Gentry & Gentry, 1978*a,b*; Potts, 1988) with high predator densities (Potts, 1988) and likely had quite altricial infants (Stanley, 1992). Thus, the transport and processing of food at places like FLK Zinj suggest that early *Homo* may have been subject to the same ecological and biological forcing mechanisms as some extant social carnivores. If so, early *Homo* may have employed a similar dual-unit foraging strategy. Certain group members (adults and non-caregivers) may have foraged together and brought animal food back to refuge areas in order to avoid predators. Mothers of altricial infants and other care-givers are likely to have foraged in closer proximity to a central-place due to the infants' vulnerability to predation and the increased energy demands placed on the mother. Processed animal remains at places like FLK Zinj may not represent provisioning or food-sharing *per se*, but the intersection of two anti-predator risk-reduction strategies. Once initiated, however, the positive feedback created by nutrient-rich feeding of altricial young and their caretakers may have set transport, provisioning, and perhaps the sharing of food in the hominid behavioral repertoire.

Sites like FLK *Zinjanthropus* probably do not represent campsites as we understand them based on modern hunter-gatherer behavior, but many behavioral pieces, as well as important ecological and biological conditions, for many of the social behaviors envisioned by (Isaac 1978a) were apparently in place by the Plio-Pleistocene.

Acknowledgements

This research was conducted with funding from the National Science Foundation, the Wenner-Gren Foundation for Anthropological Research, and the University of Wisconsin Graduate School. Permission to study the Olduvai material was provided by the Tanzanian Antiquities Department. I thank Meave Leakey, the Paleontology Section and the National Museum of Kenya for providing laboratory space. I thank H. T. Bunn for providing the FLK *Zinjanthropus* taxa and element identification data and insights into the assemblage. Administrative support and office space used during preparation of this paper at the Illinois State Museum were kindly provided by R. B. McMillan, Director, B. W. Styles, Director of the Sciences, and M. Wiant, Chairperson of the Anthropology Section. This paper has benefited from discussions with numerous individuals particularly H. T. Bunn, R. B. Potts, R. J. Blumenschine, B. W. Styles and R. W. Graham. I thank P. Shipman, two anonymous reviewers and my co-editors whose many substantive comments and criticisms helped clarify large parts of this paper. All errors and faulty logic are mine alone.

References

Altmann, J. (1980). *Baboon Mothers and Infants*. Cambridge: Harvard University Press.

Andrews, P. (1983). Small mammal faunal diversity of Olduvai Gorge, Tanzania. In (J. Clutton-Brock & C. Grigson, Eds) *Animals and Archaeology: Hunters and their Prey*, pp. 77–85. BAR International Series, 163.

Behrensmeyer, A. K. (1993). The bones of Amboseli. *National Geographic Research & Exploration* **9**, 402–421.

Behrensmeyer, A. K., Gordon, K. D. & Yanagi, G. T. (1986). Trampling as a cause of bone surface damage and pseudo-cutmarks. *Nature* **319**, 768–771.

Behrensmeyer, A. K., Gordon, K. D. & Yanagi, G. T. (1989). Nonhuman bone modification in Miocene fossils from Pakistan. In (R. Bonnichsen & M. Sorg, Eds) *Bone Modification*, pp. 99–120. Orono, Maine: Center for the Study of the First Americans.

Binford, L. R. (1981). *Bones: Ancient Men and Modern Myths*. New York: Academic Press.

Binford, L. R. (1988). Fact and fiction about the *Zinjanthropus* floor: data, arguments and interpretations. *Curr. Anthrop.* **29**, 123–135.

Blumenschine, R. J. (1986). Early hominid scavenging opportunities: implications of carcass availability in the Serengeti and Ngorongoro Ecosystems. *BAR International Series* No. **283**.

Blumenschine, R. J. (1987). Characteristics of an early hominid scavenging niche. *Curr. Anthrop.* **28**, 383–407.

Blumenschine, R. J. & Marean, C. W. (1993). A carnivore's view of archaeological bone assemblages. In (J. Hudson, Ed.) *From Bones to Behavior: Ethnoarchaeological and Experimental Contributions to the Interpretation of Faunal Remains*, pp. 273–300. CAI Occasional Paper No. 21. Southern Illinois University at Carbondale.

Blumenschine, R. J. & Selvaggio, M. M. (1988). Percussion marks on bone surfaces as a new diagnostic of hominid behavior. *Nature* **333**, 763–765.

Bonnefille, R. & Riollet, G. (1980). Palynologie, végétation, et climats de Bed I et Bed II à Olduvai, Tanzanie. In (R. E. Leakey & B. A. Ogot, Eds) *Proc. 8th Pan-Afr. Congr. Prehist. Quat. Studies*, pp. 123–127. Nairobi: The International Louis Leakey Memorial Institute for African Prehistory.

Bonnichsen, R. (1979). Pleistocene bone technology in the Berengian Refugium. *Archaeol. Survey Canada Paper* No. **89**. National Museum of Canada, Ottawa.

Bunn, H. T. (1981). Archaeological evidence for meat-eating by Plio-Pleistocene hominids from Koobi Fora and Olduvai Gorge. *Nature* **291**, 574–577.

Bunn, H. T. (1982). *Meat-eating and Human Evolution: Studies on the Diet and Subsistence Patterns of Plio-Pleistocene Hominids in East Africa*. Ph.D. dissertation. Department of Anthropology, University of California, Berkeley.

Bunn, H. T. (1983). Evidence on diet and subsistence patterns of Plio-Pleistocene hominids at Koobi Fora, Kenya, and Olduvai Gorge, Tanzania. In (J. Clutton-Brock & C. Grigson, Eds) *Animals and Archaeology: Hunters and their Prey*, pp. 21–30. BAR International Series, 163.

Bunn, H. T. (1986). Patterns of skeletal representation and hominid subsistence activities at Olduvai Gorge, Tanzania, and Koobi Fora, Kenya. *J. hum. Evol.* **15,** 673–690.

Bunn, H. T. (1989). Diagnosing Plio-Pleistocene hominid activity with bone fracture evidence. In (R. Bonnichsen & M. Sorg, Eds) *Bone Modification*, pp. 299–316. Orono, Maine: Center for the Study of the First Americans.

Bunn, H. T. (1991). A taphonomic perspective on the archaeology of human origins. *Ann. Rev. Anthrop.* **20,** 433–467.

Bunn, H. T., Harris, J. W. K., Isaac, G., Kaufulu, Z., Kroll, E., Schick, K., Toth, N. & Behrensmeyer, A. K. (1980). FxJj50: an early Pleistocene site in Northern Kenya. *World Archaeol.* **12,** 109–136.

Bunn, H. T. & Kroll, E. M. (1986). Systematic butchery by Plio/Pleistocene hominids at Olduvai Gorge, Tanzania. *Curr. Anthrop.* **27,** 431–452.

Cavallo, J. A. & Blumenschine, R. J. (1989). Tree-stored leopard kills: expanding the hominid scavenging niche. *J. hum. Evol.* **18,** 393–399.

Clutton-Brock, T. H. (1991). *The Evolution of Parental Care.* Princeton, N.J.: Princeton University Press.

Eisenberg, J. F. (1981). *The Mammalian Radiations: An Analysis of Trends in Evolution, Adaptation and Behavior.* Chicago: University of Chicago Press.

Ewer, R. F. (1973). *The Carnivores.* Ithaca: Cornell University Press.

Fiorillo, A. R. (1989). An experimental study of trampling: implications for the fossil record. In (R. Bonnichsen & M. Sorg, Eds) *Bone Modification*, pp. 61–71. Orono, Maine: Center for the Study of the First Americans.

Foley, R. A. (1987). *Another Unique Species: Patterns in Human Evolutionary Ecology.* New York: Longman Scientific and Technical.

Foley, R. A. (1989). The evolution of hominid social behavior. In (V. Standen & R. A. Foley, Eds) *Comparative Socioecology: The Behavioral Ecology of Humans and of Other Animals*, pp. 473–494. Oxford: Blackwell Scientific Publications.

Frame, L. H., Malcolm, J. R., Frame, G. W. & van Lawick, H. (1979). Social organization of African wild dogs (*Lycaon pictus*) on the Serengeti Plains. *Z. Tierspychol.* **50,** 225–249.

Gentry, A. W. & Gentry, A. (1978*a*). Fossil Bovidae (Mammalia) of Olduvai Gorge, Tanzania. Part I. *Bull. Brit. Mus. (Nat. Hist.) Geol.* Series **29,** 289–446.

Gentry, A. W. & Gentry, A. (1978*b*). Fossil Bovidae (Mammalia) of Olduvai Gorge, Tanzania. Part II. *Bull. Brit. Mus. (Nat. Hist.) Geol.* Series **30,** 1–83.

Gittleman, J. L. (1989). Carnivore groups living: comparative trends. In (J. L. Gittleman, Ed.) *Carnivore Behavior, Ecology and Evolution*, pp. 183–207. Ithaca: Cornell University Press.

Hay, R. L. (1976*a*). *Geology of the Olduvai Gorge.* Berkeley: University of California Press.

Hay, R. L. (1976*b*). Environmental setting of hominid activities in Bed I: Olduvai Gorge. In (G. Ll. Isaac & E. R. McCown, Eds) *Human Origins: Louis Leakey and the East African Evidence*, pp. 208–225. Menlo Park, Ca.: W.A. Benjamin, Inc.

Haynes, G. (1980). Prey bones and predators: Potential ecologic information from analysis of bone sites. *Ossa* **7,** 75–97.

Haynes, G. (1981). Bone modifications and skeletal disturbances by natural agencies. Unpublished Ph.D. dissertation, Catholic University of America, Washington, D.C.

Haynes, G. (1982). Utilization and skeletal disturbances of North American prey carcasses. *Arctic* **35,** 266–281.

Haynes, G. (1983). Frequencies of spiral and green-bone fractures on ungulate limb bones in modern surface assemblages. *Am. Antiquity* **48,** 102–114.

Hill, A. (1975). Taphonomy of contemporary and Late Cenozoic East African Vertebrates. Unpublished Ph.D. dissertation, University of London.

Hill, A. (1989). Bone modification by Spotted Hyaenas. In (R. Bonnichsen & M. Sorg, Eds) *Bone Modification*, pp. 169–178. Orono, Maine: Center for the Study of the First Americans.

Isaac, G. Ll. (1978*a*). The food sharing behavior of protohuman hominids. *Sci. Am.* **238,** 90–108.

Isaac, G. Ll. (1978*b*). Food-sharing and human evolution: archaeological evidence from the Plio-Pleistocene of East Africa. *J. Anthrop. Res.* **34,** 311–325.

Isaac, G. Ll. (1984). The archaeology of human origins: studies of the Lower Pleistocene in East Africa 1971–1981. In (F. Wendorf & A. Close, Eds) *Advances in World Archaeology*, pp. 1–87. New York: Academic Press.

Johnson, E. (1985). Current developments in bone technology. In (M. Schiffer, Ed.) *Advances in Archaeological Method and Theory*, Vol. 5, pp. 157–235. New York: Academic Press.

Kappleman, J. (1984). Plio-Pleistocene environments of Bed I and lower Bed II, Olduvai Gorge, Tanzania, *Paleogeog., Paleoclimat., Paleoecol.* **48,** 171–196.

Kingdon, J. (1977). *East African Mammals: An Atlas of Evolution in Africa. Volume IIIA, Carnivores.* Chicago: University of Chicago Press.

Kroll, E. & Isaac, G. Ll. (1984). Configurations of artifacts and bones at early Pleistocene sites in East Africa. In (H. Hietala, Ed.) *Intrasite Spatial Analysis in Archaeology*, pp. 4–31. Cambridge: Cambridge University Press.

Kruuk, H. (1972). *The Spotted Hyena: A Study of Predation and Social Behavior.* Chicago: University of Chicago Press.

Lamprecht, J. (1981). The function of social hunting in larger terrestrial carnivores. *Mammol. Rev.* **11,** 169–179.

Leakey, M. D. (1971). *Olduvai Gorge*, Volume 3. Cambridge: Cambridge University Press.

Lovejoy, C. O. (1981). The origin of man. *Science* **211,** 341–350.

Lyman, R. L. (1989). Taphonomy of cervids killed by the May 18, 1980, volcanic eruption of Mount St. Helens, Washington, U.S.A. In (R. Bonnichsen & M. H. Sorg, Eds), *Bone Modifications*, pp. 149–167. Orono, Maine: Center for the Study of the First Americans.

Marean, C. W. (1989). Sabertooth cats and their relevance for early hominid diet and evolution. *J. hum. Evol.* **18,** 559–582.

Martin, L. D. (1989). Fossil history of the terrestrial carnivora. In (J. L. Gittleman, Ed.) *Carnivore Behavior, Ecology and Evolution*, pp. 536–568. Ithaca: Cornell University Press.

Mech, L. D. (1970). *The Wolf: The Ecology and Behavior of an Endangered Species*. Minneapolis: University of Minnesota Press.

Moehlman, P. D. (1989). Intraspecific variation in canid social systems. In (J. L. Gittleman, Ed.) *Carnivore Behavior, Ecology and Evolution*, pp. 143–163. Ithaca: Cornell University Press.

Oliver, J. S. (1986). The taphonomy and paleoecology of Shield Trap Cave (24CB91), Carbon County, Montana, U.S.A. Unpublished MS. thesis. Institute for Quaternary Studies, University of Maine at Orono.

Oliver, J. S. (1989). Analogues and site context: bone damages from Shield Trap Cave (24CB91), Carbon County, Montana, U.S.A. In (R. Bonnichsen & M. Sorg, Eds) *Bone Modification*, pp. 73–98. Orono, Maine: Center for the Study of the First Americans.

Oliver, J. S. (1993). Carcass processing by the Hadza: bone breakage from butchery to consumption. In (J. Hudson, Ed.) *From Bones to Behavior: Ethnoarchaeological and Experimental Contributions to the Interpretation of Faunal Remains*, pp. 200–227. CAI, Occasional Paper No. 21. Southern Illinois University of Carbondale.

Oliver, J. S. (n.d.). Unpublished Ph.D. dissertation. Department of Anthropology, University of Wisconsin, Madison.

Olsen, S. L. & Shipman, P. (1988). Surface modification on bone: trampling versus butchery. *J. Archaeol. Sci.* **15,** 535–553.

Plummer, T. & Bishop, L. (1994). Hominid paleoecology at Olduvai Gorge, Tanzania as indicated by antelope remains. *J. hum. Evol.* **27,** 47–75.

Potts, R. B. (1982). *Lower Pleistocene Site Formation and Hominid Activities at Olduvai Gorge, Tanzania*. Ph.D. dissertation, Department of Anthropology, Harvard University.

Potts, R. B. (1988). *Early Hominid Activities at Olduvai*. New York: Aldine de Gruyter.

Potts, R. B. & Shipman, P. (1981). Cutmarks made by stone tools from Olduvai Gorge, Tanzania. *Nature* **291,** 577–588.

Rasa, A. (1984). *Mongoose Watch: A Family Observed*. London: John Murray.

Rood, J. P. (1986). Ecology and social evolution in the mongooses. In (D. I. Rubenstein & R. W. Wrangham, Eds) *Ecological Aspects of Social Evolution*, pp. 131–152. Princeton, N.J.: Princeton University Press.

Schaller, G. B. (1972). *The Serengeti Lion: A Study of Predator-prey Relations*. Chicago: University of Chicago Press.

Schaller, G. B. & Lowther, G. R. (1969). The relevance of carnivore behavior to the study of early hominids. *Southwest. J. Anthrop.* **25,** 307–341.

Semken, H. A. (1980). Holocene climatic reconstructions derived from the three micromammal bearing cultural horizons of the Cherokee Sewer Site, northwestern Iowa. In (D. C. Anderson & H. A. Semken, Eds) *The Cherokee Excavations: Holocene Ecology and Human Adaptation in Northwestern Iowa*, pp. 67–99. New York: Academic Press.

Shipman, P. (1981). Applications of scanning electron microscopy to taphonomic problems. In (A.-M. E. Cantwell, J. B. Griffin & N. A. Rothschild, Eds) *The Research Potential of Anthropological Museum Collections*, Volume 376, pp. 357–386. New York: New York Academy of Sciences.

Shipman, P. (1986a). Scavenging or hunting in early hominids: theoretical framework and tests. *Am. Anthrop.* **88,** 27–43.

Shipman, P. (1986b). Studies of hominid-faunal interactions of Olduvai Gorge. *J. hum. Evol.* **15,** 691–706.

Shipman, P. & Harris, J. M. (1988). Habitat preference and paleoecology of *Australopithecus boisei* in eastern Africa. In (F. Grine, Ed.) *Evolutionary History of the "Robust" Australopithecines*, pp. 343–381. New York: Aldine de Gruyter.

Shipman, P. & Rose, J. (1983). Early hominid hunting, butchering, and carcass processing: approaches to the fossil record. *J. Anthrop. Archaeol.* **2,** 57–98.

Shipman, P. & Walker, A. (1989). The costs of becoming a predator. *J. hum. Evol.* **18,** 373–392.

Sikes, N. (1994). Early hominid habitat preferences in East Africa: paleosol carbon isotope evidence. *J. hum. Evol.* **27,** 25–45.

Sikes, N. & Ambrose, S. (1993). Modeling hominid home range size using dental anatomy to estimate trophic level. Paper presented at 2nd Annual Paleoanthropology Meeting, Toronto.

Stanley, S. M. (1992). An ecological theory for the origin of *Homo*. *Paleobiol.* **18**(3), 237–257.

Susman, R. L. (1988). Hand of *Parantropus robustus* from Member I, Swartkrans: fossil evidence for tool behavior. *Science* **240,** 781–784.

Teleki, G. (1973). *The Predatory Behavior of Chimpanzees*. Lewisburg, Pa: Bucknell University Press.

Thorson, R. M. & Guthrie, D. (1984). River ice as a taphonomic agent: an alternative hypothesis for bone "artifacts". *Quat. Res.* **22,** 172–188.

van Lawick, H. & van Lawick, J. (1971). *Innocent Killers*. Boston: Houghton Mifflin.

Walker, R. C., Manega, P. C., Hay, R. L., Drake, R. E. & Curtis, G. H. (1991). Laser-fusion 40Ar/39Ar dating of Bed I, Olduvai Gorge, Tanzania. *Nature* **354,** 145–149.

Jeanne M. Sept
Anthropology Department, Indiana University, Bloomington, IN 47405, U.S.A.

Received 21 March 1994
Revision received 5 April 1994
and accepted 5 April 1994

Keywords: early hominid diet, plant food, early Stone Age, archaeology, foraging theory, actualistic studies.

Beyond bones: archaeological sites, early hominid subsistence, and the costs and benefits of exploiting wild plant foods in east African riverine landscapes

Plant remains are rare at early archaeological sites in Africa, and consequently the contribution of plant foods to the subsistence behaviors of early hominids has been difficult to investigate archaeologically. Research on the quality, abundance, and distribution of plant foods in modern African habitats can contribute to reconstructions of the paleoenvironmental context of early archaeological sites. Results of fieldwork near the upper Semliki River, Parc National des Virunga, eastern Zaire demonstrate some ecological parameters of primate plant food availability that would have constrained early hominids foraging for plant foods. Such actualistic studies provide a frame of reference from which to interpret variations in the composition of early assemblages of artifacts and fossil bones, and help focus inferences about the likely adaptive significance of plant foods in early hominid diets.

Journal of Human Evolution (1994) **27**, 295–320

Introduction

It is widely acknowledged that plant foods played an important role in the diet and subsistence practices of early hominids in Africa. Analogy to the feeding patterns of contemporary human foragers and non-human primates is one line of inference that suggests plants would have been significant foods for early hominids (e.g., Peters & O'Brien, 1981; Stahl, 1984; Sept, 1986), and fossils provide a record of relative qualities and quantities of plant food eaten by different early hominid species (cf. Walker, 1981; Gordon, 1987; Kay & Grine, 1988; Sillen, 1992). Yet, because plant food remains are so rare at Oldowan and early Acheulian archaeological sites, plant foods have become subsistence strategy after-thoughts in most archaeological reconstructions of early hominid behavior (Sept, 1992b).

The issue of plant food diet is typically treated as archaeologically uninvestigatable for Plio-Pleistocene sites. Archaeologists have argued that because little evidence of early hominid plant foods exists, we must focus our attention on what *is* preserved, the stones and bones (e.g., Potts, 1988:232–233). As other papers in this volume illustrate, recent advances in faunal analyses from early archaeological sites have demonstrated that hominid toolmakers fed from animal foods, and this has excited a resurgence of debate about the strategies which hominids might have used to acquire meat or marrow. While these taphonomic investigations of faunal assemblages are stimulating and productive (cf. Bunn, 1991), we must still remember that they rely on nested levels of "inferential confidence" (Gifford-Gonzalez, 1991) that range from identifying agents of bone modification, to recognizing signatures of butchery practice and subsequently inferring patterns of foraging behavior. In this epistemological context, archaeological visibility does not necessarily equate with adaptive significance (cf. Sept, 1992b). Asserting from a bone-biased archaeological record that meat or marrow were regular or essential foods in the dietary repertoire of early hominids verges on tautology. In other words, zooarchaeologists have clearly documented that hominids used stone tools to acquire large mammal meat and marrow, which are novel primate foods, but have yet to demonstrate that tool use and meat eating were any more frequent, regular, or important to the subsistence

0047–2484/94/010295+26 $08.00/0

strategies of Oldowan hominids than they are to the current subsistence strategies of some populations of chimpanzee, for example (Boesch & Boesch, 1989; McGrew, 1992; Wrangham & Riss, 1990). How critical were different types of foods, including plant foods, to the survival of the stone tool makers? How did diet and subsistence strategies vary in space and time, and how does the archaeological record sample such variability? Rather than allow the boldest patterns in the faunal record to dominate the debate about early hominid subsistence, archaeologists should broaden their search for evidence to investigate these more elusive paleoecological issues (cf. Foley, 1992).

How can the issue of plant foods in early hominid diet be addressed from the archaeological record, given the absence of any direct evidence for feeding on plants? One key to answering this paleoecological question lies in context-specific interpretations of what has been preserved. In addition to examining the taphonomically biased record of what hominids ate, we can study the locations where such feeding activities took place. The more we learn about the local habitats and paleolandscapes in which sets of artifacts and associated remains were buried, the more effectively we will be able to study the variability in these assemblages, and eventually model the types of subsistence choices that early hominids made. No primate is a pure carnivore, and even if meat and marrow emerge as critical foods in the dietary repertoire of various species of early hominid, we can best understand their adaptive value and evolutionary significance in relation to the likely availability of prehistoric plant foods and the costs and benefits of foraging for them.

Binford argues that archaeological materials must be compared in relation to "frames of reference" which would have conditioned prehistoric behavior (Binford, 1987). The plant resource base was integral to the paleoecological context of archaeological sites, and has the potential to be an important frame of reference for the interpretation of assemblage composition. For example, Speth (1987) suggested that if bone assemblages accumulated seasonally at some of the early sites, such patterns should be interpreted in terms of the relative availability of plant foods during those seasons. Many authors have also mentioned the distribution of plant foods, as well as shade trees or nesting trees, as important habitat variables which probably influenced the formation of stone tool assemblages (e.g., Kroll & Isaac, 1984; Binford, 1984; Schick, 1987; Sept, 1992a). To move from generalized suggestions such as these, to archaeologically useful criteria for inferring the plant food context of archaeological sites, we must seek to build explicit, relational analogies with living ecological systems (Gifford-Gonzalez, 1991). Such analogical reasoning is implicit in all interpretations of site context, and Gifford-Gonzalez reminds us (1991:234) that effective hypotheses about site context are based on forming sets of "relational analogies which circumscribe the variety and thus reduce the ambiguity of inferences" based on archaeologically recovered artifact and faunal assemblages.

In this essay I advocate an actualistic approach to improving the relational analogies we use to interpret the vegetation context of early archaeological sites, focusing on plant foods. This approach is founded on principles of evolutionary ecology (cf. Pianka, 1983; Krebs & Davies, 1984; Foley, 1987, 1992), and on a comparative knowledge of the diets of contemporary human foragers and non-human primates. Actualistic studies of plant foods have been done for a range of environments in eastern and southern Africa (Peters & Maguire, 1981; Peters, O'Brien & Box, 1984; Vincent, 1985a; Sept, 1984, 1986, 1990; Peters & Blumenschine, 1993). However, to build effective "frames of reference" for the interpretation of archaeological sites, actualistic plant food studies should explore how the distribution and abundance of plant foods vary in modern landscapes comparable to environmental situations where early archaeological sites were preserved. Here I discuss actualistic research on the abundance, distribution and quality of plant foods in riverine environments in east Africa as a guide to paleoecological reconstruction and the interpretation of archaeological assemblages as evidence of early hominid diet and subsistence.

Paleoecological analogs

Archaeological site locations and habitats

Many of the sites preserving archaeological evidence of Plio-Pleistocene hominid behavior are buried along the margins of perennial rivers and ephemeral streams in the lowland sedimentary basins of the eastern Rift Valley system (Isaac, 1984; Toth & Schick, 1986; Harris, 1983). In particular, interdisciplinary research efforts in the Turkana Basin, Kenya (Stern, 1993; Kaufulu, 1983; Kibunjia, Roche, Brown & Leakey, 1992; Howell, Haesarts & de Heinzelin, 1987; Feibel, 1988; Brown & Feibel, 1991) and at Olduvai Gorge, Tanzania (Hay, 1976; Potts, 1988), have refined our knowledge of the paleogeographical context, sedimentary conditions and taphonomic history of several sites. It should be possible in the near future to systematically analyse assemblages in relation to details of their paleogeographical placement and vegetation setting (Harris & Feibel, 1993) and compare channel floodplain sites to those that accumulated in different types of depositional environments, such as lake margins (Potts, 1988).

Reconstructions of local vegetation patterns in these channel margin and floodplain settings must integrate interpretations of fossil and geological evidence; these interpretations rely on comparisons with modern vegetation analogs, and on uniformitarian assumptions about the ecological conditions in which different types of plants will grow (Carr, 1976; Sept, 1986). It is generally accepted that the strong floristic continuities in East African vegetation during the last three million years justify such ecological assumptions (Axelrod & Raven, 1978; Bonnefille, 1984, 1985). Paleobotanical fossils provide data on the floral composition and taxonomic associations that would have characterized ancient plant communities, but because of taphonomic biases, pollen data are better suited to reconstructing patterns of regional vegetation than local plant groupings, particularly in fluviatile settings (Bonnefille, 1984; Sept, 1986). Inferences about site vegetation physiognomy can also be based on the analysis of paleosols and trace fossils, such as root casts (Retallack, 1991) which are then compared to modern analogues, as in the case of the Omo Shungura Formation.

> Many situations in the lower Omo valley today are comparable to those archaeological occurrences related to a past meandering river environment ... at the distal edge of fluviatile levees, behind gallery (or riverine) forests, and the edge of open savanna at topographic lows on an occasionally inundated floodplain (Howell *et al.*, 1987:685).

Isotopic studies of paleosols are also emerging as a sensitive indicator of ancient patterns of C-3 and C-4 plant dominance (Ambrose & Sikes, 1991; Cerling, 1992; Sikes, 1994). This research has the exciting potential to document local ecotone boundaries and vegetation gradients between mesic and xeric plant communities in riverine floodbasins.

However, discovering that a site was probably buried in a vegetation zone such as "gallery forest" is only a first step towards paleoecological reconstruction. We still must ask what the "critical environmental variables" (Pilbeam, 1984:20) were in such habitats that may have influenced the behavior of the hominids who littered the landscape there. Actualistic studies have begun to improve our knowledge of how riverine habitats in East Africa would have varied in terms of predation risk and refuge availability (Blumenschine, 1986, 1992; Sept,

1992*a*), for example. There is also a range of opinion based on actualistic research about the quality and quantity of scavenging opportunities that would have been likely along the margins of Plio-Pleistocene rivers (e.g., compare Tappen, 1992; Blumenschine, 1986, 1992; Cavallo & Blumenschine, 1989; Sept, 1994). Similarly, as described below, a "consumer's eye view" investigation of vegetation in modern riverine environments demonstrates how plant food feeding opportunities can vary in such habitats; the challenge is to use such knowledge to improve our paleoenvironmental reconstructions of archaeological sites.

East African rivers and vegetation

The diverse riverine habitats in eastern Africa can be compared in terms of some basic, inter-related attributes that influence the patterns of vegetation and the distribution of plant food resources: climate; topographic slope and placement in the drainage order or catchment area; substrate soil chemistry; sedimentary regime; and channel capacity and strength of flow (Haslam, 1978; Sept, 1984). Because water availability is one of the major constraints on plant growth in arid and semi-arid sedimentary basins in East Africa, channel margins often support a lusher plant growth than surrounding terrain where plant growth is more dependent upon the seasonal relationship between rainfall and temperature (White, 1983; Lind & Morrison, 1974). The width of distinctive belts of riverine vegetation in dry East African floodbasins is a function of access to channel groundwater, as determined by distance from the channel, topographic position above the water table, and the volume and frequency of water flow along the channel drainage. The extent to which mesic environments are likely to support a classic ribbon of gallery forest, rather than a string of shrubs or pastures of grass, depends upon edaphic factors. For example, the deep root systems of many trees and shrubs can outcompete grasses in well-drained soils, producing arboreal galleries along the stable margins of sandy drainage channels, while silty floodplain soils support perennial grasses and forbs (Sept, 1984).

Fluvial sedimentary environments are geomorphologically dynamic, and this also shapes the structure and composition of a riverine plant community. During phases of channel incision floodplains can be relatively stable plant environments, characterized by soil development and community diversification. But during channel migration and the overbank sedimentation which buries archaeological sites, stream margins and floodplains become colonies of ephemeral or ruderal plant species, unstable communities with often patchy distributions and a relatively low species diversity (Sept, 1984). For example, a number of structural features of the Omo River forest, in Ethiopia, seem to correlate with position along a successional gradient (Carr, 1976). The dynamic properties of these riverine settings have implications both for the distribution of plant foods in riverine vegetation, and for the approach that paleoecologists must take towards trying to sample and describe the vegetation of such environments in the past.

Plant foods

Research history: comparative primate analogs. The diets and feeding behaviors of non-human primates and human hunter-gatherers form a useful base from which to begin thinking about the types of plant foods that might have been eaten by early hominids, "omnivorous primates" (Harding & Teleki, 1981). Prior to the 1980s, few long-term field studies had been done that focused on the details of human or non-human primate plant food diet in Africa, so analogies to early hominid plant food diet were limited to generalized comparisons. Studies of open-country primate subsistence, for example, led to assertions that the plant foods available to hominids in savanna settings would have been restricted to seeds or tough fruits (e.g. Suzuki, 1969; Dunbar, 1976; McGrew et al., 1981).

In the first systematic attempt to define a modern analog for early hominid plant food consumers, Peters and O'Brien (1981) defined an "early hominid plant food niche" of wild plant genera and edible parts eaten by baboons, chimpanzees and humans in eastern and southern Africa that were likely to have been eaten by early hominids as well. This survey concluded that early hominid habitats would have contained a broad variety of edible plant foods, and that young leaves and fleshy fruits would have been preferred foods, with fruits having the most potential competition from other animals. In another influential synthesis Stahl (1984) formally compared classes of different types of wild plant foods eaten by primates and humans in terms of their macronutrient balance, secondary compound content, and other factors inhibiting their digestibility. In addition to focusing on how food processing technology such as fire could have significantly expanded the diet breadth of prehistoric hominids, she stressed the constraints that relative digestibility and handling costs place on primate plant food selection. She ranked the potential "edibility" of different plant parts on this basis, as a heuristic device. Fleshy fruits, flowers and young leaves and shoots ranked the highest, followed by dicot seeds, monocot seeds, and various bulbs, corms and tubers. Leaves, stems and roots high in structural carbohydrates would have been the lowest ranking foods according to this scheme.

During the last decade several long-term field projects have begun to provide a wealth of new detail about the diet and foraging strategies of "omnivorous primates" (Harding & Teleki, 1981) in Africa. Although a review of this rapidly growing body of knowledge is outside the scope of this essay, recent dietary syntheses from the three groups frequently compared to early hominids include: baboons (e.g., Altmann et al., 1987; Whiten et al., 1992; Barton et al., 1992; Norton et al., 1987); African apes (e.g., Badrian & Malenky, 1984; Collins & McGrew, 1988; Boesch & Boesch, 1989; Tutin et al., 1992; Rogers et al., 1992); and human foragers (e.g., Vincent, 1985a, 1985b; Hawkes et al., 1991; Dietz et al., 1989; Blurton Jones, 1993; Bailey, 1993).

Analogs for interpreting hominid fossil evidence for plant food diet. The arena of edibility is difficult to define for early hominids, species whose tooth and jaw morphology suggest that they ate very different mixes of foods than any living primate (e.g., Walker, 1981; Gordon, 1987; Peters & O'Brien, 1981). While the fossilized teeth and jaws of hominid individuals can be studied to compare the likely sizes and textures of foods which hominids bit and chewed (e.g., Walker, 1981; Kay & Grine, 1988; Ryan & Johanson, 1989), such morphological and toothwear studies reveal little about the dietary balance, or the different types of soft foods consumed by hominids (Gordon, 1987). As analytical techniques improve, the trace element and isotopic composition of fossil bones may potentially provide evidence of the plant/animal mix in early hominid diets (Lee-Thorp et al., 1989; Sillen, 1992). For example, recent bone chemistry analyses suggest that the "megadont" robust australopithecines from South Africa might have been more omnivorous than their tooth morphology and wear would indicate (Sillen, 1992; Sillen & Lee-Thorp, 1993).

Arguments about the balance of plant to animal foods in early hominid diets have also been based on more indirect inferences from the fossil evidence. For example, several authors have used fossil evidence to argue that the metabolic requirements of larger-bodied, larger-brained, precocial early *Homo* would have placed strong selection pressures on *Homo* to acquire "higher quality" (nutrient value/cost) foods than other hominoids, including species of *Australopithecus* or *Paranthropus* (Martin, 1983; Leonard & Robertson, 1992; Foley, 1992). It is often assumed that meat was this key "high quality" food that fueled the divergence of the *Homo* lineage, but the likely animal/plant balance and tool-use necessary to achieve such a "high quality" strategy can be debated. Similarly, while the physique of early *Homo erectus* was adapted to the thermoregulatory stresses of diurnal activity and travel in open environments (Wheeler, 1992, 1993; Walker, 1993), it cannot be determined whether this was an adaptation to hunting, scavenging, searching for widely dispersed, energy-rich plant foods (cf. Rose, 1984), or some other set of socio-ecological factors. So, in some ways we can use the fossil evidence to help define the physiological parameters that would have constrained early hominid dietary strategies, but interpreting fossil evidence in terms of hominid subsistence is dependent upon dietary analogies with living primates.

Consider fruit and leaves, two common primate foods, for example. While early australopithecines and species of *Homo* retained the primitive, broad incisors typical of frugivorous anthropoids (Kay, 1984), microwear studies suggests several hominid species had unique biting and chewing patterns (Ryan & Johanson, 1989; Kay & Grine, 1988). Still, it seems likely that all early hominid species would have eaten high quality fruits and soft seeds when they were available, even as sympatric chimpanzees and lowland gorillas do today (Tutin *et al.*, 1992). However, no early hominid had molars comparable to those of the surviving African apes, whose cusp morphologies are arguably adapted to shredding pith and herbaceous foods as well as chewing soft fruits and seeds (Wrangham *et al.*, 1992; Tutin *et al.*, 1992). It thus seems unlikely that hominid subsistence strategies were focused on the exploitation of herbaceous vegetation, though these large bodied omnivores may occasionally have eaten tender shoots, as many primates do.

Arguments that megadont australopithecines fed on "hard food objects" (Kay & Grine, 1988) raises the question of whether early hominid species could have exploited ripe seeds from legume species such as acacias. While primates such as baboons typically feed on the seeds of these species when they are still green and soft, many ripe legumes can be difficult to digest raw (Stahl, 1984). Other seeds and nuts from woody species pose fewer chemical problems, and more mechanical ones (Peters, 1987).

Or perhaps the ambiguous wear on early hominid teeth was tuber-wear (Hatley & Kappelman, 1980; Vincent, 1985*a*, 1985*b*)? Brain has argued that bones recovered from Swartkrans in South Africa had been used as digging tools by early hominids (Brain, 1988). Tubers are a preferred food of many human foragers in Africa, despite the labor costs involved in excavating them (Vincent, 1985*a*). Among the Hadza in Tanzania, for example, individuals with a wide range of skills and expertise can efficiently harvest several species of tubers with simple digging sticks (cf. Vincent, 1985*b*; Blurton Jones, 1993). While baboons and other primates commonly feed on shallow bulbs, corms and rhyzomes (e.g., Whiten *et al.*, 1992), they invest in tuber digging much less frequently than is typical of human foragers in eastern or southern Africa. The only report of which I am aware of apes digging deeply for tubers comes from eastern Zaire where chimpanzees in the Tongo forest excavated moist roots during a drought (Lanjouw, 1991; pers. comm.). Foraging for deeply buried tubers would have significantly increased the dietary repertoire of early hominids compared to living primates (Hatley & Kappelman, 1980), and such a dietary shift might be reflected in dental wear patterns. Eating more underground foods could have *increased* the grit in early hominid diet, leading to the high frequencies of microwear pitting that have been observed on the molars of early *Homo* (Walker, 1981). Or, alternatively, because some tubers have a larger volume to surface ratio than other roots, tuber-eating early hominids might have ingested *less* dietary grit than is commonly ingested by savanna baboons today who depend upon shallow bulbs, corms and rhyzomes, and this could have contributed to the unique patterns of early hominid dental wear.

Nutritional and economic constraints of a plant food diet. Because open country primates such as baboons (e.g., Whiten *et al.*, 1992) and some chimpanzees (e.g., Suzuki, 1969; Collins & McGrew, 1988; McGrew *et al.*, 1981; McGrew, 1992) eat such a diverse range of plant foods today, most authors assume that early hominids would have been just as opportunistic (Peters & O'Brien, 1981). Being eclectic, selective feeders from a diverse array of plant parts and different species might have been the key to surviving in the open habitats where hominids left their "stone age visiting cards" (Isaac, 1981). But which plant foods would early hominids have selected, and why?

Comparing the nutritional qualities of different types of plant foods in relation to the costs of foraging for them is one way of assessing which plant foods early hominids might have preferred. Within the field of behavioral ecology a number of empirical and theoretical studies have focused on the economic strategies animals use when foraging for foods (Stephens & Krebs, 1986). It is clear from this research, for example, that primary consumers such as primates face a more complex set of nutritional benefit/cost constraints than are normally accounted for in simple prey selection models of foraging theory. In addition to meeting their minimal caloric requirements, primates and human foragers dependent upon plant foods must consume the appropriate mix of amino acids and other nutrients from their highly variable plant food diet (e.g., Post, 1984; Winterhalder, 1987; Kaplan & Hill, 1992). Also, the relative quality of a plant food is often best defined in terms of its "packaging," rather than in terms of the absolute value of nutrients it contains, because the potential nutrients of many plant foods are sequestered inside indigestible bulk, or protected by hard shells or a screen of deleterious secondary compounds (Milton, 1984; Stahl, 1984; Waterman, 1984).

For example, amino acids are widely distributed in plant foods, whether in seeds, shoots or young foliage, and animals must eat an appropriate balance of these amino acids to fulfil their protein requirements for growth and development (Maynard *et al.*, 1979). As primary consumers, most primates avoid the potential problem of an incomplete amino acid balance in their plant foods by diversifying, either by switching between different plant parts and species, or by sampling insects, eggs or small mammals. Protein-rich plant foods can often be fibrous as well, and data suggest that baboons may select and manipulate plant food items and their diet breadth to optimize their protein/fiber ratio (Whiten *et al.*, 1992). While modern humans have lost the typical primate hind-gut capacity to ferment structural carbohydrates (Milton, 1987) this is probably a relatively recent evolutionary change, and an early hominid diet that included a mix of raw, wild plant foods was necessarily higher in fiber than the processed diets of anatomically modern humans (Stahl, 1984). From these types of observations we can conclude that if plant foods were used to meet the protein requirements of early hominids, then bulk may have been an unavoidable aspect of any early hominid diet, particularly during times of food scarcity. In this context it is significant that while australopithecines had the masticatory apparatus to chew quantities of non-siliceous plant foods, the reduction in cheek tooth surface area seen in species of early *Homo* suggests that selection pressures to mash bulky vegetable foods were significantly reduced in this lineage by the early Pleistocene (Walker, 1981).

While lipid-rich foods might have been preferred by hominids for their potential energy value, requirements for specific dietary lipids are negligible for primates and hypothetical hominids. For example, the human requirement for Essential Fatty Acids can be easily fulfilled with only 1–2% of daily calories—as little as 10–15 g of oil (National Academy of Sciences, 1980; Maynard *et al.*, 1979). In an African savanna environment this can be achieved by eating several hundred grams of various seeds or oily fruits such as *Cordia sinensis* berries, or even by eating half a kilo of some species of tuber (Vincent, 1985*b*). As Peters has emphasized, several woody species bear fruits with nut-like seeds that are excellent sources of oils and widely available in woodlands in the Sudano-Zambezian phytogeographic zones of central, eastern and southern Africa (Peters, 1987). In a recent article focusing on the importance of animal foods to early hominids, Bunn and Ezzo leave a mistaken impression that dietary requirements for lipids would have forced early hominids to acquire significant quantities of animal foods (Bunn & Ezzo, 1993). Their error stems from an apparent misunderstanding of the meaning of Recommended Daily Allowance statistical guidelines for dietary fat intake (National Academy of Sciences, 1980) which they interpret as minimal daily *requirements*; their claim that hominids would have required "50–70 g of fat per day" (Bunn & Ezzo, 1993:372) is nonsense and consequently exaggerates the necessary importance of animal fats to early hominid diet.[1] While early hominids very likely sought lipid-rich plant foods or animal foods for their energy value (e.g. Blumenschine & Madrigal, 1993), it is unlikely that they needed fat, in particular, in such massive quantities.

Early hominid omnivores were probably dependent upon plant foods as sources of energy (e.g., Peters & O'Brien, 1981; Stahl, 1984; Milton, 1987; O'Brien & Peters, 1991). This may be particularly true for the relatively larger-brained species with high metabolic requirements and prolonged infant dependency (Martin, 1983; Leonard & Robertson, 1992; Foley, 1992; Stini, 1988). Because protein is a metabolically costly source of energy (Speth, 1987), the soluble carbohydrate and lipid content of fleshy fruits, nuts and some seeds makes them preferred sources of energy for many primates (Milton, 1984; Leighton, 1993), and presumably early hominids as well. While behavioral ecologists favor the use of multivariate optimization or satisficing models to realistically portray the complex nutritional choices that face primary consumers (Stephens & Krebs, 1986), analysing the caloric value of plant foods compared to their handling costs can also be a useful primary model for predicting which plant foods early hominids might have preferred. In doing so, it is important to try to estimate the dietary caloric values of only the digestible constituents of plant foods, and not include the gross energy values of cellulose and hemi-cellulose included in the "crude fiber" calculations of proximate nutritional analyses (e.g., Milton, 1987).

A number of empirical studies have demonstrated that the costs of handling plant foods are an important factor in whether a non-human primate or human forager selects that plant to eat (e.g., Milton, 1984; Kaplan & Hill, 1992; Blurton Jones, 1993; Hawkes, 1992, 1993). The net caloric value of plant foods varies significantly between parts of plants, and plant species, because of the manipulative efforts needed to feed on small package items, or the time necessary to ingest bulky foods. For example, small, ripe fruits tend to have relatively low energy benefit/cost rankings because they are often fibrous, with high skin to pulp ratios, contain large seeds, and have higher relative handling times for primates compared to medium or large-sized fruits (Sept, 1984). All else being equal, the simple diet breadth models of foraging theory (Stephens & Krebs, 1986) predict that animals should prefer to eat foods that have a higher benefit to cost ratio. In this context a growing body of anthropological research suggests that ranking foods by energetic benefit/cost ratios can help analyse the subsistence

---

[1]See (Stahl, 1984) pp. 164–165 for a general discussion of common pitfalls in the use of RDA statistics as nutritional analogs for early hominids.

choices that face non-human primates, human foragers, and prehistoric hominids (e.g., Sept, 1984, 1986; Winterhalder, 1987; Peters, 1987; Speth, 1989; Whiten *et al.*, 1992; Blumenschine, 1992; Leighton, 1993; Blurton Jones, 1993; Hawkes *et al.*, 1991; Bailey, 1993; Kaplan & Hill, 1992). Several studies have also demonstrated the important role technology can play in reducing the handling costs of otherwise difficult-to-access foods, both for humans and non-human primates (e.g., Vincent, 1985*a*; Blurton Jones, 1993; Milton, 1987; McGrew, 1992; Hunt *et al.*, n.d.). Nuts, for example, which are unexploited by populations of eastern chimpanzee, are preferred foods of western chimpanzee populations who have learned to crack the nuts open with tools (cf. McGrew, 1992). In this context it is likely that early hominids who used stone hammers to flake stone could also have efficiently exploited many of the nut-bearing species common across the continent (Peters, 1987). Similarly, even simple digging tools would have made exploiting deeply buried tubers profitable for early hominids (Vincent, 1985*a*; Sept, 1984, 1986; Brain, 1988).

Plant food distribution. The density and spacing of different types of food, and the productivity of any single feeding patch, also influence the choices animals make when foraging (Pianka, 1983; Krebs & Davies, 1984; Stephens & Krebs, 1986). In fact, a number of authors have argued that there is a strong causal relationship between the relative quality and spatio-temporal distribution of plant foods and different aspects of primate behavior, particularly their territoriality and the structure of their social groupings (e.g., Wrangham & Smuts, 1980; Wrangham, 1986; Rodman, 1984). Some studies are beginning to elucidate the complex relationships that can exist between age, sex and status differences both within and between populations of non-human primates and human foragers in relation to foraging economics and the distribution of foods (e.g., Hawkes, 1992; Blurton Jones, 1993; Hunt *et al.*, n.d.). The "patchiness" of food distribution is therefore seen as a key ecological variable that would have influenced early hominid foraging behavior (Foley, 1987). However, defining and measuring the patchiness of feeding opportunities for living primates or human hunter-gatherers has been problematic (Cashdan, 1992). Even for animals as large and wide-ranging as chimpanzees, definitions of food patch range from individual fruiting trees to contiguous zones of vegetation.

Given the potential importance of the distribution patterns of different types of foods, what approaches can be taken to inferring the distribution of plant foods in early hominid habitats? Archaeologists can start by trying to define the context of their site in terms of what White (1983:27) called landscape terrain types, "a framework of land classes based on physiography, within which the different soil and vegetation units are subordinated." We already describe the context of archaeological sites in terms of small, definable sedimentological lenses, such as "sandy levee" or "proximal floodplain" deposit, and an earlier section described how researchers typically use different types of evidence to associate such geological units with generalized vegetation types, such as "gallery forest." The next step is to develop sedimentary-unit-specific hypotheses about the patterns of plant food availability and quality that would likely have occurred on these samples of paleolandscape.

There are several reasons why these site-specific plant resource hypotheses must be probabalistic: (1) the distributions of plant foods vary between habitats of similar vegetation types today (Sept, 1986); (2) plant groupings and vegetation structure can change rapidly in the unstable sedimentary environments in which archaeological sites are typically buried (cf. Carr, 1976; Western & Van Praet, 1973); (3) much of the earliest archaeological record is time-averaged (cf. Stern, 1993). Plant resource models can try to take into account the hierarchical nature of both the temporal and the spatial patterns of landscapes (Urban *et al.*, 1987). To use the terminology of landscape ecology, they are most likely to be able to resolve probable vegetation patterns at intermediate scales of space-time domains, between decades and hundreds of years, and between tens of meters and hectares (Urban *et al.*, 1987:120). To be of practical value as "frames of reference" for archaeological interpretation, hypotheses about the vegetational context of sites should aim to answer the following types of questions. Are the boundaries of sedimentary zones likely to correspond to edges or gradients in plant food distribution? If a gradient of C-3/C-4 vegetation can be defined for a site paleosol, would expected return rates of foraging for different types of plant foods be associated with this gradient? Can the richness of preferred plant foods in a site's immediate surroundings be predicted in terms of the relative variety of plant food types and their likely seasonal rhythm, compared to the likely average plant food diversity of the region as a whole? How abundant were the most profitable plant foods likely to have been in different vegetation zones? The remainder of this paper focuses on how actualistic studies of plant density and seasonal productivity in riverine habitats today can be used to address some of these questions.

Actualistic research on plant foods: Semliki River case study

I began a program of fieldwork in the 1980s, focused on studying the properties of plant foods that grow in modern East African environments comparable to river floodbasin contexts where early archaeological sites were preserved. The study sites in a series of national parks in northern and southern Kenya (Sept, 1984, 1986) and eastern Zaire (Sept, 1990, 1992*a*, 1994) were chosen because they contained relatively protected floral and faunal communities in a range of environments analogous to situations in which early sites were preserved, in terms of local climate, topography, soil chemistry and sedimentary context. These studies have sampled rivers in a range of arid and semi-arid eco-climatic zones (Pratt & Gwynne, 1977) with bimodal rainfall regimes of different seasonal amplitude.

Here new data are reported on trees and shrubs that bear edible fruit from the Semliki River region; these data augment previously published material (Sept, 1990), and provide evidence of the abundance and distribution of plant foods in savanna riverine environments (Sept, 1986). However, I present aspects of the Semliki River research here primarily as a case study to illustrate different actualistic approaches that can be used to model plant food distributions in ancient habitats.

Study area

As described previously (Sept, 1990) vegetation survey and plant food sampling studies were undertaken in June–August 1986, and subsequently continued in 1988, along the upper Semliki Valley, Parc National des Virunga, in eastern Zaire (Figure 1). At 900 meters elevation the Semliki River is a large, permanent river that drains Lake Rutanzige (ex-Edward) and flows north, cutting a shallow trough through Quaternary sediments and terraces on the floor of the western Rift Valley. The valley floor is undulating terrain, and a moderate annual precipitation averaging 750 mm in two rainy seasons supports open woodlands and wooded grasslands growing on plateau soils derived from Holocene deposits of volcanic ash. Erosion has cut steep gullies from the plateau down to the narrower portions of the river's incised floodbasin, and these ravines support dense, woody thicket vegetation. Where the river margins slope more gently, closed, bushy thickets grade upslope to wooded grasslands and more open habitats.

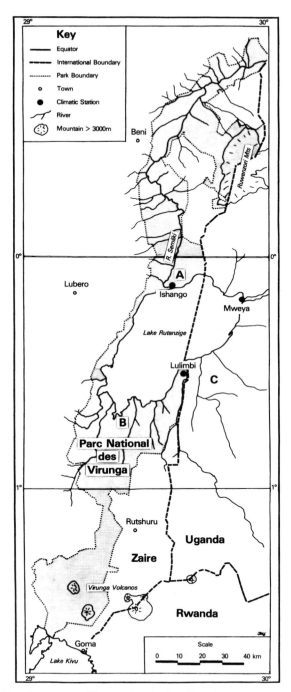

Figure 1. Map of Parc National des Virunga showing locations of the vegetation studies in the region. A=Semliki study region described in text; B=Rwindi-Rutshuru plains studied by LeBrun, 1947; C=Rwenzori National Park, studied by Lock, 1977. Vegetation of the entire Parc National des Virunga was last described by Robyns, 1943–1948.

The Semliki region provides a unique opportunity to study vegetation in a park which has been protected from the extremes of human land-use for almost 70 years (Robyns, 1943–1948). Researchers have a rare half-century's ecological perspective on vegetation change within the park, following quantitative studies of plant communities to the south of the lake done in the 1930s (LeBrun, 1947). Comparisons between these studies and vegetation descriptions done just to the east of the Semliki Valley, in Uganda (Lock, 1977) suggest that the vegetation associations described below have been characteristic of the region since it was abandoned by farming communities at the turn of the century, and probably represent relatively stable, climax communities (White, 1983). It is important to note that the woody cover in the park has fluctuated significantly in response to changing population densities of elephants (Bourliere, 1965), a historical pattern common to the dynamics of vegetation change in many parks in East Africa (e.g., Western & Van Praet, 1973).

Methods
While a number of methods were used in this study to estimate the productivity of different species of plant, and sample the abundance of different types of plant food, only information relevant to establishing the density of woody vegetation species and the relative frequency of plants with edible fleshy fruits (e.g., drupes and pomes) is presented here. Trees (stems >5 cm diameter at breast height), shrubs, vines and non-woody perennials were sampled by counting plants that were rooted within half hectare circular sample plots, located in the 40 square kilometer area directly east of the river through pacing and using compass bearings generated from a random numbers table. Sample locations were stratified within topographical zones: river margin sediments, slopes, ravines and plateau soils at different distances from the lake (Sept, 1990). A total of 124 plots were sampled during the two field seasons. For this analysis these plots have been grouped into samples of stratified canopy "forest" (*n*=6), "woodland" (*n*=54), "bushland" (*n*=30, which includes some bushed grassland), and "grassland" (*n*=34, which includes grassland with scattered trees as well as open grassland) (following Lind & Morrison, 1974; Pratt & Gwynne, 1977). Fruits of woody plants were considered "edible" for this study if they had been recorded in the literature as eaten by omnivorous or open-country monkeys or African apes (Sept, 1984, 1986). Some harvesting experiments were conducted to help estimate the fruiting productivity of different species and different vegetation zones. Different individuals picked a variety of edible fruits from a designated sample region, and their harvests were timed, counted and weighed. References for the identification of plant species in the study area can be found in an earlier report (Sept, 1990).

Results
Many species in the Semliki study area bear fleshy fruit that can be eaten by primates (Sept, 1990). Table 1 lists the most common woody species and vines bearing fleshy fruit that occurred in the study area, and the sampled frequency of these species is presented in Figure 2. Note that bushland samples have both the highest frequencies of woody species with edible fruit, and the most variety of fruiting species, followed closely by the sampled woodlands. The forest sample plots had a very low diversity of these edible fruiting species (though six samples of this type is an inadequate sample size), and were dominated instead by trees of three species: *Acacia sieberiana* (Mimosaceae), *Euphorbia calycina* (Euphorbiaceae) and *Turraea robusta* (Meliaceae). Grasslands were not empty of these fruiting plants, but had very low frequencies of trees and shrubs by definition. (Note that no nut species grew in the region, and potentially edible seeds and pods of species such as *Acacia seyal* common in the samples were not in season.)

Table 1   Species of common plant bearing edible fruit from Semliki River actualistic vegetation survey

Plant taxa	Growth form	Fruit type	Fruit size	Consumer
Apocynacea				
Carissa edulia	S	D	S	H
Boraginacea				
Cordia ovalis	T	D		M, H
Capparacea				
Capparis fascicularis	S	B	M	H
C. tomentosa	S	B	M-L	M, H
Maerua edulis	S	B	M	M, H
Cucurbitacea				
Coccinea sp.	V	B	M	M, H
Cucumis aculeatus	V	BP	L	M
Euphorbiaceae				
Securinega virosa	S	D	S	M, H
Oleaceae				
Olea africana	T	D	M	M, H
Rubiaceae				
Pavetta molundensis	S	D	S	M
Tarenna graveolens	T	D	S	M
Salvadoraceae				
Azima tetracantha	S	D	S	M
Tiliaceae				
Grewia similis	S	D	S	M, C, H
Vitaceae				
Cissus petiolata	V	D	S	C
Cissus rotundifolia	V	D	S	M

Growth form: T=tree, S=shrub, V=herbaceous or woody vine.
Fruit type: B=berry, BP=pepo; D=drupe.
Fruit size: S=small (<1 cm diameter); M=medium (1–2 cm diameter); L=large (>2 cm diameter).
Consumers: M=cercopithecine monkey; C=chimpanzee; H=human (see Sept, 1984, 1990 for consumer references).

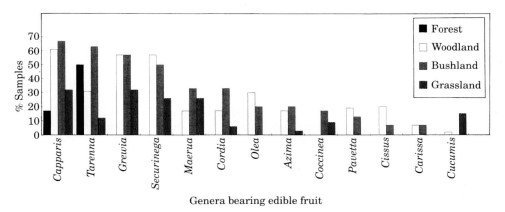

Figure 2. Frequency of fruiting taxa rooted in sample plots (half hectare samples), classified by vegetation type.

Plants of two of the most common edible fruit species grew in association; the densities of *Capparis tomentosa* and *Grewia sinensis* in the sampled plots are illustrated in Figure 3. These species often clumped in three meter high tangles, separated only by the grazing lanes and

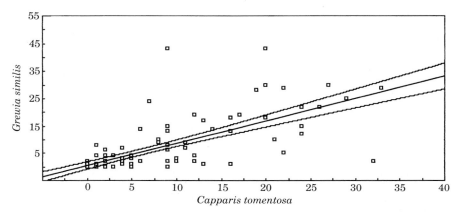

Figure 3. Comparisons of the density of *Capparis tomentosa* plants and *Grewia similis* plants in 124 sample plots (includes all vegetation types). Linear regression $R^2 = 0.54451$, $p < 0.01$.

passageways of hippopotamus. To move along the upper Semliki River in some areas of dense bush meant struggling past (or crawling through) thickets of edible species such as *C. tomentosa* and *Azima tetracantha* protected from fruit predators by thorns and spines—challenging "patches" for a potential forager.

In fact the bushland margins of the Semliki River might seem like a vast potential smorgasborg for a primate looking for plant foods until phenology and productivity patterns are considered. During the 1986 and 1988 study seasons the fruits of only a few of these shrub species were actually available. In July and early August fruit were beginning to ripen on approximately 60% of *Capparis tomentosa*, 30% of *Grewia similis*, and 10% of *Carissa edulis*. Fruit were encountered sporadically on vines and on isolated individual trees of *Cordia sinensis*. Comparing observed fruiting patterns to published herbaria records of flowering and fruiting seasons for these species (cf. Sept, 1990) demonstrated that the period during and after the March–April rains was a relatively bad time to be foraging for fruit in these habitats. As suggested in Figure 4, the greatest variety of species would have been bearing edible fruit in the area at the end of the long dry season and during the rains of November and December.

On the other hand, two of the species that were fruiting in July dominated the vegetation near the river. *Capparis tomentosa* comprised 10.85% of the more than 2000 woody plants sampled in bushland plots, and 6.84% were *Grewia similis*. Could the river margin bushland be categorized as a good July foraging patch after all? Answering this question required information about the fruiting productivity of these species.

Based on notes taken on the individual plants sampled in each plot, and some experimental harvests of the fruit on individual plants, estimates of the average number of edible fruits per plant were calculated for the most common plant species. For example, because many *Grewia* shrubs were barren, picking all the ripe fruits from 142 plants in seven of the plots yielded an average of only 8.4 fruit per plant, even though some individual shrubs yielded hundreds of the small fruit. Ripe *Capparis* fruit are few and far between, and usually eaten quickly by competitive foragers—I rarely saw more than one or two whole ripe fruit a day during observations of hundreds of shrubs. However, because almost-mature *Capparis* fruits ripen within a day or two of being picked, I estimated the productivity of *Capparis* for a hypothetical

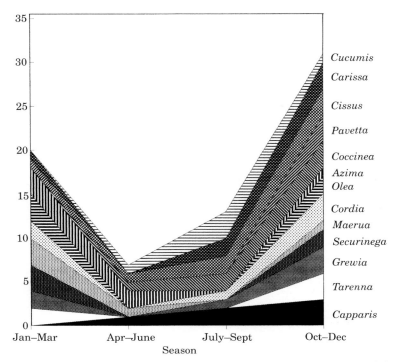

Figure 4. Relative seasonal availability of different edible fruit species. Y axis represents cumulative total of each species' availability, classified on a 4-point scale: 0=not available; 1=low availability; 2=moderate availability; 3=high availability.

hominid collector of semi-ripe fruits. Delaying consumption improves the productivity of *Capparis* habitat by an order of magnitude. The 213 *Capparis* shrubs sampled in bushland plots averaged 10·8 ± 17·7 ripening fruit per plant, and 181 woodland *Capparis* shrubs were even more variable but potentially productive, averaging 17·0 ± 40·7 fruit each. This is the one fruit sampled in this study where simple technology (e.g., carrying device) could significantly improve harvesting efficiency.

I then calculated the average fruit yield per sample plot for each species as the product of (average fruit per plant) × (species density per plot). Figures 5 and 6 show examples of the products of these calculations summarized as fruit density per hectare for *Capparis* and *Grewia*, grouping plots by vegetation type. These calculations illustrate the relatively good chances of encountering fruit feeding opportunities from these two species in bushland and woodland habitats along the Semliki. Bushlands cluster in thickets along the margins and slopes of the Semliki River, longitudinal zones of foraging opportunites for these fruits. Woodlands are more unevenly distributed across the Semliki plateau and on some terrace slopes, interspersed with tracts of more open vegetation.

Would it be worthwhile for a hungry hominid to forage for fruit in these patches of woodland or bushland? That would depend not upon the totals of harvestable fruit in each habitat, as just calculated, but on the relative foraging return rates that a hominid could achieve harvesting preferred fruit within these patches, compared to the average returns expected for other foods from the habitat at large (Stephens & Krebs, 1986). The classic "diet

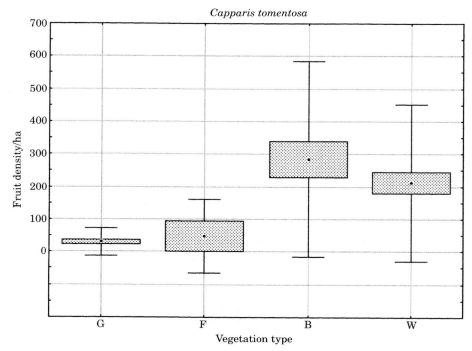

Figure 5. Average plot sample density of available fruit per hectare of *Capparis tomentosa*, samples grouped by vegetation type. G=Grassland, F=Forest, B=Bushland, W=Woodland. Boxes indicate mean ± standard error; bars indicate ± 1 standard deviation.

breadth" and "patch choice" models of foraging theory would predict that, all else being equal, hominids should prefer highly ranked foods, but broaden their diet when necessary, and forage in patches with above average return rates.

Harvesting experiments done during the 1986 and 1988 field seasons produced the following caloric-return/handling-time rankings for the major Semliki summer fruits: *Cucumis* spp (2000 kcal/hour); *Coccinea* spp (1500 kcal/hour); *Cordia africana* (1200 kcal/hour); *Capparis tomentosa* (1000 kcal/hour); *Carissa edulis* (800 kcal/hour); *Grewia similis* (490 kcal/hour); *Cissus* spp (320 kcal/hour). The lowest-ranking fruits are small, and either very watery (*Cissus* and *Carissa*) or very fibrous (*Grewia*). Fruits of *Cucumis* and *Coccinea* are large, easy to harvest and good sources of carbohydrates. *Cordia* berries require some climbing ability to acquire, but are oily and high in protein. *Capparis* seeds are laced with toxic alkaloid, but the fruit flesh is easily consumed by primates, including people, who avoid chewing the seeds (Sept, 1990).

Figure 7 illustrates the average "caloric density" of all the major fruit species in the Semliki samples, calculated by transforming fruit density data for all species into cumulative caloric yield (based on edible portions of fruit only) per plot using nutritional data from previous studies (Sept, 1984, 1990). Eighteen experimental "feed as you go" foraging bouts of one hour each in random areas of woodland and bushland yielded total returns of ripe fruits that ranged between 38–536 kcal/hour, averaging about 400 kcal/hour of edible fruit flesh, not including green *Capparis*.

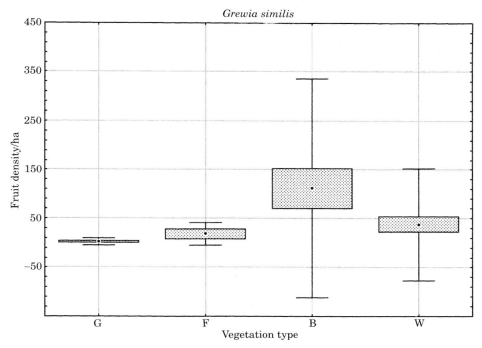

Figure 6. Average plot sample density of available fruit per hectare of *Grewia similis*, samples grouped by vegetation type. See Figure 5 for key.

While these are very crude estimates of seasonal fruit-picking return rates, it helps put the data in Figure 7 in perspective. Five or six hours of fruit feeding along the bushy margins of the Semliki River could fuel an early hominid, although these particular fruit species, eaten on a daily basis, would provide an inadequate supply of protein without supplementation from other forms of plant or animal food. Given an approximate area of 500 bushland hectares in the sampled region, if hectares averaged 5000 kcal of edible fruit (based on a "collecting" strategy that includes semi-ripe *Capparis* fruit), these borders of the Semliki could hypothetically support 1000 days of fruit foraging (at 2500 kcal/hominid/day), something like 20 adult hominids during July and August. If only ripe *Capparis* fruit were included as potential food items (a typical primate strategy), then the hypothetical "fruit-feeder carrying capacity" of the river-margin zone would drop significantly.

If feeding alternatives were available, the relatively low return rates for such a patch could be profitably abandoned. The ripening of more diverse nutritious fruits (such as *Securinega virosa*) and edible seeds (such as *Acacia seyal*) in the woodlands and wooded grasslands would likely lure plant food foragers away from the river in other seasons. Similarly, no tubers were located in the sampled plots, but if edible tubers occur in the region they would be expected to be most densely concentrated away from the river, in well-drained soils (Vincent, 1985*b*; Sept, 1984). In the process of conducting a bone taphonomy study of the region, Tappen noted scavenging opportunities occurring several times a month in open habitats away from the Semliki margins from July through November 1988 (Tappen, 1992). Predicting the frequency with which omnivorous foragers might be expected to exploit the margins of a

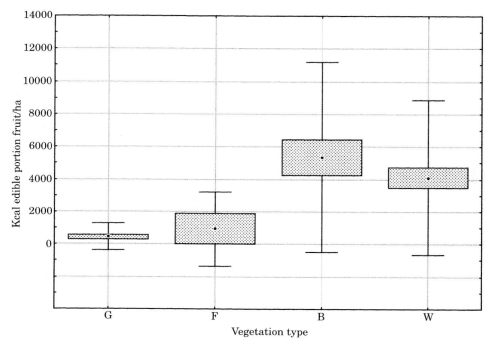

Figure 7. Average energy content (kcal) of edible portions of fruit per hectare calculated for each plot sample, grouped by vegetation type. See Figure 5 for key.

riverine habitat like the Semliki in different seasons cannot be effectively judged, however, until return rates for feeding alternatives are estimated.

Discussion and conclusions

Examining the distribution and qualities of plant foods available seasonally in the Semliki study area provides an interesting case with which to examine the costs and benefits of foraging for plant foods in savanna riverine habitats. As an incising river with little current overbank deposition, the upper Semliki has a very different pattern of vegetation and plant food distribution than is typical of fluvial sedimentary environments I have studied in eastern Africa.

Rivers meandering in accretionary floodplains in semi-arid environments are often distinguished by prominent gallery forests that stand in strong contrast to the annuals and forbs that dominate poorly drained soils of the surrounding floodplains. A steep gradient of vegetation structure can be found in such situations, associated with correspondingly strong contrasts in plant food density and distribution pattern (Sept, 1984). This is true along the middle reaches of the small, perennial Ishasha River, for example, to the south of Lake Rutanzige, which is fringed by a multi-storied groundwater forest in an active floodplain (Sept, 1992a, 1994). It is characteristic of major perennial rivers like the Omo River in southern Ethiopia (Carr, 1976), and the Tana River along its lower course in southeastern Kenya (Sept, 1984). It is also typical of the small, perennial Voi River in semi-arid southern Kenya, and of a number of channels in arid northern Kenya, where the only prominent *Acacia* trees in that arid habitat cling to the sandy margins of seasonal drainage channels, associated with productive patches of shrubs bearing edible fruit (Sept, 1986). In these types of situations riverine environments are distinctive microhabitats with a diversity and patchiness to their plant food availability that contrasts abruptly with surrounding terrain.

In the case of the upper Semliki River, however, the combination of moderate rainfall and edaphic conditions have muted the contrast between river margin and hinterland; the incising river course does not destabilize the woody plant communities away from its immediate banks, and mesic associations of plants grade more gently into the surrounding terrain types, floristically, structurally, and in terms of the density of edible fruit-bearing species. A comparable pattern of low-gradient contrast in plant food availability between stable river margins and inland plant communities, was also evident along the incising Ewaso N'giro River in arid, northern Kenya (Sept, 1984). In some ways it is misleading to refer to the Semliki bush vegetation sampled in this study as a potential feeding "patch." It is actually more of a sinuous, larger-scale zone of predictable plant food availability. Moving away from the Semliki River vegetation groupings grade into mixtures of woodland and bushed grassland. And while microhabitats such as forested ravines recur in the study region, overall variation in plant food distribution occurs in a more homogenous pattern; it grades more gently and perhaps at a larger scale than the other examples mentioned above.

The salience of riverine paleoenvironments as resource patches, or zones for hominid exploitation, probably varied considerably. During periods of river incision, riverine environments were probably characterized by relatively gentle gradients of vegetation structure and plant food availability. Mesic-xeric vegetation contrasts would have been more apparent, and river margin foraging opportunities for many plant foods more diverse and patchy, during periods of active overbank deposition, particularly in semi-arid or arid settings. This pattern may have been evident not only for plant food abundance, but for the localisation of scavenging and refuge opportunities as well (Sept, 1992a, 1994; Tappen, 1992). As a frame of reference for interpreting archaeological site assemblages, such resource salience varies in relation to a multivariate edaphic/climatic context that should be detectable through fossil and geological evidence. Feibel's recent work (Feibel, 1988; Harris & Feibel, 1993) detailing the varying fluvial environments in which archaeological sites formed in the Turkana basin is a good example of how paleogeographical work must form the foundation of such analyses.

Prehistoric vegetation patterns similar to those of the upper Semliki today are likely to be sampled in the sedimentary record only as diachronic horizons at disconformities between a stable soil horizon and a new depositional facies, associated with archaeological materials that are significantly time-averaged lag assemblages (cf. Stern, 1993), as illustrated in Figure 8. For example, while early archaeological sites in the Turkana Basin are associated with channel settings of various configurations (Harris & Feibel, 1993), pedogenic facies account for only a small fraction of the facies associations in the Koobi Fora Formation (Feibel, 1988). Many of the sites buried under overbank sediments in the relatively arid Turkana basin were more likely to have been formed in settings with strong mesic-xeric contrasts in both vegetation structure and plant food availability. Innovative paleoecological work with paleosols, such as that being done by Sikes (1994), has the potential to test such an hypothesis, and narrow the scale of vegetation reconstructions at such sites in both space and time, potentially until they are assemblage-specific.

While the Semliki River research is not directly relevant for reconstructing paleovegetation patterns at many of the Koobi Fora sites, as an actualistic case study, this research illustrates how I advocate formulating models of the structure and composition of plant communities for

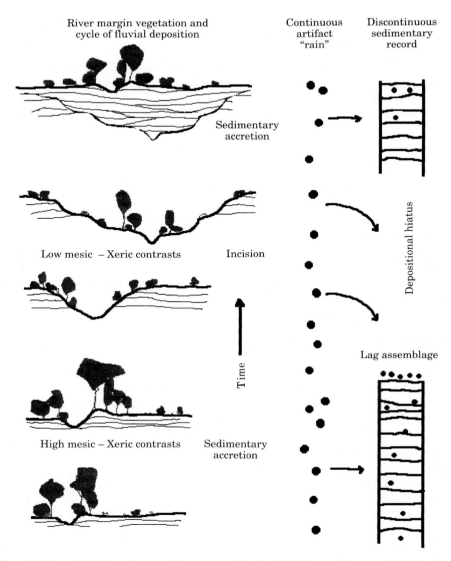

Figure 8. A model of how the vegetation context of archaeological materials accumulating in fluvial sediments might change with the depositional cycle. If a semi-arid climate is held constant, assemblages in overbank or proximal floodplain deposits are more likely to have been associated with a high contrast mesic/xeric vegetation gradient and gallery forest, while materials that accumulate as lags on pedogenic surfaces during phases of channel incision are more likely to have been associated with a lower mesic/xeric gradient and less likely to have formed near prominant gallery forest.

that subset of ancient landscapes sampled by the archaeological record. As illustrated in Figure 9, hypotheses about the probable patterns of plant food abundance and distribution at specific archaeological sites can be seen epistemologically as an intermediate step between environmental reconstructions of archaeological sites, and assemblage-specific hypotheses about early hominid behavior. Good paleoenvironmental research lays the groundwork for reconstructing

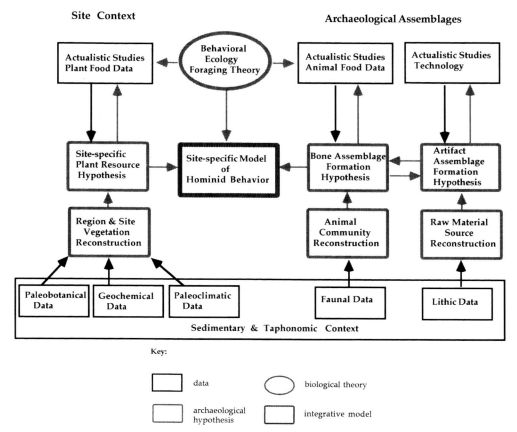

Figure 9. Diagram showing stages of inference through which behavioral interpretations of specific archaeological sites are derived from different types of evidence. The chart highlights some of the major methodological and theoretical links between empirical data recovered from archaeological sites, actualistic studies used to interpret these archaeological data, and the hypotheses through which these data are integrated. However, only a selection of the suite of analogical inferences that structure such interpretations are included. See (Gifford-Gonzalez, 1991).

site contexts, but actualistic studies help us define the paleoecological variables which we need to reconstruct.

Actualistic plant food research has two main goals. First, I think we must improve our ability to reconstruct the vegetation context of specific archaeological assemblages in terms of attributes that would have influenced hominid behavior. To reconstruct vegetation patterns as early hominid habitat we might think about paleoenvironments from the perspective of a refuging forager (Sept, 1984, 1992a), a large-bodied, bipedal primate with a taste for meat, but a dependence upon plant foods. Plant foods, in omnivorous scenarios of hominid evolution, play the role of low-risk, dependable foods with predictable locations and known processing efforts. By formally estimating the probabilities that specific archaeological assemblages formed in different plant food contexts, archaeologists may be able to elucidate aspects of assemblage patterning that have so far been overlooked. Second, if regular associations can be revealed between attributes of the bone and stone assemblages at different sites, and their plant food context, this may lead archaeologists to inferences about likely plant food components of early hominid stone tool maker diets that could potentially be tested with independent lines of evidence, such as the wear patterns on stone tools (Keeley & Toth, 1981), or the chemical composition of hominid skeletal remains (Sillen, 1992).

In summary, while plant foods are poorly preserved at early archaeological sites in East Africa, various lines of fossil evidence and analogies with living primates and human foragers suggest that plant foods would have provided a major source of calories for early hominids. Many early archaeological sites have been preserved in stream margin sediments, suggesting that more effort needs to be made to reconstruct the vegetation and plant food context of these sites. This can be done by comparing evidence from the fossil and geological records with actualistic studies of vegetation and plant communities in similar modern sedimentary environments. An example of actualistic plant food research along the upper Semliki River, in eastern Zaire, is presented to illustrate the important parameters of plant food quality, abundance, and distribution that should be modeled to build realistic paleoecological models of the paleoenvironmental contexts of early archaeological sites.

Acknowledgements

I thank the editors for their persistence in inviting me to contribute a paper to this volume. I appreciate the useful and critical comments by Ann Stahl, Martha Holder, and anonymous reviewers on an initial draft of the manuscript. I would also like to thank the Republic of Zaire, Departement de l'Environnement, Conservation de la Nature et Tourisme, for permission to conduct the Semliki River research, along with thanks to N. T. Boaz, J. W. K. Harris, and A. S. Brooks for the invitation to join the Semliki Research Expedition. My research was made possible by grants from the National Science Foundation and the L. S. B. Leakey Foundation. Special thanks are also due to Dr Mugangu Trinto Enama and Dr Nyakabwa Mutambana for their research assistance in the field.

References

Altmann, S. A., Post, D. G. & Klein, D. F. (1987). Nutrients and toxins of plants in Amboseli, Kenya. *Afr. J. Ecol.* **26,** 279–293.
Ambrose, S. H. & Sikes, N. E. (1991). Soil organic isotopic evidence for vegetation change in the Kenya Rift Valley. *Science* **253,** 1402–1405.
Axelrod, D. I. & Raven, P. H. (1978). Late Cretaceous and Tertiary history of Africa. In (M. Werger, Ed.) *Biogeography and Ecology of Southern Africa*, pp. 77–130. The Hague: Junk. Monographiae Biologicae.
Badrian, N. & Malenky, R. K. (1984). Feeding ecology of *Pan paniscus* in the Lomako forest, Zaire. In (R. L. Susman, Ed.) *The Pygmy Chimpanzee. Evolutionary Biology and Behavior*, pp. 275–299. New York: Plenum Press.
Bailey, R. (1993). *The Behavioral Ecology of Efe Pygmy Men*. Ann Arbor, Mich.: Museum of Anthropology, University of Michigan Anthropological papers/Museum of Anthropology, University of Michigan; no. 86.
Barton, R. A., Whiten, A., Strum, S. C. & Byrne, R. W. (1992). Habitat use and resource availability in baboons. *Anim. Behav.* **43,** 831–844.
Binford, L. R. (1984). *Faunal Remains from Klasies River Mouth*. Orlando, FL: Academic Press.
Binford, L. R. (1987). Researching ambiguity: frames of reference and site structure. In (S. Kent, Ed.) *Method and Theory for Activity Area Research: an Ethnoarchaeological Approach*, pp. 449–512. New York: Columbia U.P.
Blumenschine, R. J. (1986). *Early Hominid Scavenging Opportunities: Implications of Carcass Availability in the Serengeti and Ngorongoro Ecosystems*. Oxford: British Archaeological Reports.
Blumenschine, R. J. (1992). Hominid carnivory and foraging strategies, and the socio-economic function of early archaeological sites. In (A. Whiten & E. M. Widdowson, Eds) *Foraging Strategies and Natural Diet of Monkeys, Apes and Humans. Proceedings of a Royal Society Discussion Meeting held on 30 and 31 May 1991*, pp. 211–221. Oxford: Clarendon Press.

Blumenschine, R. J. & Madrigal, T. C. (1993). Variability in long bone marrow yields of East African ungulates and its zooarchaeological implications. *J. Archaeol. Sci.* **20**, 555–587.

Blurton Jones, N. (1993). The lives of hunter-gatherer children: effects of parental behavior and parental reproductive strategy. In (M. E. Pereira & L. A. Fairbanks, Eds) *Juvenile Primates. Life History Development, and Behavior*, pp. 309–326. Oxford: Oxford University Press.

Boesch, C. & Boesch, H. (1989). Hunting behaviors of wild chimpanzees in the Tai National Park. *Am. J. phys. Anthrop.* **78**, 547–573.

Bonnefille, R. (1984). Cenozoic vegetation and environments of early hominids in East Africa. In (R. O. Whyte, Ed.) *The Evolution of the East Asian Environment, Vol II: Paleobotany, Paleozoology and Palaeoanthropology*, pp. 579–612. Hong Kong: University of Hong Kong Center of Asian Studies.

Bonnefille, R. (1985). Evolution of continental vegetation: the paleobotancial record from East Africa. *S. Afr. J. Sci.* **81**, 267–270.

Bourliere, F. (1965). Densities and biomasses of some ungulate populations in eastern Congo and Rwanda, with notes on populations structure and lion/ungulate ratios. *Zoologica Afr.* **1**(1), 199–207.

Brain, C. K. (1988). New information from the Swartkrans Cave of relevance to "robust" australopithecines. In (F. E. Grine, Ed.) *Evolutionary History of the "Robust" Australopithecines*, pp. 311–316. New York: Aldine de Gruyter.

Brown, F. H. & Feibel, C. S. (1991). Stratigraphy, depositional environments and paleogeography of the Koobi Fora Formation. In (J. M. Harris, Ed.) *Koobi Fora Research Project, Volume 3, The Fossil Ungulates: Geology, Fossil Artiodactyls and Paleoenvironments*, pp. 1–30. Oxford: Clarendon.

Bunn, H. T. (1991). A taphonomic perspective in the archaeology of human origins. *Ann. Rev. Anthrop.* **20**, 433–467.

Bunn, H. T. & Ezzo, J. A. (1993). Hunting and scavenging by Plio-Pleistocene hominids: nutritional constraints, archaeological patterns, and behavioral implications. *J. Archaeol. Sci.* **20**, 365–398.

Carr, C. J. (1976). Plant ecological variation and pattern in the lower Omo basin. In (Y. Coppens, F. C. Howell, G. Ll. Isaac & R. E. Leakey, Eds) *Earliest Man and Environment in the Lake Rudolf Basin*, pp. 432–470. Chicago: Chicago U.P.

Cashdan, E. (1992). Spatial organization and habitat use. In (E. R. Smith & B. Winterhalder, Eds) *Evolutionary Ecology and Human Behavior*, pp. 237–268. New York: Aldine de Gruyter.

Cavallo, J. A. & Blumenschine, R. J. (1989). Tree-stored leopard kills: expanding the hominid scavenging niche. *J. hum. Evol.* **18**, 393–399.

Cerling, T. E. (1992). Development of grasslands and savannas in East Africa during the Neogene. *Palaeogeog. Palaeoclim. Palaeoecol.* **97**, 241–247.

Collins, D. A. & McGrew, W. C. (1988). Habitats of three groups of chimpanzees (*Pan troglodytes*) in western Tanzania compared. *J. hum. Evol.* **17**, 533–574.

Dietz, W. H., Marino, B. & Peacock, N. R. (1989). Nutritional status of Efe pygmies and Lese horticulturalists. *Am. J. phys. Anthrop.* **78**/4, 509.

Dunbar, R. I. M. (1976). Australopithecine diet based on a baboon analogy. *J. hum. Evol.* **5**, 161–167.

Feibel, C. S. (1988). Paleoenvironments of the Koobi Fora Formation, Northern Kenya. PhD Dissertation, University of Utah, Salt Lake.

Foley, R. (1987). *Another Unique Species. Patterns in Human Evolutionary Ecology.* New York: Longman Scientific & Technical/John Wiley & Sons.

Foley R. A. (1992). Evolutionary ecology of fossil hominids. In (E. A. Smith & B. Winterhalder, Eds) *Evolutionary Ecology and Human Behavior*, pp. 131–166. New York: Aldine de Gruyter.

Gifford-Gonzalez, D. (1991). Bones are not enough: analogues, knowledge, and interpretive strategies in zooarchaeology. *J. Anthrop. Archaeol.* **10** (3), 215–254.

Gordon, K. D. (1987). Evolutionary perspectives on human diet. In (F. E. Johnston, Ed.) *Nutritional Anthropology*, pp. 3–39. New York: Alan R. Liss.

Harding, R. S. O. & Teleki, G. (Eds) (1981). *Omnivorous Primates: Gathering and Hunting in Human Evolution.* New York: Columbia U.P.

Harris, J. W. K. (1983). Cultural beginnings: Plio-Pleistocene archaeological occurrences from the Afar, Ethiopia. *Afr. Archaeol. Rev.* **1**, 3–31.

Harris, J. W. K. & Feibel, C. (1993). Changing patterns of land use by Plio-Pleistocene hominids in the Lake Turkana Basin. Paper presented at *Four Million Years of Hominid Evolution in Africa. An International Congress in Honour of Dr Mary Douglas Leakey's Outstanding Contribution in Palaeoanthroplogy*. Arusha, Tanzania: August 1993.

Haslam, S. M. (1978). *River Plants. The Macrophytic Vegetation of Watercourses.* Cambridge: Cambridge University Press.

Hatley, T. & Kappelman, J. (1980). Bears, pigs, and Plio-Pleistocene hominids: a case for the exploitation of below ground food resources. *Hum. Ecol.* **8**, 371–387.

Hawkes, K. (1992). Sharing and collective action. In (E. A. Smith & B. Winterhalder, Eds) *Evolutionary Ecology and Human Behavior*, pp. 269–300. Hawthorne, NY: Aldine de Gruyter.

Hawkes, K. (1993). Why hunter-gatherers work: an ancient version of the problem of public goods. *Curr. Anthrop.* **34**, 341–362.

Hawkes, K., O'Connell, J. F. & Blurton Jones, N. (1991). Hunting income patterns among the Hadza: big game, common goods, foraging goals, and the evolution of the human diet. *Phil. Trans. R. Soc., Lond.* B **334**, 243–251.

Hay, R. L. (1976). *The Geology of Olduvai Gorge*. Berkeley and Los Angeles: University of California Press.

Howell, F. C., Haesarts, P. & de Heinzelin, J. (1987). Depositional environments, archeological occurrences and hominids from Members E and F of the Shungura Formation (Omo basin, Ethiopia). *J. hum. Evol.* **16** (7/8), 665–700.

Hunt, K. D., Nishida, T. & Wrangham, R. W. (n.d.). Sex differences in chimpanzee positional behavior, activity budget and diet: relative contributions of rank, reproductive-demands and body size, and implications for tool-use. *Am. J. phys. Anthrop.* (submitted).

Isaac, G. Ll. (1981). Stone age visiting cards: approaches to the study of early land-use patterns. In (I. Hodder, G. Ll. Isaac & N. Hammond, Eds) *Patterns in the Past*, pp. 37–103. Cambridge: Cambridge University Press.

Isaac, G. Ll. (1984). The archaeology of human origins: studies of the lower Pleistocene in East Africa 1971–1981. In (F. Wendorf, Ed.) *Advances in World Archaeology*, pp. 1–87. New York: Academic Press.

Kaplan, H. & Hill, K. (1992). The evolutionary ecology of food acquisition. In (E. A. Smith & B. Winterhalder, Eds) *Evolutionary Ecology and Human Behavior*, pp. 167–202. Hawthorne, NY: Aldine de Gruyter.

Kaufulu, Z. (1983). The geological context of some early archaeological sites in Kenya, Malawi, and Tanzania. PhD Dissertation, University of California, Berkeley.

Kay, R. F. (1984). On the use of anatomical features to infer foraging behavior in extinct primates. In (P. S. Rodman & J. G. H. Cant, Eds) *Adaptations for Foraging in Nonhuman Primates*, pp. 21–53. New York: Columbia University Press.

Kay, R. F. & Grine, F. E. (1988). Tooth morphology, wear and diet in Australopithecus and Paranthropus from southern Africa. In (F. E. Grine, Ed.) *Evolutionary History of the "Robust" Australopithecines*, pp. 427–448. New York: Aldine de Gruyter.

Keeley, L. K. & Toth, N. (1981). Microwear polishes on early stone tools from Koobi Fora, Kenya. *Nature* **293**, 464–465.

Kibunjia, M., Roche, H., Brown, F. H. & Leakey, R. E. (1992). Pliocene and Pleistocene archaeological sites west of Lake Turkana, Kenya. *J. hum. Evol.* **23**, 431–438.

Krebs, J. R. & Davies, N. B. (Eds) (1984). *Behavioral Ecology. An Evolutionary Approach* (2nd ed.). Sunderland, MA: Sinauer Associates Inc.

Kroll, E. M. & Isaac, G. L. (1984). Configurations of artifacts and bones at early Pleistocene sites in East Africa. In (H. J. Hietala, Ed.) *Intrasite Spatial Analysis in Archaeology*, pp. 4–31. Cambridge: Cambridge University Press.

Lanjouw, A. (1991). An experiment in long-term conservation. Tongo chimpanzee conservation project, Zaire. In (P. G. Heltne, Ed.) *Understanding Chimpanzees: Diversity and Survival*, 15, pp. 32–33. Chicago: Bulletin of the Chicago Academy of Sciences.

LeBrun, J. (1947). *La Vegetation de la Plaine Alluviale au Sud du Lac Edouard*. Bruxelles: Patrimoine de l'Institut royal des sciences naturelles.

Lee-Thorp, J. A., Van der Merwe, N. & Brain, C. K. (1989). Isotopic evidence for dietary differences between two extinct baboon species from Swartkrans. *J. hum. Evol.* **18**, 183–190.

Leighton, M. (1993). Modeling dietary selectivity by Bornean orangutans: evidence for integration of multiple criteria in fruit selection. *Int. J. Primatol.* **14**, 257–313.

Leonard, W. R. & Robertson, M. L. (1992). Nutritional requirements and human evolution: a bioenergetics model. *Am. J. hum. Biol.* **4**, 179–195.

Lind, E. M. & Morrison, M. E. S. (1974). *East African Vegetation*. Bristol: Longman Group Ltd.

Lock, J. M. (1977). The vegetation of Rwenzori National Park. *Bot. Jahrb. Syst.* **93**, 372–448.

Martin, R. D. (1983). Human brain evolution in an ecological context. In *52nd James Arthur Lecture on the Evolution of the Brain*. American Museum of Natural History.

Maynard, L. A., Loosli, J. K., Hintz, H. F. & Warner, R. G. (1979). *Animal Nutrition* (7th edn). New York: McGraw-Hill.

McGrew, W. C. (1992). *Chimpanzee Material Culture. Implications for Human Evolution*. Cambridge: Cambridge University Press.

McGrew, W. C., Baldwin, P. J. & Tutin, C. E. J. (1981). Chimpanzees in a hot, dry, and open habitat: Mt Assirik, Senegal, West Africa. *J. hum. Evol.* **10**, 227–244.

Milton, K. (1984). The role of food-processing factors in primate food choice. In (P. S. Rodman & J. G. H. Cant, Eds) *Adaptations for Foraging in Nonhuman Primates*, pp. 249–279. New York: Columbia University Press.

Milton, K. (1984). The role of food-processing factors in primate food choice. In (P. S. Rodman & J. G. H. Cant, Eds) *Adaptations for Foraging in Nonhuman Primates*, pp. 249–279. New York: Columbia University Press.

Milton, K. (1987). Primate diets and gut morphology: implications for hominid evolution. In (M. Harris & E. Ross, Eds) *Food and Evolution: Towards a Theory of Human Food Habits*, pp. 93–115. Temple University Press.

National Academy of Sciences (1980). *Recommended Dietary Allowances* (9th edn). Washington D.C.: National Academy Press.

Norton, G. W., Rhine, R. J., Wynn, G. W. & Wynn, R. D. (1987). Baboon diet: a five-year study of stability and variability in the plant feeding and habitat of the yellow baboons (*Papio cynocephalus*) of Mikumi National Park, Tanzania. *Folia Primatol.* **48**, 78–120.

O'Brien, E. M. & Peters, C. R. (1991). Ecobotanical contexts for African hominids. In (J. D. Clark, Ed.) *Cultural Beginnings: Approaches to Understanding Early Hominid Life-Ways in the African Savanna*, pp. 1–15. Bon, R. Habelt Monographien/Romisch-Germanisches Zentralmuseum, Forschungsinstitut fur Vor- und Fruhgeschichte, Bd 19.

Peters, C. R. (1987). Nut-like oil seeds: food for monkeys, chimpanzees, humans, and probably ape-men. *Am. J. phys. Anthrop.* **73**, 333–363.

Peters, C. R. & Blumenschine, R. J. (1993). Modeling resource distributions and hominid land use in the Olduvai Basin during lowermost Bed II times. In *Four Million Years of Hominid Evolution in Africa. An International Congress in Honour of Dr Mary Douglas Leakey's Outstanding Contribution in Palaeoanthroplogy*. Arusha, Tanzania.

Peters, C. R. & Maguire, B. (1981). Wild plant foods in the Makapansgat area: a modern ecosystem analogue for Australopithecus africanus adaptations. *J. hum. Evol.* **10**, 565–583.

Peters, C. R. & O'Brien, E. M. (1981). The early hominid plant-food niche: insights from an analysis of plant exploitation by *Homo*, *Pan*, and *Papio* in eastern and southern Africa. *Curr. Anthrop.* **22**, 127–140.

Peters, C. R., O'Brien, E. M. & Box, E. D. (1984). Plant types and seasonality of wild plant foods, Tanzania to south-western Africa: resources for models of the natural environment. *J. hum. Evol.* **13**, 397–414.

Pianka, E. R. (1983). *Evolutionary Ecology* (3rd edn). New York: Harper & Row.

Pilbeam, D. (1984). Reflections on human ancestors. *J. Anthrop. Res.* **40**, 14–22.

Post, D. G. (1984). Is optimisation the optimal approach to primate foraging? In (P. S. Rodman & J. G. H. Cant, Eds) *Adaptations for Foraging in Nonhuman Primates*, pp. 280–303. New York: Columbia University Press.

Potts, R. (1988). *Early Hominid Activities at Olduvai*. New York: Aldine de Gruyter.

Pratt, D. J. & Gwynne, M. D. (1977). *Rangeland Management and Ecology in East Africa*. Sevenoaks, England: Hodder and Stoughton Educational.

Retallack, G. J. (1991). *Miocene Paleosols and Ape Habitats of Pakistan and Kenya*. Oxford: Oxford University Press.

Robyns, W. (1943–1948). *Flore des Spermatophytes du Parc National Albert*. Bruxelles: Institut du parc national Congo-Belge.

Rodman, P. S. (1984). Foraging and social systems of orangutans and chimpanzees. In (P. S. Rodman & J. G. H. Cant, Eds) *Adaptations for Foraging in Nonhuman Primates*, pp. 134–160. New York: Columbia University Press.

Rogers, M. E., Maisels, F., Williamson, E. A., Tutin, C. E. G. & Fernandez, M. (1992). Nutritional aspects of gorilla food choice in the Lope Reserve, Gabon. In (N. Itoigawa, Y. Sugiyama, G. P. Sackett & R. K. R. Thompson, Eds) *Topics in Primatology Volume 2. Behavior, Ecology, Conservation*, pp. 255–266. Tokyo: University of Tokyo Press.

Rose, M. D. (1984). Food acquisition and the evolution of positional behavior: the case of bipedalism. In (D. J. Chivers, B. A. Wood & A. Bilsborough, Eds) *Food Acquisition and Processing in Primates*, pp. 509–524. New York: Plenum Press.

Ryan, A. S. & Johanson, D. C. (1989). Anterior dental microwear in Australopithecus afarensis: comparison with human and non human primates. *J. hum. Evol.* **18**, 235–268.

Schick, K. D. (1987). Modeling the formation of Early Stone Age artifact concentrations. *J. hum. Evol.* **16**, 789–808.

Sept, J. M. (1984). Plants and early hominids in east Africa: a study of vegetation in situations comparable to early archaeological site locations. PhD dissertation, University of California, Berkeley.

Sept, J. M. (1986). Plant foods and early hominids at site FxJj50, Koobi Fora, Kenya. *J. hum. Evol.* **15**, 751–770.

Sept, J. M. (1990). Vegetation studies in the Semliki Valley, Zaire as a guide to paleoanthropological research. In (N. Boaz, Ed.) *Virginia Mus. Nat. Hist. Mem.* **1**, 95–121.

Sept, J. M. (1992a). Was there no place like home? A new perspective on early hominid sites from the mapping of chimpanzee nests. *Curr. Anthrop.* **33** (2), 187–207.

Sept, J. M. (1992b). Archaeological evidence and ecological perspectives for reconstructing early hominid subsistence behavior. In (M. B. Schiffer, Ed.) *Archaeological Method and Theory*, pp. 1–56. Tucson: University of Arizona Press.

Sept, J. M. (1994). Bone distribution in a semi-arid chimpanzee habitat in eastern Zaire: implications for the interpretation of faunal assemblages at early archaeological sites. *J. Archaeol. Sci.* **21**, in press.

Sikes, N. (1994). Early hominid habitat preferences in East Africa: paleosol carbon isotope evidence. *J. hum. Evol.* **27**, 25–45.

Sillen, A. (1992). Strontium-calcium ratios (Sr/Ca) of *Australopithecus robustus* and associated fauna from Swartkrans. *J. hum. Evol.* **23**, 495–516.

Sillen, A. & Lee-Thorp, J. A. (1993). Correspondence: Diet of Australopithecus robustus. *S. Afr. J. Sci.* **89**(4), 174.

Speth, J. D. (1987). Early hominid subsistence strategies in seasonal habitats. *J. Archaeol. Sci.* **14**, 13–29.

Speth, J. D. (1989). Early hominid hunting and scavenging: the role of meat as an energy source. *J. hum. Evol.* **18**, 329–343.

Stahl, A. (1984). Hominid dietary selection before fire. *Curr. Anthrop.* **25**, 151–168.

Stephens, D. W. & Krebs, J. R. (1986). *Foraging Theory*. Princeton N.J.: Princeton U.P.

Stini, W. A. (1988). Food, seasonality and human evolution. In (I. de Garin & G. A. Harris, Eds) *Coping with Uncertainty in Food Supply*, pp. 32–51. Oxford: Oxford University Press.

Stern, N. (1993). The structure of the lower Pleistocene archaeological record: a case study from the Koobi Fora Formation. *Curr. Anthrop.* **34**, 201–226.

Suzuki, A. (1969). An ecological study of chimpanzees in a savanna woodland. *Primates* **10**, 103–148.

Tappen, M. J. (1992). Taphonomy of a central African savanna: natural bone deposition in Parc National des Virunga, Zaire. PhD dissertation, Harvard University.

Toth, N. & Schick, K. (1986). The first million years: the archaeology of human origins. In (M. B. Schiffer, Ed.) *Advances in Archaeological Method and Theory*, pp. 1–96. New York: Academic Press.

Tutin, C. E. G., Fernandez, M., Rogers, E. M., Williamson, E. A. & McGrew, W. C. (1992). Foraging profiles of sympatric lowland gorillas and chimpanzees in the Lope Reserve, Gabon. In (A. Whiten & E. M. Widdowson, Eds) *Foraging Strategies and Natural Diet of Monkeys, Apes and Humans*, pp. 19–24. Oxford: Clarendon Press.

Urban, D. L., O'Neill, R. V. & Shugart Jr, H. H. (1987). Landscape ecology. *Bioscience* **37** (2), 119–127.

Vincent, A. (1985*a*). Plant foods in savanna environments: a preliminary report of tubers eaten by the Hadza of Northern Tanzania. *World Archaeol.* **17**, 1–14.

Vincent, A. (1985*b*). Underground plant foods and subsistence in human evolution. PhD dissertation, University of California, Berkeley.

Walker, A. (1981). Dietary hypotheses and human evolution. *Phil. Trans. R. S. Lond.* **292**, 57–64.

Walker, A. (1993). Perspectives on the Nariokotome discovery. In (A. Walker & R. Leakey, Eds) *The Nariokotome* Homo erectus *Skeleton*, pp. 411–430. Cambridge, MA: Harvard University Press.

Waterman, P. G. (1984). Food acquisition and processing as a function of plant chemistry. In (D. J. Chivers, B. A. Wood & A. Bilsborough, Eds) *Food Acquisition and Processing in Primates*, pp. 177–211. New York: Plenum Press.

Western, D. & Van Praet, C. (1973). Cyclical changes in the habitat and climate of an east African ecosystem. *Nature* **241**, 104–106.

Wheeler, P. E. (1992). The thermoregulator advantages of large body size for hominids foraging in savannah environments. *J. hum. Evol.* **23**, 351–362.

Wheeler, P. E. (1993). The influence of stature and body form on hominid energy and water budgets: a comparison of Australopithecus and early Homo physiques. *J. hum. Evol.* **24**, 13–28.

White, F. (1983). *The Vegetation of Africa*. La Chaux-de-Fonds: UNESCO.

Whiten, A., Byrne, R. W., Barton, R. A., Waterman, P. G. & Henzi, S. P. (1992). Dietary and forging strategies of baboons. In (A. Whiten & E. M. Widdowson, Eds) *Foraging Strategies and Natural Diet of Monkeys, Apes and Humans*. Oxford: Clarendon Press.

Winterhalder, B. (1987). The analysis of hunter-gatherer diets: stalking an optimal foraging model. In (M. Harris & E. B. Ross, Eds) *Food and Evolution*, pp. 311–339. Philadelphia: Temple University Press.

Wrangham, R. W. (1986). Ecology and social relationships in two species of chimpanzee. In (D. I. Rubenstein & R. W. Wrangham, Eds) *Ecological Aspects of Social Evolution*, pp. 352–378. Princeton: Princeton University Press.

Wrangham, R. W. & Riss, E. V. Z. B. (1990). Rates of predation on mammals by Gombe chimpanzees. *Primates* **31** (2), 157.

Wrangham, R. W. & Smuts, B. (1980). Sex differences in the behavioral ecology of chimpanzees in the Gombe National Park, Tanzania. *J. Reprod. Fertility, Supplement* **28**, 13–31.

Wrangham, R. W., Conklin, N. L., Chapman, C. W. & Hunt, K. D. (1992). The significance of fibrous foods for Kibale Forest chimpanzees. In (A. Whiten & E. M. Widdowson, Eds) *Foraging Strategies and Natural Diet of Monkeys, Apes and Humans*, pp. 11–17. Oxford: Clarendon Press.

Andrew Hill

Department of Anthropology, Yale University, Box 208277, New Haven, CT 06520, U.S.A.

Received 31 January 1994
Revision received 3 March 1994
and accepted 31 March 1994

Keywords: hominid, behavioural ecology, methods.

Early hominid behavioural ecology: a personal postscript

Recent work in the field of early hominid behavioural ecology is discussed, and some currently significant issues outlined. Present methods are evaluated from the point of view of approaches formerly employed.

Journal of Human Evolution (1994) **27**, 321–328

Introduction

I have been invited by the editors of this volume to comment briefly on the state of investigations into early hominid behavioural ecology today, principally as demonstrated by the papers gathered here, in contrast to behavioural ecology carried out in the past. Consequently I would like to attempt to tie together what appears to be a very diverse set of contributions by concentrating mainly on an element that necessarily runs through them all, and that is method. This does not aim to be a fully comprehensive treatment, but I want to mention some of the things I have felt were problems in earlier attempts to examine paleoecological matters, and see whether, and how, these pitfalls have been avoided in this volume. Obviously this results in a very personal view. I am not claiming any special insights, and certainly I am not attempting to formulate a prescription for how behavioural paleoecology should be carried out.

What are currently seen as important issues in early hominid behavioural ecology? One set of problems is concerned with the origins of particular characteristic human behaviours. Other key issues are early hominid habitat preference and land use, the procurement of food and lithic materials, processing behaviours, the use of fire, competitive interactions with other carnivores and hominids, social organization and cognitive skills. Another constant problem that should be addressed is whether what we see as problems at any one time are really the interesting or relevant ones.

The following comments are grouped under loosely constrained headings that reflect some of these important issues, or areas I have felt were problems, and which are also treated to some extent in the papers forming this volume. Origins are important, both of hominoid taxa and of innovative behaviours, and the first section discusses some aspects of this. How work in paleoecology should respond to methods applied to neoecology is a fruitful area of discussion, as is the rather vexed issue of the ways in which analogy should be applied to past situations. Another section considers the effects of emphasis in research on a single site versus the attempt to make inter-site comparisons. The former stance is perhaps more conducive to the formulation of single hypotheses. This leads to a brief consideration of the idea of simplicity in hypotheses, and how rival ideas might be evaluated. The determination of synchronicity of events in the past is an important element of early hominid behavioural ecology, and a consideration of time resolution forms another short section. Bones and stones are the obvious elements of early sites. The next section alludes to the less obvious and usually less considered components of sites, such as fish remains. Finally, when considering the early Pleistocene of eastern Africa, a time and region where we know there are about four contemporary hominid species, how do we know which did what? A few concluding lines are given to this large question.

0047–2484/94/010321+08 $08.00/0

Origins

One group of problems that are probably always relevant and interesting includes those connected with the origins of taxa or behaviours. There are sometimes attempts (e.g., Conkey, 1991) to criticize and belittle the concern with origins, to explain it as lust for intellectual power, pandering to a public fascination with the oldest, and a corresponding search for popular fame. However, this attitude betrays a lack of understanding of evolutionary processes. The origin of Hominidae, or of *Homo*, for example, are interesting and important evolutionary events. If we are ever to understand the reasons for these events, then it helps to know when and where they occurred, and to understand the environmental and ecological matrix in which they took place. It is likely that evolutionary innovation is associated with environmental change at some level, be it global or more locally. We need to know the timings of such evolutionary origins as accurately as possible, to be able to correlate them with various indicators of climatic and environmental change. This is why the determination of origins is important, and it is also worth saying that it is an activity that always requires field work. These origins are taxonomic ones, correlated with morphological changes that presumably reflect behavioural change. In addition, apart from phyletic origins, there are more exclusively behavioural origins to consider, ones that are not, or not so obviously connected with morphological change. Some of these are clearly important in that they mark the acquisition of characteristics that demarcate at least some hominids behaviourally from other primates.

One of these traditionally distinctive characteristics is the manufacture of stone tools (but see, for example, McGrew, 1992). The origin and widespread use of flaked stone tools, however, is clearly a very important step in human evolution, and Kibunjia (1994) provides important information about one of the earliest occurrences, west of Lake Turkana, Kenya. Dated at about 2·36 Ma this is only slightly younger than the earliest evidence of genus *Homo* dated at 2·42 Ma (Hill *et al.*, 1992). This coincidence might lead some to suggest that *Homo* is responsible for the innovation of flaked tool production, but given the density of data we have available it is not a necessary conclusion at present. Similarly, and more broadly, the origin of *Homo* and stone tools has been correlated with a sharp phase of aridity at 2·5 Ma (Vrba *et al.*, 1989, for example). Here again the quality of the present record does not lead to overwhelming confidence in the proposal, particularly as the 2·5 Ma "event" appears to be simply part of a general cooling curve starting at 3 Ma and extending to younger than 2 Ma. Kibunjia himself also points out the ameliorating local effects of large lake basins and river systems.

Another important behavioural innovation in human evolution was the use of fire, and at least since the discovery of *Australopithecus prometheus* the detection of the evidence of controlled fire has been recognized as an elusive matter. Bellomo (1994) apparently demonstrates the context and controlled use of fire at 1·6 Ma. This has very important implications for how we interpret the abilities of humans at that time. Competence with fire obviously confers a variety of practical advantages, such as warmth, defense, the ability to modify food. Also, once we have camp fires we have a literal focus for social activity.

Neoecological methods

One general problem in early human behavioural ecology concerns the difficulty, and indeed the inappropriateness, of applying neoecological techniques. The earliest attempts in paleo-ecology seemed to treat the past as if it were the present, and people accepted and tried to implement the techniques of neoecology. For example, contemporary ecologists count the relative numbers of different species as an aid to understanding community dynamics. Early attempts at paleoecology tried to do the same, essentially by pretending that bone scraps were living animals. And even when taphonomy reared its ugly head it was initially geared to the attempt to correct for the biases in fossil samples so that modern ecological techniques could be applied.

I think it has taken us a relatively long time to fully realize the obvious; that fossil assemblages are quite different from modern living communities and require radically different treatment. This situation demands that we develop original and innovative techniques to overcome the limitations of the fossil record. Clearly we need to tackle the fossil and archaeological record on its own terms; to use information that actually is preserved in order to find out what we need to know, rather than reading into the record what we would like to see there. This in turn often means that we need to examine the modern world from the point of view of the fossil record, rather than *vice versa*. This was a point also made by Blumenschine *et al.* (1994). We should look at the present from what we know of the past, rather than at the past with spectacles tinted with the preconceptions of our own, or other scientists', modern knowledge.

Many of the papers in this volume avoid this neoecological difficulty by using relatively innovative techniques. One obvious set of examples are those papers that discuss detailed damage to bone, such as those by Selvaggio (1994), Blumenschine *et al.* (1994), Oliver (1994) and Bunn (1994). Sikes (1994) is one of a number of people now who are employing isotopic methods to assess vegetation that is impossible to observe directly, or even from fossils. In tropical areas trees and shrubs use a different photosynthetic pathway from open country grasses, and a signature of this is incorporated in fossil soil organic matter and carbonates, enabling the nature of fossil plant biomass to be assessed. In a slightly different sense Bellomo's investigation of the existence and nature of fire falls into this category (Bellomo, 1994), where he has developed techniques to distinguish camp fires from other kinds of fire that could be due to non-human causes.

Another example is the paper given by Plummer and Bishop (1994). They use a method for determining the environmental preferences of bovids that involves looking at aspects of bovid anatomy ignored by modern ecologists because they have better indicators of environment, namely direct observation. Most of these techniques require reference to the present in terms that are dictated by our knowledge of the past.

Analogy

Analogy to me has always been a difficult area as well (e.g., Hill, 1984). It is rather insidious because its problems operate on many levels. Obviously we need to make comparisons with what we know about the present, but crudely applied and simple analogies can lead us into erroneous and, worse still, very dull ideas. At the level of the species, for example, pretending that the very early hominids are almost exactly like modern chimpanzees, or any other particular animal, seems to me a dead end. Work is then devoted to confirming this view of the past, and this practice prohibits the detection of differences, which are probably more likely, and much more interesting. A similar problem involves the simple ethnographic analogies that were incorporated into the home base model (e.g., Isaac, 1978).

McHenry (1994), however, has a sophisticated appreciation of analogy in his investigation of early hominid body size and encephalization. Data from modern primates are used to detect basic ecological parameters and principles. These principles then ought to be as applicable to the past as they are to modern situations. His conclusions are important as body size relates to many significant life-history parameters. For example, there are implications for the mating systems and foraging patterns of different hominids. As McHenry suggests, problems still exist, primarily because of shaky fossil data, but it is an illuminating avenue to pursue. Similarly, Oliver (1994) sensibly looks for analogies not among the phylogenetically close, but with other animals that share some particularly similar life history characteristics or inferred basic behaviours. He explores the socioecological correlates of transport behaviour, based on a range of modern carnivores, and of having altricial young. This method employs more fundamental principles of ecology and behaviour that have a greater chance of being general at any time or place.

Plummer & Bishop (1994) again avoid making simple comparisons, and display a refined use of analogy in their taxon-free approach of investigating the functional anatomy of extinct bovid metapodials. This study throws light on the range of habitats exploited by hominids at Olduvai. It also seems to show that the antilopini-alcelaphini criterion for assessing habitat developed by Vrba (1980) may not exactly apply. Vrba's idea was a very good one, and she did justify the taxonomic criterion with some functional suggestions, but ultimately it suffers from the assumption that bovid tribes were adapted to similar environments in the past as now.

A more general and worrying point is that it is easy to make analogies from fossil sites at an environmental level to particular modern African environments. These modern African environments are mostly represented by data from game parks, which are almost the only places where animals can be observed today in their more or less natural states. But tourists do not like closed habitats. They cannot see the animals. As a result, our present samples of the so-called natural world in Africa are biased towards open environments. This consideration applies not just to animals, but also to vegetation. The popular prominence of the Serengeti as a tourist archetype of what an African habitat is like has no doubt given rise to the feeling that savanna grasslands are the typical present day east African environment, when of course they are not, as Sikes (1994) documents.

Single sites: single hypotheses

A recurring problem has been that investigators have often been tethered, you might say, to a single site. And this tethering has applied not just to the investigation, but to the methods that were used and the ideas that resulted. Often methods were developed specifically for the particular local problems posed by a particular site. That, of course, is quite legitimate, and some wonderful studies have resulted. Brain's work at Swartkrans is a fine example (e.g., Brain, 1981, 1993). But sometimes this practice led to the development of separate vocabularies and to a consequent lack of comparability of data collected. We have evidence in a number of papers in this volume that this problem is disappearing. Sites are extensively compared, and important insights come from exploring the variety of information different sites provide. Potts (1994), for example, contrasts aspects of Olorgesailie and Olduvai in this way, and many other authors share a concern that their information can be compared with that from other investigators' sites.

Perhaps the earlier focus on single sites also led to separate and isolated hypotheses to explain what went on there. The development and defense of a single hypothesis, sometimes labelled as *the* working hypothesis, was a practice that I feel was limiting. Separate workers took unreasonable pains to protect their separate hypotheses of site interpretation, hominid land use practices, and so on. Today there seems to be less polarization. Obviously we all have pet theories, and we all know our colleagues are fools and charlatans, but at least there is more of an attempt to integrate ideas, or to indicate common elements of them, for example as in Potts' paper (Potts, 1994). Potts indicates what I think is a very beneficial move away from the "single best reconstruction". I have suggested (Hill, 1984) that there is an opposition between the most probable narrative and the most falsifiable theory. The theory that seems most probable in terms of what is already known is not the most falsifiable, nor is it the most simple. Other contributors, like Bunn (1994), are ready to set up several hypotheses and indicate ways in which they may be evaluated and tested.

A benefit of multiple hypotheses is that they stimulate the production of increasingly subtle data in order to respond to them. Observations are made of matters that were not previously considered to be data. Things formerly thought obvious are now questioned, and examined much more carefully. I think in this context particularly of the increasingly discriminating information about bone damage that is recorded, as in Selvaggio's and Oliver's papers (Selvaggio, 1994; Oliver, 1994). These are aimed at, for example, distinguishing between scavenging and hunting, disentangling carnivore involvement and the work of humans, assessing the timing of access to carcasses, and examining the degree of competition between humans and other animals. Finer distinctions are also made of vegetation, as Sikes (1994) shows. Bellomo (1994) and Kroll (1994) provide very detailed spatial information and analyses of particular sites. The paleoenvironment is interpreted in more intricate and less simple ways.

There is another way in which we are no longer tied to sites, and which explores variation in behaviour. That is the relatively recent attempt to find out what prehistoric people were doing in the landscape as a whole. Potts' work at Olorgesailie is an example (Potts, 1994), as is Bunn's investigation at Koobi Fora (Bunn, 1994). Sikes' Olduvai work, in association with Blumenschine, is also relevant here (Sikes, 1994), as are Kroll's questions about the definition of sites (Kroll, 1994).

Simplicity

There is a related issue involving simplicity (Hill, 1979, 1984). A former practice was to defend a hypothesis by asserting that it was the simplest interpretation of the available data. But we are involved in a subject where we certainly know that much of the data we would like to have is unavailable, and the world is a very complex place. It is ourselves who are simple, not the world outside us. Focussing on simplicity in our hypotheses is probably a good research strategy. Simple hypotheses are probably more testable. However, there is no reason to believe that the simplest of a set of hypotheses is the most likely to be true. This is something that has to be recognized in taxonomy and cladistics as well. I was very pleased to find that in these papers no one was invoking simplicity as a defence of their ideas. This additionally manifests itself as a greater tolerance of a variety of ideas concerning what hominids were doing in the past.

It is likely that early hominids were much more complex than the data show, or can ever show. One of the important comments in this context of Blumenschine et al. (1994) is that if we are fixed and narrow in our ideas about the behavioural repertoire of hominids then we prevent ourselves from being able to detect variability itself in hominid behaviour. Yet a flexible behavioural response may be a significant ability that was selected for in evolution.

Time resolution

One general problem made particularly acute in this area is that of time resolution. Stern (1994) provides an important warning for landscape archaeologists, of something already acknowledged by many geologists and paleontologists, that we cannot recognize instantaneous time planes. That in the early Pleistocene at East Turkana, and probably elsewhere, "contemporaneous" means "within about 80 ka years" at best. The palimpsestuous nature of the record still does not receive enough general attention, though Kroll (1994) and Bellomo (1994) note its importance. I have pointed out in the larger context of mammalian evolution in the Neogene (Hill, 1987), that we need to match the resolution of our questions with the resolution of the available data. Discovering the resolution of the record, in time or in various other features, is an important first step to any interpretation. Also, although we may now realize that fossil assemblages are quite different from communities of animals, we may not yet fully recognize that the various planes that can be detected in sediments are not exactly similar to landscapes.

Seeing more than the obvious

Paradoxically, avoiding a preoccupation with the obvious has been another problem. Stone tools and mammal bones are the most conspicuous items in the record, and consequently a great deal of attention is paid to them. But as plant-oriented archaeologists have indicated before, we may be overemphasizing the use of mammalian meat as a resource. Stewart (1994) indicates the same regarding fish. Fish may indeed be a major food item, and have been particularly important seasonally in periods of stress when mammalian meat could not be fully utilized. Up to a point this relative neglect also applies to bird and small mammal fossils, as in general fine-meshed screens have not been routinely used at most sites.

The relative importance of plants and animals as food is a question in human behavioural ecology that has great impact, as it impinges upon a consideration of hunting *versus* gathering, upon implied gender roles and how the social structure of early hominid groups was organized. Unfortunately plant fossils only very rarely exist. This provides an example of a more general methodological problem. Due to the nature of the fossil and archaeological record there are many factors of hominid life that on general grounds must have been important, but for which we have no direct evidence. Despite this lack of evidence, and therefore the possibility that hypotheses cannot be falsified, these factors have to be considered and evaluated more obliquely. In this light Sikes (1994) presents a useful review of work on the modern distribution of the kinds of plants that may have been a food resource to early hominids. She also provides information from her own work on carbon isotopes regarding floral microhabitats at Olduvai.

Who did what?

A further important issue, not really fully confronted by anyone in this volume, except tangentially by Oliver (1994) and Potts (1994), is that of hominid taxonomic diversity. We know that during the time periods and places most of the contributors to this issue are discussing there were at least three or four species of sympatric hominid in eastern Africa. Who was responsible for what? What were the interactions among them? I have already hinted at the problem of identifying the makers of the earliest stone tools, as does Sikes (1994) and Oliver (1994). Oliver (1994) acknowledges Susman's suggestion (Susman, 1988,

1991) that *Paranthropus* was probably capable of making stone tools, but justifies an assumption that early *Homo* is responsible for the early sites, at Olduvai and elsewhere in east Africa. However, this still leaves us with three contemporary species to choose from, *Homo habilis*, *H. rudolfensis* and *H. ergaster* (or *H. erectus*). To give another concrete example involving this issue: the Olduvai "*Zinjanthropus*" site (FLK) is probably the most intensely studied locality in the world. We now know a lot about carnivore-hominid interactions at the site, and many other aspects of site formation. But what was Zinj doing there? What was the role of *Paranthropus boisei*? Was this species responsible for the site, as Louis Leakey originally thought (Leakey, 1959)? Or is *P. boisei* part of the food remains of a species of *Homo*? Which species? Has it been brought to the site by some other carnivore? Different answers to these questions have quite different implications for how we interpret early hominid behavioural ecology.

Conclusions

In McHenry's delicate phrase: ". . . the dance between precision and uncertainty enlivens paleoanthropology . . ." (McHenry, 1994). If there is a lesson from these "new looks at old questions" presented in this volume it is perhaps that the clue to progress in human behavioural ecology is realism, rather than a naive optimism or a paralyzing pessimism. We have to be as precise as possible in those situations where it is possible to be precise. But we must acknowledge uncertainty clearly and firmly in the many situations where it exists, and may always exist. It is always worth reminding ourselves that it is really extremely remarkable that we know anything about our remote ancestry. When animals die, they rot, and usually little remains. The evidence of our distant past consists of relatively few bones and stones that have been preserved and discovered simply because of a few essentially random geological accidents in the east and south of Africa. It should not be surprising that there will always remain many interesting things about early hominids that we will never know. But it is clear from these papers that much progress is being made. There are interesting new ideas, well connected with ecological theory, and good suggestions for how the fossil and archaeological record should be examined to determine their relevance. These hypotheses are being examined much more critically and correspondingly refined data are being collected to evaluate them with a much greater degree of scientific rigor than has been the case in the past. A few years ago it seemed that work in this field was just geared to confirming our preconceptions about the behavioural ecology of early hominids. Now it seems more likely to surprise us with a greater range of possibilities.

Acknowledgements

I want to thank the organizers of the symposium on Early Hominid Behavioural Ecology at the American Association of Physical Anthropologists meeting in Toronto in 1993, Jim Oliver, Nancy Sikes and Kathy Stewart, for having invited me to discuss such an eclectic and fascinating collection of presentations on that occasion, and for subsequently inviting me to add my comments to this volume. Also, of course, I thank the individual contributors for having provided such stimulating papers. Tom Gundling, Sally McBrearty, and nameless reviewers gave me very helpful comments on the manuscript.

End note
This paper discusses only those papers which were presented in the 1993 AAPA symposium.

References

Bellomo, R. V. (1994). Methods of determining early hominid behavioral activities associated with the controlled use of fire at FxJj20 Main, Koobi Fora. *J. hum. Evol.* **27**, 173–195.

Blumenschine, R. J., Cavallo, J. A. & Capaldo, S. D. (1994). Competition for carcasses and early hominid behavioral ecology: a case study and a conceptual framework. *J. hum. Evol.* **27**, 197–213.

Bunn, H. T. (1994). Early Pleistocene hominid foraging strategies along the ancestral Omo River at Koobi Fora. *J. hum. Evol.* **27**, 247–266.

Brain, C. K. (1981). *The Hunters or the Hunted? An Introduction to African Cave Taphonomy*. Chicago: University of Chicago Press.

Brain, C. K. (1993). *Swartkrans: A Cave's Chronicle of Early Man*. Transvaal Museum Monograph #8. Transvaal Museum: Pretoria.

Conkey, M. W. (1991). Original narratives: the political economy of gender in archaeology. In (M. di Leonardo, Ed.) *Gender at the Crossroads of Knowledge: Feminist Anthropology in the Postmodern Era*, pp. 102–139. Berkeley: University of California Press.

Hill, A. (1979). Ockham bearded. *New Scientist* 6 December: 810.

Hill, A. (1984). Hyaenas and hominids: taphonomy and hypothesis testing. In (R. Foley, Ed.) *Hominid Ecology and Community Ecology: Prehistoric Human Adaptation in Biological Perspective*, pp. 111–128. London: Academic Press.

Hill, A. (1987). Causes of perceived faunal change in the later Neogene of East Africa. *J. hum. Evol.* **16**, 583–596.

Hill, A., Ward, S., Deino, A., Curtis, G. & Drake, R. (1992). Earliest *Homo*. *Nature* **335**, 719–722.

Isaac, G. L. (1978). The food-sharing behavior of protohuman hominids. *Sci. Am.* **238**, 90–108.

Kibunjia, M. (1994). Pliocene archaeological occurrences in the Lake Turkana Basin. *J. hum. Evol.* **27**, 159–171.

Kroll, E. M. (1994). Behavioral implications of Plio-Pleistocene archaeological site structure. *J. hum. Evol.* **27**, 107–138.

Leakey, L. S. B. (1959). A new fossil skull from Olduvai. *Nature* **184**, 491–493.

McGrew, W. C. (1992). *Chimpanzee Material Culture*. Cambridge: Cambridge University Press.

McHenry, H. M. (1994). Behavioral ecological implications of early hominid body size. *J. hum. Evol.* **27**, 77–87.

Oliver, J. (1994). Estimates of hominid and carnivore involvement in the FLK *Zinjanthropus* fossil assemblage: some socioecological implications. *J. hum. Evol.* **27**, 267–294.

Potts, R. (1994). Variables versus models of early Pleistocene hominid land use. *J. hum. Evol.* **27**, 7–24.

Plummer, T. W. & Bishop, L. C. (1994). Hominid paleoecology at Olduvai Gorge, Tanzania, as indicated by antelope remains. *J. hum. Evol.* **27**, 47–75.

Selvaggio, M. M. (1994). Carnivore tooth marks and stone tool butchery marks on scavenged bone: archaeological implications. *J. hum. Evol.* **27**, 215–227.

Sikes, N. E. (1994). Early hominid habitat preferences in East Africa: paleosol carbon isotope evidence. *J. hum. Evol.* **27**, 25–45.

Stern, N. (1994). The implications of time-averaging for reconstructing the land-use patterns of early tool-using hominids. *J. hum. Evol.* **27**, 89–105.

Stewart, K. (1994). Early hominid utilisation of fish resources and implications for seasonality and behavior. *J. hum. Evol.* **27**, 229–245.

Susman, R. L. (1988). Hand of *Paranthropus robustus* from Member 1, Swartkrans: fossil evidence for tool behavior. *Science* **240**, 781–784.

Susman, R. L. (1991). Who made the Oldowan tools? Fossil evidence for tool behavior in Plio-pleistocene hominids. *J. Anthrop. Res.* **47**, 129–151.

Vrba, E. S. (1980). The significance of bovid remains as an indicator of environment and predation patterns. In (A. K. Behrensmeyer & A. Hill, Eds) *Fossils in the Making: Vertebrate Taphonomy and Paleoecology*, pp. 247–272. Chicago: University of Chicago Press.

Vrba, E. S., Denton, G. H. & Prentice, M. L. (1989). Climatic influences on early hominid behavior. *Ossa* **14**, 127–156.